SUSTAINABLE FACILITY MANAGEMENT

OPERATIONAL STRATEGIES FOR TODAY

John P. Fennimore, MSc, P.E.

PEARSON

Boston Columbus Indianapolis New York San Francisco Upper Saddle River
Amsterdam Cape Town Dubai London Madrid Milan Munich Paris Montreal Toronto
Delhi Mexico City São Paulo Sydney Hong Kong Seoul Singapore Taipei Tokyo

Editorial Director: Vernon R. Anthony
Senior Acquisitions Editor: Lindsey Gill
Editorial Assistant: Nancy Kesterson
Director of Marketing: David Gesell
Senior Marketing Coordinator: Alicia Wozniak
Marketing Assistant: Les Roberts
Senior Managing Editor: JoEllen Gohr
Program Manager: Maren L. Miller
Operations Specialist: Deidra Skahill
Development Editor: Leslie Lahr

Art Director: Jayne Conte
Cover Designer: Suzanne Duda
Manager, Image Asset Services: Mike Lackey
Cover Art: The Port Authority of NY and NJ
Full-Service Project Management: Nitin Agarwal, Aptara®, Inc.
Composition: Aptara®, Inc.
Printer/Binder: Edwards Brothers Malloy
Cover Printer: Lehigh-Phoenix Color/Hagerstown
Text Font: Minion

Credits and acknowledgments borrowed from other sources and reproduced, with permission, in this textbook appear on the appropriate page within the text.

The author and Pearson Education, Inc., do not assume and hereby disclaim any liability for any loss or damage caused by errors or omissions resulting from negligence, accident, or other causes.

Microsoft® and Windows® are registered trademarks of the Microsoft Corporation in the U.S.A. and other countries. Screen shots and icons reprinted with permission from the Microsoft Corporation. This book is not sponsored or endorsed by or affiliated with the Microsoft Corporation.

Many of the designations by manufacturers and sellers to distinguish their products are claimed as trademarks. Where those designations appear in this book, and the publisher was aware of a trademark claim, the designations have been printed in initial caps or all caps.

Library of Congress Cataloging-in-Publication Data is available from the Publisher upon request.

10 9 8 7 6 5 4 3 2 1

ISBN 10:   0-13-255651-0
ISBN 13: 978-0-13-255651-4

This edition of **Sustainable Facility Management: Operational Strategies for Today** *is dedicated to facility managers and their staffs across the globe—the unsung heroes who make our daily lives better through their diligence, hard work, and persistence in dealing with all matters of the built environment.*

ABOUT THE AUTHOR

John P. Fennimore has followed a broad and robust career path that includes engineering, military duty, seamanship, commercial property management, and education. He found many similarities among military duty, ship management, and facility management. All of his experiences culminated in the writing of this textbook.

John began his professional career after receiving a bachelor of engineering degree in marine engineering from State University of New York Maritime College, followed by a commission as ensign in the United States Naval Reserve. Subsequently, John worked for several years as chief engineering officer aboard commercial ships plying the Great Lakes. During the height of the cold war, he provided his marine engineering expertise to the U.S. government by traveling overseas to manage the in-port maintenance work aboard high-technology submarine surveillance vessels.

Returning home, John then became director of facilities at Ramapo State College of New Jersey and later at Alverno College in Wisconsin. Opportunity for personal development presented itself again in the form of commercial property management. John took on the task of managing the facility operation of a portfolio of high-rise office buildings, shopping centers, and industrial properties for RREEF Funds (now RREEF Real Estate). Seeking continued professional and personal growth, John obtained a state license as a professional engineer and finally a master's degree in facility management from Leeds Metropolitan University in the United Kingdom. These achievements led to positions as a facilities engineering consultant in the field and also as a lecturer at the Milwaukee School of Engineering. Today, John instructs a program of facilities maintenance technology at Waukesha County Technical College.

PREFACE

Sustainable Facility Management: Operational Strategies for Today is written for those who not only want to obtain a thorough understanding of the methods, skills, and equipment used in the art of facility management, but also seek firsthand knowledge about how to manage effectively in a changing world. The book covers multiple subject areas and can easily be used as the main text for a capstone course in a facility management degree program. For facility managers, the all-encompassing nature of the work makes it an excellent how-to guide for sustainable practice. The text's emphasis on the practical operational aspects of sustainable facility management also makes it well suited for an operational course in a certificate in sustainable management program. The book contains topics that are absolutely essential for sustainable facilities, and some of these topics are not covered in any other textbook. Chapter questions, assignments, and numerous photos and diagrams make it a unique choice for facility management instruction. Also unique to the book are the questions and assignments that conclude each chapter. These features challenge students and reinforce their learning.

Here are some chapter-by-chapter highlights of the text:

Chapter 1, "Facility Types and Management Methods," analyzes the organizational structure and cultural aspects of properties and facilities. It delves into green rating systems that have become an essential part of facility design and operation. Job titles covered under facility management and new job titles due to the pursuit of sustainability are defined. There is a discussion of the many job opportunities available to those entering the field, which provides an opening for a section on how to manage. Information on various management methods are used, which leads to a section on the important questions every new facility manager needs to ask.

Chapter 2, "Sustainable Maintenance Operations," discusses work-order systems, types of work orders, maintenance formats, computerized maintenance management systems, and property condition surveys. The proper maintenance of facilities using advanced methods is essential to sustainable operations.

Chapter 3, "Managing Outsourced Services," covers the many elements of successful outsourcing through discussions on contract management, benchmarking, and service-level agreements. It tackles the difficult topic of outsourcing facility management. The chapter ends with a look at outsourcing in the United Kingdom.

Chapter 4, "Financial Management and Control," handles the basics of budgeting and utilizing funds. The chapter covers different types of budgets and how to budget for financial control. It also delves into using financial information to discover hidden problems at a facility, and the importance of proper budgeting to aid in green initiatives.

Chapter 5, "Construction Management and Sustainable Design," is a topic that every facility manager must perform, either as an owner's representative or as an independent general contractor. The chapter begins with the steps in construction, site development, types of construction contracts, and the importance of choosing the right construction contract for the desired outcome. Sustainable construction methods and sustainable design are covered thoroughly. Sustainable design covers building information modeling and the use of laser scanning to create interactive 3D models of existing construction. Computerization in construction is a major step toward sustainable construction. The text discusses network diagrams used in construction planning and shows the importance of the critical path method to control project duration. The chapter ends with a discussion of recycling construction materials and postconstruction issues.

Chapter 6, "Fire and Security Systems and Disaster Prevention," covers vital topics. Disasters can strike at any time. Successful emergency management is essential, but its success hinges on performing risk management beforehand and developing the organization, equipment, and procedures to handle the threats in relation to the likelihood of their happening. A review of firefighting systems is provided, along with security management, access control, and cyberattack.

Chapter 7, "Facility and Global Environmental Management," explains why a misstep in environmental management can cost the facility millions of dollars, ruin their reputation, lead to extensive litigation, and harm the environment. Phase I all appropriate inquiry (AAI) environmental site assessments are an important step in protecting the facility from environmental issues. Tenant operations that may harm the facility are also discussed. Additional environmental concerns such as Legionnaires' disease, cell phone tower electromagnetic radiation, radon, asbestos, lead and lead-based paint, and underground storage tanks are discussed. The chapter then talks about global environmental concerns, ozone, greenhouse gases, and carbon footprint.

Chapter 8, "Building Systems and Controls" describes why managers need to understand sophisticated heating, ventilation, and air conditioning (HVAC) systems and controls, which exist in even small buildings today, and the operational strategies that can provide energy savings if they are followed correctly. This chapter concentrates on distribution systems and terminal devices. Heat reclamation, chilled beam, demand-controlled ventilation (DCV), and other ventilation control systems are covered.

Chapter 9, "Major Building Equipment Systems and Subsystems," showcases various types of main system commercial HVAC equipment and their applications. Boiler basics and power plant and air-conditioning plant operations are detailed. Methods and equipment that achieve energy savings are covered in detail. Specific types of equipment, such as ice storage equipment, condensing boilers, and cooling towers, are major topics. Reducing both boiler system and cooling system reliance on toxic chemicals is important for sustainability. The chapter focuses on various methods of nonchemical water treatment.

Chapter 10, "Energy Management and Renewable Energy," discusses a major cost for facilities: the energy they consume. As energy resources become scarcer, their cost to the facility will continue to rise. Managing resources while pursuing renewable alternatives is the major goal of energy management. This chapter discusses the many sustainable options available: wind, solar, photovoltaic, geothermal, and hydropower, and the fuel cell technology used at the new One World Trade Center. Commercial rate structures are discussed, along with how to reduce costs. The chapter also points out important aspects of alternating current power systems that need to be addressed in order to reduce costs. Lighting is a major part of electrical consumption, and the chapter discusses different types of lighting, and the advantages and costs of each.

Chapter 11, "Building Site Interior and Personnel Management," focuses on water, the most precious commodity. It is becoming less available for drinking due to pollution and climate change. The ability to reduce the amount of water entering storm sewers is a significant factor in creating a sustainable environment and helps ensure that there will be plentiful clean water in the future. The chapter discusses landscaping techniques, including xeriscaping to reduce water use, and a method of rainwater treatment that takes storm runoff and converts it to water for landscaping use. The chapter then changes focus and discusses green housekeeping and safe practices for housekeepers. Sustainable training and productivity measurement follows. The chapter ends with recycling, waste management, and restroom water conservation.

Chapter 12, "Green Building Construction," discusses common construction methods, which leads into a discussion of green building. Roofs are high-cost items that require frequent maintenance, and extensive coverage is given to this important topic. The chapter also discusses green roof and green wall construction, and leak detection with green roof systems. Common building problems that facility managers can expect are reviewed. The chapter then shifts focus to interior construction. Phase change materials, which are used in interior wall construction and provide a passive form of air conditioning, are introduced. The chapter then discusses important green aspects of elevator and escalator systems, and ends with interior finish considerations that include green carpet and hard-surface flooring.

Chapter 13, "Strategic Planning and Project Financial Analysis," emphasizes that strategic planning requires an evaluation of the available options. Discovering these options is achieved through SWOT analysis, brainstorming, and mind maps. Project evaluation, which is essential for approval by upper management, consists of financial methods such as life-cycle costing analysis, simple payback, and return on investment. The chapter then discusses churn management and space planning. The chapter (and hence the book) ends by showcasing the world-famous sustainable design of Council House 2 in Melbourne, Australia.

Sustainability is a quest rather than an absolute. It is a goal that we, as facility managers, should be striving toward for the good of the planet.

Instructor's Resources

To access supplementary materials online, including a Solutions Manual and Image Bank, instructors need to request an instructor access code. Go to **www.pearsonhighered.com/irc** to register for an instructor access code. Within forty-eight hours of registering, you will receive a confirming e-mail that includes an instructor access code. Once you have received your code, locate your text in the online catalog and click on the Instructor Resources button on the left side of the catalog product page. Select a supplement, and a login page appears. Once you have logged in, you can access instructor material for all Prentice Hall textbooks. If you have any difficulties accessing the site or downloading a supplement, please contact Customer Service at **http://247pearsoned.custhelp.com/**.

Acknowledgments

First, I would like to thank my family who supported this effort in ways too numerous to count. Then I would like to thank the book's reviewers: Dr. Joseph O. Arumala, University of Maryland Eastern Shore; Michael Bobker, City College of New York, CUNY; Chris Garbett, Leeds Metropolitan University, United Kingdom; Sarel Leibovich-Lavy, Texas A&M University; Angela Lewis, University of Reading; Kathy Roper, Georgia Institute of Technology; and Claude Villiers, Florida Gulf Coast University. They selflessly gave their time and used their extensive knowledge and experience to make many valuable suggestions, helping to improve the book's content. I would also very much like to acknowledge the great contribution and unflinching dedication of all the editors, with special thanks to Leslie Lahr and Marianne L'Abbate, both of whom worked tirelessly on this project. Finally, I would like to thank all those who made direct material donations of photos, diagrams, and information. In the world of today, little is as simple as it seems. I thank all those persons who "jumped through the corporate hoops" to provide valuable material for the book. Without this essential help, the book would not be able to contribute to the goal of making facility management sustainable through education.

John P. Fennimore

CONTENTS

INTRODUCTION

What makes a great facility manager (FM)?

The complexity and varied nature of the demands of the job makes the answer to this question a difficult one. Facility managers must possess an extremely varied skill set. Therefore, the answer to the question, What makes a great facility manager? must account for the type of facility and its corporate culture, status in its industry, and financial goals. A facility manager might be considered great at one facility, but his or her ability might be judged as less than great at another. An example is the facility manager of a small school district whose building and grounds operation functions in what could be describe as the janitorial mode. The facility workers are referred to as custodians or janitors (these job titles are somewhat interchangeable and are not used consistently throughout industry). Their work consists of mostly cleaning and nontechnical interior and exterior grounds maintenance tasks. The custodians/janitors also perform a variety of minor jobs, such as unlocking doors and gates, putting up signs, rearranging room furniture, and the like. The district's technical maintenance work is contracted by employing outsourced specialty trades such as electrical; plumbing; and heating, ventilation, and air conditioning (HVAC). In this case, the facility manager is most focused on the day-to-day operation, managing his or her workers, overseeing the activities of outsourced technical staff, planning for school events, monitoring regulatory requirements, and controlling the maintenance budget.

A facility manager with a baccalaureate degree in civil or mechanical engineering, architecture, or construction management may not be a good fit for a situation where the janitorial mode of operation is employed. But for a growing commercial facility operation that is purchasing existing properties requiring substantial mechanical system upgrades or for a facility that is undertaking the construction of new buildings, a facility manager with engineering qualifications and experience is a great asset. The surprise for many school districts operating in the janitorial mode is that many of them will eventually undertake large renovation or new construction projects. During these heavy construction periods, they would benefit from having someone on staff with broad construction management skills to manage these projects successfully.

One facility management position is not necessarily better than another. The facility manager working in the janitorial mode obtains great experience operating an educational facility, dealing with the massive housekeeping issues, scheduling HVAC equipment to meet the constantly varying needs for after-school programs and events, providing the high level of landscaping necessary to maintain the quality and safety of athletic fields, and working with a unionized staff. The facility manager/engineer in a commercial facility, on the other hand, has an outsourced janitorial staff and an in-house technically skilled maintenance staff. The facility manager/engineer doesn't deal with operational housekeeping issues; however, he or she must manage the many strategic facility planning and construction issues. The facility manager/engineer must consider the competitive financial aspects of all facility decisions. What can be inferred from making FM position comparisons is that *the manager who meets the needs of the facility is the best manager for the facility*.

In the pursuit of greatness, note the local market for real estate. If the facility manager is in a small stable market, any increases in her or his educational credentials, professional licenses, and certifications might actually make the facility manager overqualified for the local market demand for top-level managers. In this case, the facility manager might face the difficult decision between leaving a cherished home location or staying in spite of being overqualified for the local market. Making the move to a larger, more vital market can exercise the skills of a facility manager to the fullest, which in turn allows that facility manager to be great.

Technology, energy, environmental concerns, and the issue of terrorism are all driving change in the way facilities are constructed, maintained, and operated. The site called Trump International Hotel and Tower Chicago is an example of the necessity of rethinking building design. The original design called for the tower to be the world's tallest residential structure, but the events of 9/11 necessitated a reduction in the building's height to make it less of a target for terrorists (9/11 almost forced the project to be scrapped entirely). The project was delayed for about two years to accommodate design changes. As it turned out, building height was the problem, but it also was the solution! Through the use of reinforced concrete construction rather than steel I-beam frame construction, the height between each floor was reduced, resulting in more floors in a shorter building. A significant gain in rentable square feet was achieved, making the building more profitable, and the reduced height also resulted in a safer building.

A facility manager's job requirements and the talents necessary to perform them are changing. The result is a shift in thinking by the highest levels of corporate management and persons charged with hiring facility managers. Organizations want facility managers who are professionals with the requisite qualifications. Therefore, the way to begin a career as a facility manager is by obtaining a solid educational background in the disciplines required by the facilities management profession. Obtaining a facility management credential tells the employer that the employee candidate has current knowledge and professionally developed management skills. This is very important in a field that is rapidly changing because it can tip the scales in favor of a facility management job candidate who is competing with an experienced manager who lacks a credential.

A program of lifelong learning makes the standard FM into a sustainable FM. Lifelong learning allows the facility manager to acquire the knowledge and skills needed for responding effectively to the problems of today and to be well prepared for handling the inevitable demands of the future. The combination of knowledge about sustainability and the skills to achieve sustainability provides for continued greatness as a facility manager, no matter what the facility type or the difficulty level of the challenges presented.

FACILITY TYPES AND MANAGEMENT METHODS

INSIDE THIS CHAPTER

What Are the Elements of a Successful Career in Facility Management?

The two major employment areas for the facility manager (FM) are in properties and facilities. These workplaces are discussed in their cultural context in order to illustrate why they operate differently. The employment opportunities and ways of advancing in the profession are discussed, along with the titles for the different positions that offer a variety of opportunities. The various types of properties and facilities and their respective definitions of quality are explained. The relatively new, green rating systems launch the discussion of sustainable facilities and sustainable operations. We close the chapter with a review of effective facility management methods and a valuable checklist for new managers.

1.1 POSITION OF FACILITY MANAGER

The U.S. Department of Labor, Bureau of Labor Statistics (BLS), places facility managers under the heading of Administrative Services Managers in their *Occupational Outlook Handbook,* 2010–11 edition. The BLS describes facility managers specifically as those who plan, design, and manage buildings, grounds, equipment, and supplies. Increasingly, FMs develop and implement plans that incorporate energy efficiency into a facility's operations and structures, hence the addition of the term *sustainable* to their title. All of their duties require integrating the principles of business administration, information technology, architecture, and engineering.

The BLS also points out that the tasks assigned to a facility manager can vary greatly. Depending on the organization being served, their responsibilities may fall into wide-ranging categories such as operations and maintenance, real estate, project planning and management communication, finance, facility function, technology integration, and an awareness of environmental issues. Tasks within these broad categories may include space and workplace planning, budgeting, purchase and sale of real estate, lease management, renovations, and architectural planning and design.

Work Locations

Places where facility managers work fall into two broad categories: properties and facilities. These two divisions are not set in stone, and these terms are often used interchangeably, especially when discussing the actual buildings themselves. Frequently, the various aspects of building construction exhibit no tangible difference between a facility and a property.

The organizational term *property* is the common identifier for commercial buildings and implies the right of ownership. Properties are buildings such as shopping malls, office buildings, corporate headquarters, apartment and condominium communities, and industrial structures. A common thread among property organizations is ownership for profit. A person or a corporation who owns the property has title to the property and a deed. Properties produce income when a tenant signs a lease, occupies the space, and pays rent, unless, of course, the property is owned and occupied by the same company as is the case with some industrial complexes; in this situation, it is the company's operations that produce the income.

Most large office tower complexes and many other types of commercial properties are owned by pension funds and various forms of real estate investment vehicles such as a real estate investment trust

(REIT). These investments all aim to reduce risk through diversification by purchasing several properties in different geographic areas. Pension funds frequently hire management and investment companies to operate the properties, purchase and dispose of properties, handle financial issues and leasing, and provide a return to the fund participants.

The term *facilities* relates to a variety of organizations. Examples are government buildings, military bases, public airports, marine terminals, local and state educational, religious facilities (churches and schools), and hospital campuses, both private and public. In the case of government facilities, funding comes from tax revenues. Other facilities may be funded through donations and fees, and generally operate as nonprofits. Facilities frequently have a greater number of in-house staff than property operations. But even with their larger in-house staffs, most facilities rely on some outsourced service providers for specialized work that is either beyond their capability, or can simply be done more cost effectively by **outsourcing**.

Position Titles

The varied nature of the responsibilities placed on the facility managers has led to numerous job titles. These titles tend to vary depending on whether the manager is at a property or a facility.

A facility manager at a property may be given the title operations manager, building manager, general manager, or, of course, facility manager.

How these titles are used is as follows:

Operations manager and *building manager* are the common titles for FMs who work at an office property. The title of operations manager may be bestowed on an FM whose responsibilities may include a portfolio of properties comprising retail, office, residential, and industrial properties, or any combination or part thereof.

General manager is a title used at hospitality properties such as hotels and motels. This title is appropriate because it recognizes that, in addition to managing the facility, this facility manager must perform many tasks associated with "running the store," or maintaining the operation. Management of this nature sounds simple, but it is actually quite a massive task that includes supervision of food service operations, marketing, strategic planning, supervision of hotel staff, solving customer problems, maintenance of financial records, and a multitude of other duties involving the day-to-day maintenance of the organization.

The facility manager of a large vacation resort property may be called the director of facilities, even though the resort is a private property. This job title reflects the multiple services and large number of staff under the director's supervision. The situation is similar to a large facility like a college or university. If the company owns several resort properties, the person in charge of all of them is the *regional director of facilities*; he or she may have several facility managers reporting to him or her.

A facility manager at a facility may be given one of the following titles: director of facilities, director of the physical plant, building superintendent, manager of buildings and grounds, plant operations manager, or vice president of facility operations.

Director of facilities is a common title used today at many facilities such as hospitals and universities.

Director of the physical plant is an older title that reflects the days when it was fairly common to have a separate plant building containing boilers and mechanical equipment, which served the needs of several buildings on a campus.

Building superintendent is another old title from the early days of apartment management. The title was also adopted by some K–12 educational facilities. For these facilities, the title *manager of buildings and grounds* is more common today.

Plant operations manager is a newer title frequently used for a manager whose duties include the plant operation. At a very large healthcare facility, for example, the *plant operations manager* may actually work for the *vice president of facility operations*. The vice president of facility operations would report to the top management of the healthcare facility.

1.2 CULTURAL DISTINCTIONS AND ORGANIZATIONAL CULTURE

The most important differences between different facilities or properties are created by cultural distinctions, which are known collectively as *corporate culture*. Cultural distinctions influence an organization's management structure, management philosophy, policies, goals, and definition of organizational success. The effect of corporate culture might be revealed by the amount of time allocated to an FM to develop a schematic design for a construction project.

Example: Renovation

An FM may be given the responsibility to renovate a light industrial/warehouse building, reconfigure its interior, and change its use to office/retail. In one organization's culture, a manager who operates quickly and efficiently is the cultural norm. This organization gives the FM ninety days to develop plans and specifications.

Another organization may emphasize the sustainability of a project and therefore requires a detailed analysis of not only the site and the building in question, but also how the new building will function around the existing buildings and how it will fit into the community. In this case, it would be reasonable to spend up to two years in the design phase.

Example: Worker Evaluation

In another example, a nonprofit facility operated by a charitable organization might define facility worker management success in terms of keeping workers employed and developing their skill level. Rather than defining an entirely successful facility maintenance service by the number of maintenance jobs completed in a day, the maintenance department is considered successful when management has found a way to keep workers employed who would otherwise find it difficult to keep a position at another organization. As workers develop professionally in this type of cultural environment, they can take on increasingly more complex tasks and grow professionally with the facility. A specific type of maintenance work is issued to the worker who is capable of performing the required task, but management would not necessarily penalize another worker who should be able to do that same job but can't. This may be true even when the maintenance worker's job title specifically includes tasks that the worker finds difficult to complete without help.

Alternatively, a profit-motivated office property operation defines employee operational success in terms of a competitive, cost-effective operation that hires only fully trained, independent workers with a proven track record of experience. Management in this type of environment might penalize a worker for an inability to perform a task completely. This operation might also utilize outside contractors (referred to as *outsourcing*) to perform work in the most cost-effective manner, with staff sizes kept to a minimum. The profit-motivated property operation might be reluctant to send workers for additional training, fearing they may lose their investment in training should the workers decide to take their new skills elsewhere. On the other hand, workers may be reimbursed or offered bonus pay for taking the initiative and obtaining the training that benefits the operation without the cost implications to the property should they leave.

Organizational Culture

An effective facility manager needs to understand organizational culture and its effect on facility operations. For instance, the cultural values of an organization based on an altruistic model, focusing on trust and a philanthropic approach (which would likely be the case at a charitable organization), must be considered when making management decisions. An FM in this situation might find that employing the more profit-motivated management style discussed earlier would be viewed highly unfavorably by the organization and would be in direct conflict with its culture. An organization in a highly competitive, profit-driven environment, however, might praise a manager for undertaking the difficult and unpleasant task of dismissing incompetent workers and replacing them with an efficient facility team.

An FM might discover that an organization needs to review and modify some of its values to adjust to current realities. Values out of step with the realities of a changing marketplace may actually harm the organization, putting it at a competitive disadvantage by having expenditures that are too great.

Example: Modification of Organizational Culture

The charitable organization already mentioned may have begun hiring facility workers years ago when wages and benefit costs were comparatively low. Over time, relatively small salary increases have compounded their value, one increase upon another, as with interest on a bank account. The facility is now faced with a low-skilled, older workforce that is paid top dollar in the labor marketplace.

The facility manager in this situation would have good reason to request a modification of the organizational culture in order to save the organization. The FM might recommend that a training program to update staff skill levels be instituted to meet the changing needs of the facility. Persons refusing or unable to increase their skill level would suffer demotion and/or termination. This is harsh medicine and a bitter pill to swallow for the worker who has become comfortable with the prior job situation.

As previously stated, the facility manager must first understand the culture of the organization when seeking change. The FM must then consider various other factors in staff and operational planning, some of which are:

- **Market for the services provided by the organization**—Is the market growing, stable, or shrinking in size?
- **Worker talent pool**—What are the strengths and weaknesses of the current staff?
- **Training availability**—Are there programs in place to train staff in new methods easily and affordably?
- **The current budget and overall financial condition of the organization**—Is the organization solvent?
- **Facility condition**—What are the current maintenance needs?
- **Technology**—What technology is available that could aid in operational decision making?
- **Immediate needs versus future goals of the organization**—What is a hazard and what are the long-term improvements under discussion?
- **Risk**—What are the risks of either action or inaction?

Cultural distinctions also affect the manager's job title, job requirements, authority to make decisions, and the definition of a successful operation. An example of this is shown in Figure 1.1 whereby a profit-driven organization is compared to a nonprofit organization. The physical facilities might be very similar, but the facility management goals and methods employed are very different. Cultural distinctions also affect all of the facility's workers in many of the same ways. The facility manager should discuss with upper management how the culture of the organization affects the facility operation overall.

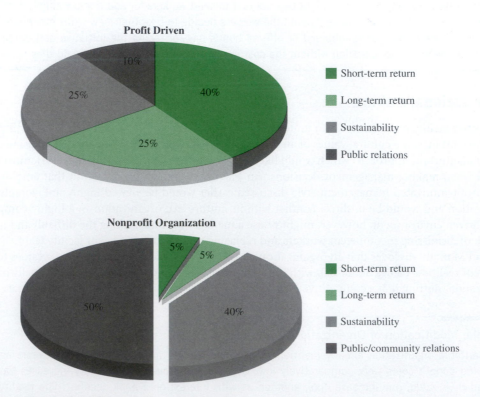

FIGURE 1.1 Comparison Between Two Organizations

1.3 ADVANCING IN THE FACILITY MANAGEMENT PROFESSION

The Bureau of Labor Statistics points out that most facility managers have an undergraduate or graduate degree in engineering, architecture, construction management, business administration, or facility management. Some managers work their way up from technical positions and move into management. Persons who are already managers advance by taking facility management positions that offer greater responsibilities. In addition, competency-based certification such as the Certified Facility Manager (CFM) and the Sustainability Facility Professional (SfP™) designations from the International Facility Management Association (IFMA) can give prospective candidates an advantage when seeking employment or advancement within an organization.

In IFMA's Profiles 2011 salary report, which tabulated the results of 4,353 facility managers from forty-five countries who responded to their survey, the combination of salary and bonus pay the average facility managers earns is $99,578 USD per year. It is interesting to note that new facility managers, with three years experience or less, earned $65,000 USD per year. The full salary report can be obtained online from the IFMA bookstore.

When an FM has an undergraduate degree in facilities management or a related field, the obvious choice is to advance that degree or attain other professional qualifications in order to seek a higher level of responsibility and authority. In addition to possessing IFMA's CFM designation, other designations, degrees, and licenses can be very attractive to an employer, especially if they relate to the facility's operational type or the company's core business. Table 1.1 offers examples of additional credentials that an FM might pursue for advancement at a particular property or facility type.

Several distance-learning offerings can lead to graduate degrees in the United States and in countries such as Great Britain, Australia, and Canada. Because not all foreign degrees are equivalent to a U.S. degree, a U.S. student should first have the degree program evaluated by a professional educational evaluation company to ensure that the degree will meet U.S. requirements for the credential sought. Once the degree is completed, the evaluation company calculates the actual number of equivalent credits awarded by each course within the degree, based on the student's course transcripts.

Several U.S. institutions of higher education, which have for decades specialized in traditional areas of study, have recently embraced facility management. In particular, state maritime colleges now offer both undergraduate and master's degrees in facility management. Maritime college students have the advantage of graduating with actual power plant operation experience in addition to a degree in engineering or management.

1.4 PROPERTIES

The built environment comprises a range of structures: from historic buildings, hundreds of years old, to modern skyscrapers employing the latest technology. The good news is that they all employ facility managers. During difficult times when tenant fortunes wane, facility managers are still needed to operate the property, even at very low property occupancy levels. This keeps FMs in demand under most economic conditions. The following property (commercial) classifications are addressed in this section to illustrate the variety of opportunities for the FM:

- Office properties
- Hotels and apartments buildings
- Industrial properties
- Retail properties

Office Properties

Buildings that rise ten stories or more in height are considered high-rise buildings by many building professionals. Figure 1.2 shows some of the high-rises belonging to the Chicago skyline. The National Fire Protection Association (NFPA) considers a building to be high-rise at a height greater than 75 feet (25 meters) for fire prevention purposes. The construction, maintenance, and operation of high-rise buildings have more stringent fire and safety requirements than buildings of less height.

The size of office buildings is measured in square feet in the United States, and by the square meter just about everywhere else in the world. Office buildings range in size from 2,000 square feet to over 1 million square feet. The area of each tower of the World Trade Center was 3.8 million square feet, and

TABLE 1.1 Credentials for Possible Professional Development in the United States By Industry Group

	Commercial Properties or Property Management Firms	State or Federal Agency	Hospital or Healthcare Facility	Industrial	Educational Facility
Licensed real estate broker	✓				
Graduate degree in real estate or business management	✓	✓			
LEED AP®	✓	✓	✓	✓	✓
Degree in engineering, architecture or construction management	✓	✓	✓	✓	✓
Sustainability Facility Professional™ (SfP)	✓	✓	✓	✓	✓
Certified Facility Manager (CFM)	✓	✓			✓
Registration as a licensed professional engineer or registered architect		✓			
OSHA certificate		✓	✓	✓	✓
Asbestos certification		✓			✓
Lead-based paint certification		✓			✓
Government Operator of High Performance Buildings (GOHPB) from the Association for Facilities Engineering (AFE)		✓			
Certified Healthcare Facility Manager (CHFM) designation from the American Hospital Association			✓		
Power plant operator's license as chief engineer			✓		
Certificates in healthcare activities: construction; medical gas; heating, ventilation, and air conditioning (HVAC); infection control; electrical safety; and underground storage tanks			✓		
Six Sigma Designation—master black belt, black belt, green belt, yellow belt, and white belt				✓	
Certificate in lean manufacturing				✓	
Certified Plant Maintenance Manager (CPMM) from the Association for Facilities Engineering (AFE)				✓	
Advanced degree in management					✓
Power plant operator's license					✓
Certified Educational Facilities Professional (CEFP) from the Association of Higher Education Facilities Officers (APPA)					✓

the area of the Willis Tower (formerly the Sears Tower) in Chicago is 4.56 million gross square feet and 3.81 million rentable square feet. **Gross square feet** includes areas that are needed for access and mechanical spaces; examples include corridors, stairways, elevators, and lobbies.

Most of the space that is not rentable falls under the space category known as common area. **Common areas** are spaces that are available to all tenants and visitors; the expenses related to such areas may be billed to the tenants as a **common area maintenance** charge. Common areas are not simply limited to

FIGURE 1.2 Part of the Chicago Skyline

Source: © pongsakorn1/Fotolia

corridors, restrooms, and lobbies. Areas such as teleconferencing rooms, meeting rooms, lunch rooms, handball courts, and exercise rooms are all considered common areas. The exterior walks, parking areas, and landscaping also fall under the heading of common area.

Office Property Classification

The quality of modern real estate is indicated by class ratings. Office buildings are commonly rated as Class A, Class B, and Class C. Classification level is a direct reflection of the design, equipment, architecture, operational efficiency, management structure, maintenance level, indoor environmental quality, amenities, location, historical significance, transportation efficiency, and synergistic relationship to surrounding businesses. Properties scoring high in these areas receive a Class A designation.

Class A Office Buildings   Class A office buildings have the most desirable location, a modern or classic design, high-end interior design components, the latest mechanical equipment, excellent maintenance, and a top-notch management team. Class A buildings periodically undergo major renovations and mechanical system upgrades that seek to improve energy efficiency and comfort to maintain their Class A status.

Class B Office Buildings   Class B office buildings are in a desirable location, but are older and may be showing some wear and tear. Maintenance and management are performed well, but to a lesser degree. Renovations are smaller in scope, and more emphasis is placed on providing a good value to the tenants for their lease dollar.

Class C Office Buildings   Class C office buildings have lower quality or older interior fixtures and mechanical equipment. These buildings may have some comfort issues with regard to the heating and cooling systems and are located in less desirable areas. Class C office buildings may have been created through redevelopment of buildings previously designed for other uses such as factory buildings or warehouse structures. At this building class, maintenance is focused on keeping equipment running as long as it can. Replacement of equipment occurs only when the equipment is well past the end of its useful life.

Exceptions to the Classification Rule   Some factors may take precedence over the rules on office building classifications. For instance, a building might be rather old, but it may have architectural and historical significance that boosts its class rather than detracting from it.

    The importance of office building class for facility managers is that class tends to guide the level and quality of maintenance.

Example: Maintenance by Building Class

In a Class A office building, cracked, marble floor tile would be replaced with tile reserved for that purpose from the original installation of the floor. This reserve allows the tile installer to provide a perfect match. Tile with the same color, shade, and marbling pattern as the original can only come from the original quarry; it must be mined from stone at a particular depth and location in the quarry or variances in color and pattern will occur. To handle this type of future maintenance issue, Class A buildings have storage areas designated for replacement materials. Class A office buildings have their lobby tile floor on a strict maintenance schedule of cleaning and polishing, performed by outside experts who are contracted specifically for this task. Subsequently, no deliveries are allowed to be moved across the lobby floor; grit or small stones stuck in the tires of a hand truck might scratch the floor and detract from its appearance.

Tile maintenance at a Class B office building is different. If the lobby tile is still firmly attached, a crack is repaired with an epoxy grout that closely matches the color of the tile. Maintenance on the floor is performed on an unscheduled basis, defined by the condition of the floor and whether the building is in leasing mode (where new tenants are sought). Much of the floor is covered with mats for a good portion of the year to minimize the maintenance required.

In a Class C building, maintenance of a minor crack is deferred, and the crack might remain in that condition for several years before further deterioration occurs. If the lobby floor tiles deteriorate on a large portion of the floor, all the tiles are taken up and replaced with vinyl composition tile or another product, such as a ceramic tile more in keeping with a Class C status.

It is clear to see how building class affects when maintenance is performed, who performs the maintenance, the quality of maintenance, the facility operation, and the maintenance costs. It is also very important to consider building class in all major construction or mechanical system upgrading decisions. A Class C building surrounded by other Class C buildings might actually be hurt financially by spending too much money on lobbies and amenities for the tenants. Such spending would reduce the return on investment and put pressure on the building owner to raise rents. Tenants who look for Class C space do so because they are looking for low rents. Quality, comfort, and beauty are not their first priority.

Hotels and Apartment Buildings

Hotels and apartment buildings are described by the number of hotel rooms for rent or apartments for lease. Hotels are rated by the number of stars they are given by the organization performing the review. A four- or five-star hotel has the same quality level of interior finish as the common areas found in a Class A office building. There is no comparable quality indicator for apartments other than vague advertised classifications, such as *luxury* or *deluxe*. Websites are available to provide apartment ratings based on the opinions of prior tenants.

Industrial Properties

Factories, corporate campuses, and light industrial/warehouse buildings compose some of the different industrial property types. Facility managers at factories take care of the building along with the various industrial production support duties. Production managers are responsible for the actual factory production. Corporate campuses are more akin to office complexes and can be found on property owned or leased by the corporation and also inside research parks with other corporations. Facility managers at a corporate campus may provide support for a variety of employee activities for both professional and leisure reasons.

Light industrial/warehouse buildings are structures that have a front office area, a back production area, and a loading dock or overhead door area. See Figure 1.3 for an example of such a structure. The buildings are generally either steel sided and steel framed, concrete block, or concrete tilt-up construction. Industrial properties may have a single tenant, or the building may be divided into a multitenant arrangement with several businesses occupying *demised spaces*. Demised tenant spaces are separate tenant spaces requiring a fire-rated wall between them that goes from the floor to the deck above. These buildings range from small, 20,000 square-foot buildings to mammoth, 250,000 square-foot buildings with a height as high as 36 feet. These buildings are often found in industrial or business parks. Facility managers usually don't manage individual buildings; rather, they work for the owner of the industrial park and manage it in its entirety.

FIGURE 1.3 Light Industrial Warehouse
Source: © hansenn/Fotolia

Retail Properties

Strip malls and shopping malls make up the majority of retail properties. Strip malls are generally composed of buildings one or two stories in height, with exterior entrances directly attached to the walkways and parking lots. Strip malls generally do not have onsite management but instead are frequently managed and maintained by a third-party contractor. Some regional shopping centers (larger versions of strip malls) and large outlet malls are managed onsite. Regional shopping centers and malls are above 400,000 square feet in area and may be owned and managed by the same company. Tenants are frequently responsible for maintaining their HVAC equipment, storefront, signs, and interior space. The management firm takes care of all base building issues (e.g., roofs, utility service, building exterior), common areas, sidewalks, landscaping, and parking areas.

A shopping mall is one large, enclosed building (see Figure 1.4) or a series of large buildings connected by enclosed or elevated walkways. The mall manager or general manager is in charge of the mall; her or his position is very similar to a property manager in an office building. Mall managers handle the leasing and financial issues for the property. A maintenance supervisor, operation manager, or (at a large mall) a facility manager handles the building issues and reports directly to the mall manager.

1.5 FACILITIES

Facilities are generally considered to operate as nonprofit organizations. This statement is completely true for federal government buildings, state and local educational facilities, government-operated prisons, and military bases. However, there are exceptions. Hospitals, private educational facilities, privately run prisons, and airports must remain competitive and profit-motivated in order to remain in operation.

Facilities include noncommercial buildings such as the following:

- Healthcare facilities
- Educational facilities
- Federal government buildings
- Military bases, prisons, and airports

FIGURE 1.4 Enclosed Shopping Mall
Source: © Pavel Losevsky/Fotolia

Healthcare Facilities

The size of a hospital is determined by the number of hospital beds. An approximate representation follows: A hospital with less than 150 beds is considered small; 150 to 350 beds, medium; and 350 or more beds, large. To define the amount of maintenance required at a hospital, a combination of building age, square footage, type of care being administered, and daily patient load are more accurate indicators of size and the subsequent volume of maintenance required, rather than just the number of beds.

In the United States, all aspects of hospital maintenance and maintenance record keeping are inspected and reviewed by an organization called The Joint Commission (TJC), previously known as the Joint Commission on the Accreditation of Healthcare Organizations (JCAHO). Loss of accreditation has a severe impact on the status of the hospital and may result in a loss of income from Medicare and Medicaid patient reimbursement. In addition to satisfying Joint Commission requirements, effective hospital facility managers must have systems in place for accomplishing the required maintenance. Maintenance must be documented; the data made accessible; and a system be in place to ensure that future maintenance will be correctly defined, of consistent quality, and scheduled at the appropriate time.

The construction quality of hospitals and the level of hospital maintenance can vary, but neither may drop below governmental standards. There are no formal class levels (e.g., A, B, and C) for hospitals as there are for office buildings because hospitals and schools are not looked upon as commodities for sale or lease. It is well recognized, however, that hospitals do compete with one another. Healthcare management professionals have learned that they must provide an attractive and efficient facility in order to garner revenues from patients who have a choice between hospitals. Figure 1.5 shows a typical, well-kept hospital emergency room entrance. To be competitive, hospitals must update their facility interior and exterior architecture, beautify their grounds, upgrade their facility comfort-control systems, improve services, and provide high levels of maintenance and housekeeping. Figure 1.6 shows a liquid oxygen heating unit, one of the important and unique pieces of equipment that facility managers encounter in a hospital operation.

Educational Facilities

The size of a postsecondary educational facility is derived from the number of enrolled students. Small colleges have 2,000 or fewer students. Large universities may have 40,000 or more students. According to *High School Guide's* article "Top 100 Largest High Schools in America," (August 26, 2011), the largest high school has an enrollment of 8,076 students. There are no direct, quality, building class ratings of college and university buildings. The author's experience is that, in general, higher education facilities equate to a Class B office building. A notable exception to this is when university campuses include buildings that are centerpieces for the institution. These buildings may have historic value or are

FIGURE 1.5 Hospital Emergency Room Entrance

Source: © Sean Pavone Photo/Fotolia

FIGURE 1.6 Liquid Oxygen Heating Unit
Source: Photo courtesy of Oconomowoc Memorial Hospital.

instrumental in making a statement about the institution (for example, high-technology lab buildings and school of architecture buildings). These buildings can easily be considered Class A and are operated and maintained as such. K–12 (elementary, middle, and high school) educational facilities lacking air conditioning generally fall into an equivalent of Class C. The maintenance and management would be at a similar level as well.

The overall quality of school buildings and mechanical systems at all levels of education is improving. Many school buildings have achieved this through the improvement of indoor air quality (IAQ) and the addition of air conditioning where appropriate. (The use of air conditioning may be cost-prohibitive and reserved for assembly areas or administrative areas that operate year-round.)

Federal Government Buildings

The largest owner and operator of government facilities in the United States is the General Services Administration (GSA). The GSA owns or leases approximately 9,600 facilities and preserves 478 historic buildings. This equates to a facility asset value of $500 billion USD. This undertaking requires a full-time staff of 12,635 federal employees and a fleet of 215,000 vehicles. The GSA's budget for fiscal 2008 was $26.3 billion. The GSA undertakes a variety of initiatives that promote the advancement of the field of facility management, sustainable facility design, and environmental stewardship. GSA's website provides a variety of tools for facility managers, procurement professionals, project managers, and leasing specialists that assist anyone undertaking these initiatives.

The facility managers at these facilities have a broad range of responsibilities. The buildings are varied and range from historically significant to highly technological. The GSA tends to lead in the implementation of technology to manage and design its buildings. It has implemented a program to convert much of its traditional two-dimensional, paper, construction documents to three-dimensional, digital, interactive, building information modeling (BIM) representations.

Military Bases, Prisons, and Airports

Military bases, prisons, and airports are too varied and their operations too specialized to try to place them into one building category. The management of these facilities is frequently outsourced to companies that provide the management, maintenance, and housekeeping services under a multiyear contractual arrangement.

Working at military bases or other secure government installations may require the manager to get special government clearance by having her or his background investigated thoroughly. The manager may find that an escort will be assigned to her or him when leaving certain areas until security clearance is ascertained. Escorts are also provided at prisons, but in this case it is done for the manager's personal safety.

FIGURE 1.7 Airport Terminal Seating
Source: © tomas/Fotolia

Prisons may be managed by government personnel or by private organizations under a contract agreement. Prisons provide a myriad of services for the inmates and must do so in a secure manner.

Airports vary greatly in size and in the number and type of aircraft and passengers being serviced. Today, a greater emphasis is placed on security and the prevention of terrorism than was done in the past. The goal is to move passengers in the most safe and efficient manner and to provide for their comfort (Figure 1.7). Managers working at military bases, prisons, and airports gain a specialization in the field of facility management, making them highly sought after.

1.6   GREEN BUILDING RATING SYSTEMS

Energy efficiency, the environment, and sustainability are the main drivers of a new set of rating systems, certifications, and building requirements. Three types of organizations research, develop, and administer building quality programs:

- Governmental organization
- Nonprofit organization
- Professional society

The following are some of the green building rating systems used for buildings in the United States, Canada, Australia, and the United Kingdom:

- **Leadership in Energy and Environmental Design (LEED) Certification**—United States Green Building Council (USGBC)
- *Energy Star Label*—United States Environmental Protection Agency (USEPA)
- **Building Energy Quotient (bEQ)**—American Society of Heating, Refrigerating, and Air Conditioning Engineers (ASHRAE)
- **BREEAM®**—BRE (the Building Research Establishment Ltd.) Environmental Assessment Method (BRE Global is the company responsible for BREEAM and is owned by the BRE Trust.)
- **Green Globes®**—Green Building Initiative (GBI)
- **Green Star**—Green Building Council Australia (GBCA)

The World Green Building Council (WGBC) is the umbrella organization that provides leadership and supports member councils. According the WGBC website, in 2012 there were Green Building Councils (GBCs) in twenty-five different countries around the globe. GBCs provide building certification and certification for individuals wishing to become involved with accreditation projects or simply to become more aware of the requirements for building certification and sustainability in general.

The accredited professional (AP) is the designation that is generally used to describe someone who has received training and has passed an examination, which enables him or her to develop the scope of work, supervise the work, and evaluate green building projects. The actual construction design necessary for the green building project requires that the designer is a licensed professional engineer registered in the state where the work is to be performed. A professional engineer may be required by governing bodies for project supervision during the construction, especially if there are modifications to the building's structure. Depending on the green certification applied for, professional engineers could be required by the green building certification agency to certify the correctness of the energy calculations made by the accredited professional. These calculations are used in the application form for a green building certification. A licensed professional engineer who is also an accredited professional can, in most cases, perform the entire body of work required to achieve a green building certification.

Leed Building Certification

The U.S. Green Building Council, through a multitude of volunteer committees composed of experts in the fields covering the various areas of certification, has developed nationally recognized standards for the design, construction, and operation of new and existing buildings. This process was given the acronym LEED (for Leadership in Energy and Environmental Design). Under the category of new building construction, LEED 2009 (the current LEED version at this writing), credits were given in the following areas:

- Sustainable sites
- Innovation and design processes
- Materials and resources
- Energy and atmosphere
- Water efficiency
- Indoor environmental quality

Professionals from the various fields who have earned the LEED Accredited Professional (LEED-AP) credential review the above areas for building certification. For the latest information on LEED, contact the U.S. Green Building Council or the Green Building Certification Institute (GBCI) through their websites. Staying current with the latest developments in LEED is necessary because LEED is constantly evolving and taking on the challenges of new areas of classification. LEED requirements and administrative structure change with advances in technology and through the practical experience gained as a greater number of building construction projects achieve LEED certification.

The level of LEED certification a building achieves is frequently used as an overall building quality indicator. Organizations may compare LEED certification levels of properties when considering a possible new location for their company to lease. LEED ratings are valuable not only to the owners of a property, but to tenants as well. The certification level a building has achieved indicates a definable level of design and management professionalism, sustainability, maintainability, comfort, environmental quality, and energy efficiency. For the corporate tenant, this translates into cost savings and increased worker productivity.

The LEED system recognizes building quality by awarding points, or credits, that rate a building at four levels:

- Certified   40–49 points
- Silver   50–59 points
- Gold   60–79 points
- Platinum   The highest level and represents an achievement of 80 or more points

The LEED rating system is frequently applied to new building construction, but it can also be applied to buildings that may be decades old. This is done through the LEED for Existing Buildings: Operations & Maintenance (LEED-EBOM) credit evaluation system. It provides a way for building owners to evaluate their existing building systems and make sustainable decisions for modernization of mechanical systems and improvements in building construction. The general categories where points can be earned leading to certification are: sustainable sites (SS), water efficiency (WE), energy and atmosphere (EA), materials and resources (MR), indoor environmental quality (IEQ), innovation in operations (IO), and regional priority (RP).

At the time of this writing, LEED 2012 has not been finalized. It has also been renamed as LEED v4. Some of the changes may be additions and revisions to evaluation categories. LEED v4 will most certainly

TABLE 1.2 LEED Professional Designations

LEED Professional Designations	
LEED Green Associate	Nontechnical aspects of green building
LEED AP BD + C	Building design and construction
LEED AP O + M	Operations and maintenance
LEED AP ID + C	Interior design and construction
LEED AP ND	Neighborhood development
LEED AP Homes	Homes

add more building configurations and building usage types, as well as revisions to credit weightings for LEED points. Updated information can be obtained from the USGBC's website.

Table 1.2 lists LEED professional designations, which provide specific areas of specialization for green professionals.

Energy Star® Label*

Commercial buildings can achieve the ENERGY STAR label from the USEPA by initiating energy and water resource conservation measures. Most people in the United States are familiar with the ENERGY STAR label attached to a refrigerator, water heater, washing machine, or other energy-consuming appliance. Similarly, an ENERGY STAR labeled building must meet the EPA's standards for energy usage for the specific type of building.

To begin the process of achieving the ENERGY STAR label for an existing building, the facility manager uses the EPA's free computer program called Portfolio Manager. The FM inputs data on the building's energy and water use. The program then benchmarks the data against all buildings in its database. The program can then help the FM manage energy and water consumption, rate building energy performance, estimate a building's *carbon footprint*, set investment priorities, and verify and track the progress of energy and water improvement projects.

To achieve an ENERGY STAR label for a building, all information submitted must be reviewed and verified for accuracy by a state-licensed professional engineer. The engineer must affix his or her stamp to the EPA's form. Achieving a rating of 75 or more qualifies any one of the fifteen eligible types of buildings for the ENERGY STAR label. Buildings that are eligible, at this writing, are:

- Bank or financial institution
- Courthouse
- Data center
- Hospital (acute care and children's)
- Hotel
- House of worship
- K–12 school
- Medical office building
- Municipal wastewater treatment plant
- Office building
- Residence hall or dormitory
- Retail store
- Senior care facility
- Supermarket
- Warehouse

The EPA is also working on developing benchmarks of plant performance for auto assembly, cement, and corn-refining plants.

---

*The information in this section was provided courtesy of USEPA's Energy Star Program.

The Portfolio Manager program can also generate a statement of energy performance (SEP) for each building. It provides a summary of important energy information and building characteristics such as site and source energy intensity, carbon dioxide (CO_2) emissions, gross floor area, and number of personal computers. This information is required to apply for the ENERGY STAR label; satisfy requirements for LEED for existing buildings; support mortgage sale and lease transactions; document performance of energy service contracts; and communicate energy performance with tenants, owners, and customers.

Building Energy Quotient

Building Energy Quotient (bEQ) is a program started by ASHRAE. It provides a rating system that includes both the designed energy usage and the actual operational energy usage for all commercial building types. Using bEQ provides FMs with data on their building's energy consumption, indoor air quality (IAQ), and energy demand profiles. It provides the information necessary to achieve the goal of quantifying the energy used by the building so that energy reductions can be made. The bEQ rating system is as follows:

A+	Net-Zero Energy
A	High Performance
A−	Very Good (meets design goals)
B	Good
C	Fair
D	Poor
F	Unsatisfactory

ASHRAE has a unique record of success, with a current membership of over 50,000 worldwide and an enormous impact on the HVAC industry, setting the standards for operation and design used throughout the world.

BREEAM

BREEAM is an assessment method with a rating system that covers a building's sustainable design and environmental assessment. BREEAM was developed in the United Kingdom by BRE Global Ltd., a company who is part of the BRE Trust, which is a registered charity in England and Wales. BRE Trust's objective is to advance knowledge and innovation in all matters concerning the built environment. (BRE previously stood for Building Research Establishment, but today is the trading name of the Building Research Establishment, Limited.)

BREEAM Rating System

Unclassified	< 30
Pass	≥ 30
Good	≥ 45
Very Good	≥ 55
Excellent	≥ 70
Outstanding	≥ 85

BREEAM 2008 Assessment Areas

- Management
- Health and well-being
- Energy
- Transport
- Water
- Materials
- Waste
- Land use and ecology
- Pollution

Green Globes

Green Globes evolved from BREEAM as an online interactive assessment and rating tool provided by the Green Building Initiative (GBI) in the United States. The Green Globes rating system was published in 2010 by the American National Standards Institute (ANSI) as one of their standards. The trade name in Canada is BOMA BESt and is provided by the Building Owners and Managers Association (BOMA) of Canada.

Programs provided by Green Globes include:

- Green Globes New Construction (NC)
- Green Globes Continual Improvement of Existing Buildings (CIEB)
- Green Globes for Healthcare

Certification ratings, from highest to lowest, are:

- Four Globes
- Three Globes
- Two Globes
- One Globe

Success in the Green Globes assessment is achieved when a building reaches at least 35 percent out of a total of 1,000 possible points, thus qualifying a building for certification. To complete the process, a third-party assessor is appointed by GBI. The assessor performs an onsite building audit.

Green Star

According to the Green Building Council of Australia (GBCA), Green Star is an environmental rating system that evaluates the level of environmental design and construction of buildings. At this writing, there are 4 million square meters of Green Star–certified buildings in Australia.

All Green Star rating tools have the following categories:

- Management
- Indoor environment quality
- Energy
- Transport
- Water
- Materials
- Land use and ecology
- Emissions
- Innovation

Green Star Ratings, from Highest to Lowest

Rating	Score	Achievement Level
6 Stars	75 to 100	World Leadership
5 Stars	60 to 74	Australian Excellence
4 Stars	45 to 59	Best in Practice
3 Stars to 1 Star		Not Certified by GBCA

Green Rating Systems Summary

If there were only one quality rating system for buildings, it would be a much easier task to decide whether to undertake the process of having a building rated. The facility manager must decide which rating system best meets the needs of the facility. The major considerations are:

- Cost, time, outsourced service, and in-house personnel that will be needed to complete the project
- Prestige of the rating system that relates to marketability of the property
- Strengths of the chosen rating system must match the needed improvement areas of the facility (e.g., energy, sustainability, IAQ, carbon footprint)

- Type of building under consideration: new building or existing building needing an upgrade
- Type of rating system: specific to the type of facility or more generic in nature
- Life-cycle costs and maintainability
- Likelihood of success
- Cost to maintain the rating

It is imperative that, after achieving an ENERGY STAR label, LEED Certification, BREEAM, Green Globes, or bEQ rating, facility managers look to their facility's future. A shiny plaque bolted to a lobby wall proclaiming the achievement is a valuable marketing tool for the property and a reminder to all that the facility is serious about sustainability. But not maintaining the quality level that the designation represents renders it meaningless. Successful buildings need systems and operators that are part of a regimented plan of continuous improvement.

1.7   THE FACILITY MANAGEMENT TEAM

TWO areas will be considered: operational teams for large commercial properties and large university facilities. The positions of the team members are generally different because of the cultures, requirements, and differences in the operational realities of public and private organizations. Differences between the two reflect the entrepreneurial and profit-driven aspect of properties and the service-driven requirements of a university. Similarities are reflected in the fact that both require professionals who can plan for the future and make decisions that are sustainable.

Team Members for Large Commercial Properties

Large commercial properties are owned by institutional investors and private investment trusts. Property management is performed by companies who provide strategic financial real estate services, which generally include portfolio evaluation, property purchase, disposition, and day-to-day management. These services are provided to return the maximum profit for investors. The management company receives a fee for its services.

There are corporate managers above and below the rank of the following facility management levels. These individuals are responsible for the corporate aspects of management operation as a whole. Depending on the structure of the corporation, there may be a managing partner, partners or a CEO (chief executive officer), COO (chief operating officer), and president. Reporting to them are various financial officers and marketing and sales personnel.

The team members for large commercial office properties include the following:

- National manager of engineering
- Chief sustainability officer and sustainability manager
- Portfolio manager
- Property manager
- Facility manager

National Manager of Engineering   A title that a facility manager may aspire to above all other titles previously mentioned is national manager of engineering. Companies that manage properties across the United States generally have someone in-house who can provide direction to the entire operation from an engineering perspective. This manager directs operational policy issues, ensuring the success of the facility operation.

The engineering department is also responsible for performing *due diligence* during property acquisition and disposition, for instance. Some of the engineering staff members, such as environmental engineers, architects, civil engineers, and construction managers, possess specific professional expertise that is valuable for property acquisitions and development. Facility managers who have shown themselves skillful by having handled a variety of situations and challenges may be recruited to work in this department as well.

Chief Sustainability Officer and Sustainability Manager   When sustainability is of high importance to the organization and therefore warrants its own stand-alone department, a chief sustainability officer (CSO) is the person in charge. Several factors motivate organizations to create an entirely new department, even during poor economic times. The two main drivers are:

- Acknowledgment by the industry of the importance of sustainability
- Demands of clients who may own or rent the property (sustainable properties being widely recognized as superior)

The CSO ensures that all properties, whether owned or managed by the organization, are sustainable in all aspects. The CSO must also have a thorough knowledge of financial management and be able to present initiatives to top corporate management. The CSO has a staff but also utilizes the skills of the organization's FMs. The CSO provides the leadership and direction for the organization in the pursuit of sustainability. A major aspect of the position is to institute sustainable projects. Frequently the goal of such projects is to attain recognition as an ENERGY STAR–labeled building or as a USGBC–LEED certified property.

The sustainability manager takes direction from the CSO and ensures that sustainable projects are operating as designed. In a small organization, the sustainability manager may perform many of the same tasks as a CSO by initiating and then managing projects once completed. Small organizations may not be able to afford a new department dedicated to sustainability. In this case, the facility manager may need to act as the sustainability manager. Even facility maintenance departments with a highly skilled building engineering staff may need the direction, focus, and review of operations that a sustainability manager can provide.

Portfolio Manager   The portfolio manager is a person skilled in business administration, finance, and real estate. As the name implies, the portfolio manager is responsible for the financial success of a group of properties that may extend over a multistate area. The portfolio manager provides the strategic vision for the properties in keeping with the financial objectives of the property owners and directs the property managers, who oversee individual or small groups of properties that make up part of the portfolio. The portfolio may be composed of properties included in one or more investment funds or real estate investment trusts. For privately held companies, the portfolio manager reports to the partners who own the entire real estate organization. For public corporations, the portfolio manager reports to the top corporate officers of the real estate management company.

Property Manager   The property manager handles the financial aspects of the property's operation. The property manager reports timely financial data, which is compiled with data from other properties to create quarterly and year-end reports for investors. Property managers are generally required to have a state-issued real estate broker's license in order to negotiate lease terms with tenants, sign leases in the owner's name, and represent the interests of the property owners. It's a requirement in situations where the property manager works for a company other than the owners of the property. We refer to this as third-party management.

A property manager at an apartment complex is known as an apartment manager and must supervise the maintenance team. Office property managers must be able to design a negotiation strategy and understand the financial implications of all negotiated lease terms. They must also be able to negotiate with corporations, both domestic and multinational, because some property managers at office buildings do so on a regular basis.

Facility Manager   In large office towers or multibuilding commercial properties, the facility manager reports directly to the property manager. In addition to the obvious duties of operations, maintenance, and planning, the FM is responsible for tenant lease provisions as they relate to maintenance repair and remodeling. On some occasions, the FM is a licensed real estate broker and performs the additional function of property manager. This multiskilled person is simply referred to as the property manager or property asset manager, both of which are generally considered to be on a higher level than facility manager.

Team Members for Facilities

The facility management team frequently has more facility department positions than a property operation. This reflects the expanded needs, reporting requirements, and service levels required for a facility. There are many types of facilities and, for the sake of brevity, a discussion of the facility management team at a university is provided. Figure 1.8 illustrates the organizational structure at a typical university. Key management positions at universities include the following:

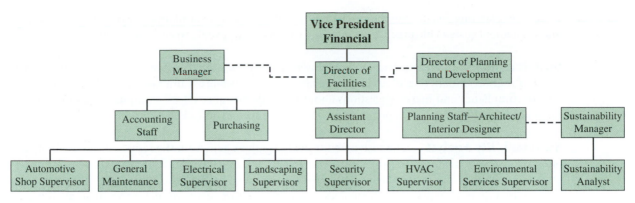

FIGURE 1.8 Organizational Chart of a Medium-size State University

- Director of facilities
- Assistant director of facilities
- Director of facility planning and development
- Vice president, financial
- Trade supervisors
- Director of sustainability and energy conservation
- Sustainability manager
- Sustainability analyst
- Environmental services manager

Director of Facilities   The director of facilities is the chief of all the trade shops. Trade shops are composed of the following tradespersons: carpenters, plumbers, HVAC technicians, electricians, general maintenance workers, power plant operators, vehicle mechanics, locksmiths, landscapers, security, and housekeepers. The director provides the strategic direction for the facility, helps develop and maintain the budget, and works with upper management during capital improvement and major maintenance projects. This position reports to the vice president, financial.

Assistant Director of Facilities   This team member aids the director of facilities and provides more direct contact with and supervision of the various building trade supervisors.

Director of Facility Planning and Development   This position may be utilized if construction projects are underway or are in the planning stage. The director of facilities planning and development will be greatly involved in the development of a university's five-year plan if the plan includes construction or remodeling. This position requires experience in construction, architecture, and/or mechanical design.

Vice President, Financial   This member of the university facility management team interacts and consults with the director of facilities on budgetary issues and on developing a five-year master plan for the university. If there isn't a director of facilities planning and development, the vice president will work directly with the director of facilities on major maintenance and construction issues. The facility manager may also work with a business manager, who handles issues dealing with the budget.

Trade Supervisors   Trade supervisors report to the assistant director for the day-to-day issues involved in running the facility, and they directly supervise the maintenance personnel. Landscape supervisors have been especially active in sustainability initiatives such as water and chemical use, composting, xeriscaping, equipment exhaust, recycling, and trash removal. *Xeriscaping* is the use of native plants that, under normal conditions, do not require any supplemental irrigation.

Director of Sustainability and Energy Conservation   The director of sustainability and energy conservation provides major program planning, develops program areas, sets goals, and provides the vision for a sustainable facility. This position works in conjunction with the director of facilities and the director of planning and development. The position also works with academic program heads to develop and implement sustainable practices and energy conservation specific to each program area.

Sustainability Manager   The sustainability manager ensures that sustainability and energy conservation program initiatives are carried out. This team member must interact with all departments to ensure compliance and make suggestions about methods for more effective implementation. The sustainability manager may have a professional designation such as an LEED-AP.

Sustainability Analyst   The sustainability analyst gathers data from the results of sustainability and energy conservation efforts. To gain control of a project, it is imperative that hard data feedback be gathered and analyzed. This is the only way to know if a program has been successful. The information can also be used to develop program efficiencies by analyzing effort versus results. Some areas analyzed are waste streams and recycling, chemical use, material conservation, electric energy and fuel use, program cost data, and water conservation.

Environmental Services Manager   The environmental services manager is the person in charge of housekeeping. This position takes on the management of green cleaning, recycling, and waste reduction initiatives. The manager works closely with facility personnel to ensure that maintenance and facility use functions will not be affected by housekeeping operations.

1.8 LEADERSHIP

Leadership is the ability to create and direct the goals of the organization, provide strategic direction, and motivate personnel to excel personally and as team members. A top-notch leader must also possess managerial skills, particularly in the areas of organization, human resources, and operational quality evaluation. These skills are necessary to keep the leader focused, functional, and effective. Great leaders excel at seeing the big picture, delegating responsibility, and motivating the team to strive toward excellence. This kind of leader creates a fruitful and productive working environment.

In discussions of leadership characteristics, a military leader is frequently used as an example. Although there is quite a bit of overlap of when discussing leadership traits, there are key differences between military leadership and corporate leadership. Table 1.3 offers a generalized comparison of military leadership versus corporate leadership traits.

TABLE 1.3 Comparison Chart Between Military and Corporate Leadership Traits

Military Leadership Traits	Corporate Leadership Traits
Leadership is part of daily life—members expected to follow without reward.	Leadership is a transaction—members receive rewards by following the leader.
Coolness under fire.	Coolness under pressure.
Leads a unit (team) where member individualism is discouraged.	Inspires followers to go above and beyond the stated mission.
Leadership is a defined hierarchy—always follows rank.	Leadership is a defined hierarchy, but leadership position does not always follow rank because members can assume a leadership position depending on needs dictated by the situation.
Authority must be visible—uniforms, saluting, formalized speech.	Authority might not be visible; it is driven by corporate title. Professional speech used.
Absolute control—orders must be followed without question.	Empowers others; stimulates creativity.
Stern personality.	Charismatic personality.
Physical punishment an available tool for the leader, as is dismissal or demotion.	Economic and status punishment—dismissal or demotion.
Establishes unconditional loyalty.	Creates a sense of purpose.
Big picture is provided by political leaders.	Creates the big picture for the organization.

1.9 MANAGEMENT METHODOLOGIES

Management is the ability to take strategic direction from the leader; put it in an organizational framework; and provide the equipment, training, supplies, and environment so that team members can work effectively toward the goals of the organization. An expert manager is able to motivate people to work diligently toward a goal without having to use formal authority. Managers make things happen, get the job done, and keep the facility running at optimum efficiency.

Management methodology provides the manager with a framework of steps to follow when managing the day-to-day operation and striving for improvement in all aspects. There are numerous management methods. We will discuss a number of methods that are often employed in facilities management.

- Sustainable facilities management
- Hands-on management
- Management by walking around (MBWA)
- Six Sigma™
- Total quality management (TQM)

Sustainable Facilities Management

Sustainable facilities management is unique in a management landscape dominated by highly specialized management positions. Its success relies on effectively blending many specialized disciplines, such as the following areas of management: human resources, construction, disaster, security, financial, building, grounds, public relations and media, contract, environmental, housekeeping, strategic, and energy. These are the multiple management specialty areas in which FMs practice on a daily basis. When executed to its fullest potential, sustainable facilities management is truly a multidisciplinary art.

Hands-on Management

Hands-on management means that the manager actively participates in the operational decisions and, at times, pitches in to help accomplish a difficult maintenance or repair task. On the other hand, hands-off management means that the manager relies on the judgment and ability of supervisory personnel to manage the day-to-day operational tasks of the facility.

Hands-on management is a frequent choice of FMs who have come up through the facility worker ranks, having built a reputation as being an excellent facilities maintenance technician or maintenance supervisor. Hands-on managers who become directly involved with project work at their facility should assess the benefits versus the costs of spending time at this level. These managers should question their true motivation in becoming so involved in day-to-day maintenance work issues. Does the manager really need to have such an active role in normal day-to-day operations, or is the manager simply trying to avoid performing difficult administrative tasks?

Hands-on managers may also unwittingly alienate supervisory personnel whose job is to manage in a hands-on fashion. Supervisors frequently work alongside FMTs to get the job done, but having a facility manager working alongside as well can be uncomfortable at times for the supervisor. Services provided by an FM come at a higher cost than if a maintenance supervisor or outsourced service person handles the task. The true value of a skilled hands-on manager is illustrated when the manager demonstrates maintenance techniques for the education of the workforce. Demonstrating techniques to technicians and supervisory staff in a hands-on fashion can provide for an excellent learning experience and enhance camaraderie on the job.

Management by Walking Around

The style of management by walking around (MBWA) means that the manager walks through the facility to observe the actual day-to-day operational activities of the facility personnel. MBWA isn't hiding in the bushes with binoculars trying to catch someone not performing the required work. MBWA is a method of management that should be planned, and the specific goals that the walk should achieve must be well defined. Notes must be taken during the walk so that, on subsequent walks, the FM can see if improvements have been made or if a particular situation has deteriorated. Notes should be dated, and locations the manager walked to should be recorded. The weather and seasonal conditions at the time of the walk and any prior conditions that may be relevant should be noted. For example, when reviewing the landscaping at a shopping center, note the amount of moisture that has fallen over the past few days. If the site is experiencing a period of drought, note that, too, so there is a fair representation of the landscape condition in comparison to the environmental conditions. Incomplete notes may actually provide information that misdirects the reader. If the poor landscape condition is noted without the drought information, a subsequent evaluation that again shows the landscape to be in poor condition might result in a conclusion that the condition stems from a lack of proper maintenance, which would be an inaccurate conclusion.

MBWA can be an effective way to manage a facility, but it is time intensive and too often degrades into something no more valuable than the so-called stroll in the park. To make the best us of MBWA, keep the following tips in mind:

1. Define the places most important for you to visit.
2. Provide supervisory personnel tasks (a sort of mini-walk) on a scheduled basis so that you don't have to walk the entire facility every time.
3. Break down walks into specific areas.
4. List the objectives that you hope to achieve during the walks for the site visited.
5. Record information and provide a rating system for every situation reviewed.
6. Inform personnel in the areas reviewed of the conclusions and obtain their input.
7. Review the information, define corrective action, and plan for implementation.
8. Follow up to see that the corrective action achieved the required result.

Example: Epitome of MBWA

A housekeeping manager at a very prestigious university jogs through the entire campus at a very quick pace every morning Monday through Friday, starting at 6:30 a.m. This allows the manager to see firsthand the efforts of his nighttime staff and write work orders for his daytime staff for any items missed. He first checks the high-profile areas that the university's decision makers would see first. The manager then flies through dormitories and other high-use areas during off hours. After that, the academic wings are inspected. Last, the manager goes through supply and equipment storage and distribution areas. The goal isn't to count supply items; an automated inventory system provides that information. The housekeeping manager is double-checking to make sure chemical containers are properly labeled, and that there are no obvious problems with the stored chemicals.

Walking through storage areas also gives the manager knowledge of how staff members keep up with organization and cleanliness of their own workspaces. The manager understands that good workers frequently direct all their efforts for the benefit of the facility and expend minimal effort on areas from which they themselves would benefit directly. He doesn't want his workers to lose the positive recognition that their hard efforts have previously achieved, based on a school administrator stumbling across a messy supply storage area or break room, for instance. The housekeeping manager's efforts are focused on evaluating the job being done, before the customers (students, teachers, visitors, and administrators) make their own evaluation.

During the rest of the day, he can take the information he has accumulated and plan for action. The plan was more than just a series of work orders or lists of needed supplies. MWBA provided the information to implement additional worker training where needed, worker wage increases, worker disciplinary action, governmental regulatory compliance, and worker safety. The data also provided him with information for self-evaluation, well in advance of any future evaluation by upper management.

Six Sigma™

Six Sigma is a management method created by Motorola in 1981 to improve its manufacturing operation. Because of its roots, Six Sigma is frequently practiced at manufacturing facilities. Six Sigma is a registered service mark and trademark of Motorola. The Six Sigma performance level is achieved when there are only 3.4 failures per million operations. Stated another way, Six Sigma translates into a 99.99966 percent success rate. This low failure rate is achieved by a DMAIC process (DMAIC stands for *d*efining, *m*easuring, *a*nalyzing, *i*mproving, and *c*ontrolling).

Six Sigma affects all operations on a continuous basis. The Six Sigma process strives to achieve predictable results consistently for any task that is undertaken. It utilizes management strategy, or review and reevaluation, and utilizes statistical tools to achieve the result level sought. Six Sigma, with its emphasis on production, manager level, and specific results that can be duplicated, differs from other methodologies in several ways. Another notable difference is that Six Sigma defines the level of competency of the participants by using terminology borrowed from martial arts. There are four Six Sigma skill levels: master black belt, black belt, green belt, and yellow belt. The process involved in gaining belt status is extensive and involves training, mentoring, teaching, coaching, experiment design, and project supervision. More information can be obtained from the American Society for Quality website.

Total Quality Management

Total quality management (TQM) is a management style with the goal of establishing a process of continuous improvement in order to generate superior results constantly. TQM's major emphasis is in improving the quality, efficiency, and cost of goods produced in a manufacturing environment. TQM is what propelled Japan to prominence in manufacturing in the 1980s. Facility managers can practice the TQM process by first discovering the specific product. For FMs at office properties, the product that is sold is usable space. The space that is sold must be comfortable, regardless of the environmental conditions outside, and it must be functional. Space functionality means that the space provides a productive and profitable environment for tenant operations. So for a manager practicing TQM in the built environment, the goal is to provide a sustainable space that meets all stakeholder needs. This is an FM's greatest challenge.

There are four main processes that must be applied when FMs want to institute TQM as a management style: *kaizen, atarimae hinshitsu, miryokuteki hinshitsu,* and *kansei.*

Kaizen: Measurable Outcomes   Kaizen is a process whereby the facility operation strives for perfection in all endeavors on a continuous basis. The process must also provide results that are measurable and justifiable improvements. Success lies not only in showing constant improvement but in being able to track the improvement, measure it, duplicate it, and improve upon it. In a facility, kaizen may be applied to everyday functions such as the elevator service, as we will see in the next example.

Example: Elevator Service

The goal of an archaic elevator management system was simply to operate with a couple of breakdowns expected in the course of operation. The old systems did not tabulate any data or record any faults. Elevator technicians underwent the painstaking process of running the elevators up and down to find hidden faults. This involved taking control of the elevators for several hours during a nighttime or early-morning period when there was little or no need for elevator service.

Years later, when elevator controls became computerized, our elevator technical team could review the data logged. They could now zero in on the necessary corrections to improve service. No longer did technicians have to guess at wait times or the number of missed stops; the data was compiled and ready for use for meaningful operational and mechanical corrections. These actual operational logs can also improve service by making elevators *smart.* This is done by programming elevators to head to floors where service will likely be needed, rather than always waiting for a call button to be depressed by a passenger waiting in the elevator lobby. More data offers more control and, with that, improved service.

Atarimae Hinshitsu: Elimination of Defects   Atarimae hinshitsu focuses on the elimination of defects so that things work as designed. This is done so a product or service operates as intended (e.g., bicycle tires rotate, buildings provide heat, pens write). This can be referred to as the make-it-work step.

Miryokuteki Hinshitsu: Make It Work Well   Miryokuteki hinshitsu stresses that, not only should things work, but the user of the product should enjoy using it. The product should be aesthetically pleasing and create an enjoyable experience. Not only does the pen write, but it looks good and feels comfortable in your hand. Not only does the building provide heat, but the visible heating vents fit nicely into the surroundings and the heat is warm and comfortable, not blasting out of an ugly, dirty grill. This is the make-it-work-well step.

The kaizen section used an example of elevator service improvement; elevator service can also be used to explain miryokuteki hinshitsu. It is why elevator music is piped into many elevators. The attempt is to make the elevator ride experience more pleasing to the user. Although better than hearing the elevator noises and the rumblings from fellow passengers, most riders would agree that using elevator music is an effort that has met with limited success.

Kansei: Actual Use and Customer Needs   Kansei is an examination of how tenants and customers actually use the facilities so improvements can be made based on use and need. The FM must understand how the customer sees service. The FM might be devoting an extensive amount of effort for the tenants and still not be meeting their needs completely.

Example: Storage Space

A tenant in an office building enjoys being a tenant in the building and wants to renew its lease. But instead the tenant has decided to leave the property at the end of its lease because the costs are just too high for its business, which is now experiencing declining revenue. The FM visits the tenant's office and discovers that the tenant has a huge amount of files in tall cabinets that occupy a large room and some open areas. The tenant also has clerical personnel occupying several small, private offices. The tenant has purchased high-value office space but is using it inefficiently by wasting it on storage and having clerical staff in private offices instead of in more space-efficient cubicles.

The FM first speaks to the property manager, then offers a solution to the tenant to downsize the current space by providing low-cost basement storage for files that are not in regular use. The FM also suggests that clerical workers can be relocated to smaller cubicles, which will save space and provide for easier interaction among staff members. This whole process is the practice of kansei. The FM understands how the customer uses the product (office space) and has offered suggestions to modify it to meet the customer's need.

To sum it up, the four quality processes of TQM are:

1. Kaizen (change for the better): continuous Improvement

   The entire process of striving for perfection.

2. Atarimae hinshitsu (first to discover, then to eliminate defects): elimination of defects

   Things must work as designed, without failures or poor operation.

3. Miryokuteki hinshitsu (quality that fascinates or enchanting quality): make it work well

   Provide a pleasant user experience. Using the product or service should be an enjoyable experience.

4. Kansei (the five senses): product's actual use

   How the product or service is used and customer's service viewpoint. Knowing how customers use the product or service and understanding how the customer views the service. Basically, kansei is akin to walking a mile in the shoes of the customer.

TQM from the Ground Up

The first step is to collect data, which is done by sending a survey to the building's tenants, customers, and contractors. Some contractors provide their services at many similar facilities and can be a good source of **benchmarking** information. The data need to be compiled in tables or graphs and analyzed. At this point, the FM is working on the process of atarimae ahinshitsu, the steps of which are first to discover and then to eliminate defects.

The survey should not be filled out anonymously, under the assumption that more honest information will be provided. This may be true, but at a minimum the FM needs to know the job title of the person filling out the survey; the option of providing the actual name of the person and contact information for additional feedback should be provided. For instance, the most valuable information would be obtained from a corporate executive if the question asked is, Does the leased space meet the corporate objectives of your organization? The FM also needs to know if the answer applies to the entire company or just to a narrow subdepartmental area in which the person filling out the form works.

The FM should target specific persons to answer the questions so that he or she obtains more meaningful answers and should ask that the form be given to the persons best able to answer the questions. The FM should also ask where the person works inside the building. Location data can be very helpful if the survey comes back indicating heating or cooling problems. It could signal a building-wide problem, a problem affecting only one floor, or a problem limited to a specific location. Once facility problems are dealt with, the FM can start working on making the experience a pleasant one (miryokuteki hinshitsu).

Example: HVAC Monitoring and Control

For many years, building management was opposed to allowing people working in the building to have any control over temperature or airflow within their space. This was prior to the advent of direct digital control (DDC) for HVAC monitoring and control. With pneumatic-controlled equipment, tenants were told to wear the appropriate clothing or bring a sweater to work. The building management would try to provide an overall 72 degree Fahrenheit space temperature, plus or minus 2 degrees. The building management couldn't allow the tenants any flexibility with regard to temperature control because no feedback was provided about actual space temperature.

If a tenant wanted his office warmer, he turned the temperature dial on the thermostat. If he turned it too high, the room would get hot. Raising the temperature setting also raised the temperature in the other rooms that were served by that thermostat. Other tenants might now be too warm. If a complaint came in that the heat was too high, a technician would have to be dispatched to the complaint location. The technician would then have to track down the location of the thermostat and test it, only to discover that the problem was caused by someone turning the thermostat temperature setting up too high. A technician could easily waste the entire day with temperature complaints resulting from people adjusting their thermostats.

Having all building thermostats set at 72 degrees made troubleshooting easy, with quick dismissal of warm or cold complaints. If a technician walked into a room that the tenant claimed was too hot and the temperature was 74 degrees, the space was within the plus or minus 2 degree tolerance and the technician could dismiss the complaint: It was a case of someone getting "overheated" due to the business events of the day and not because of an issue with the HVAC equipment.

A new building computer system based on DDC provides a central location from which to view temperature data from the spaces within the building and compare them to the thermostat's setting. Now the first step in troubleshooting is to view the temperature setting on the computer for the space in question and compare it to the thermostat's set-point temperature. If the space temperature is incorrect for the thermostat's setting, then the technician knows immediately that further investigation is necessary. With the thermostat shown on the floor plan on the computer, the technician is no longer required to go to the mechanical blueprints or to search the building for the location of the thermostat. These tools help to eliminate defects quickly, saving time and money.

Maintenance technicians might not be able to dismiss complaints so quickly with direct digital control. Instead, they can check the other parameters that affect comfort, such as air movement, humidity, and purity, because the complaint about the temperature being too warm might only be a symptom of the actual problem. DDC systems give the technician additional diagnostic tools.

TQM and New Tenants

An FM at an office or industrial property should ask to sit in on some lease negotiations. The FM should also ask permission to be in attendance when the property is shown by real estate brokers to potential clients. In this way, the FM can learn firsthand the needs of these clients. The FM should also visit a potential client's present location. The FM can then learn what the tenant's needs are by actually observing and noting its ongoing operation. The FM might be uncomfortable doing this because he or she is not an expert in the client's business. The FM can obtain expert assistance from consultants in the business field in which the client operates, or from consultants skilled in the general field, whether it is manufacturing, packaging, chemical or petroleum operations, or distribution. Should the client's operations generate hazardous waste, it is imperative that the associated risk be evaluated thoroughly.

Example: Visit to a Manufacturing Operation

An FM is visiting a client with a manufacturing operation. The FM observes the flow of materials: loading into a facility, storage location, and distribution to the processing machines. The FM scrutinizes how wastes and recyclable materials are processed and stored. Often there are obvious deficiencies in material handling and department location that the client may have learned to live with after many years at a facility. Sometimes there is a problem with how materials flow through the plant.

Problems the FM notices include the following: parts being stored in a location that is too far from the manufacturing area, a storage area for waste that is neither safe nor secure, poor ventilation and heating, a ceiling that is too low to allow for maximum storage, and a space that is too small for safe and efficient operations. The potential tenant might no longer have to live with these deficiencies if it moves into a space that is specifically designed for it at the FM's property. The FM is presented with a great opportunity to show the client the many benefits of relocating by gearing the space specifically to the client's needs.

Remember that many clients may not have construction experience or the ability to read blueprints. Therefore they might not fully understand what the FM's company can do for them, or they may fear the cost of improvements. The information about the suggested efficiency improvement can be expressed in block diagrams. Block diagrams visually define the client's operations in a sequence leading to the final delivery of the manufactured good (see Figure 1.9). A basic floor plan of the FM's property showing how the client's operations would fit into the space can garner much interest from a potential client. It is true that some companies can be fairly successful, even if they aren't operating at peak efficiency. But the FM can show the client how it can be more efficient and ultimately more profitable in a custom-designed facility.

1.10 NEW MANAGER: TAKING OVER THE PROPERTY OR FACILITY OPERATION

Once hired by an organization, the new FM must have a plan for successful transition into the new position. Along with an understanding of the type of facility and its culture, the new FM must be aware of the past status of the facility management operation as it relates to the new position. Did past events provide for a positive transition into new management or a painful one? A new manager needs to understand these factors, so she or he can better understand the reasons behind interactions with fellow employees

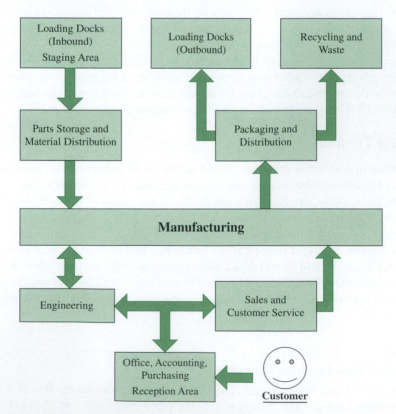

FIGURE 1.9 Workflow Block Diagram for a Tenant Engaged in Manufacturing

and customers. In addition, the new FM must understand the expectations of upper management to correct any deficiencies.

Commencement of the New Position

There are four possible paths to commence a position as the new FM: cold start, warm start, hot start, and legacy start.

Cold Start   The facility manager is brand new at the facility, or the previous manager has left the facility prior to the arrival of the new facility manager. The new FM must hit the ground running, but she or he can build a new operation based on the best practices in the industry.

Warm Start   The prior facility manager has not yet left the facility and will be spending several days with the new FM to help make the transition. The manager may have found a new and better position outside or inside the organization. This can be a great situation for the new incoming manager who can ask questions and also take time to consider a particular situation and ask more questions later.

Hot Start   The prior manager was fired outright or given the option to quit for committing a serious infraction. The option to quit might require the prior manager to spend a few transition days with the replacement manager. This undoubtedly will be an awkward and uncomfortable situation for both the incoming and outgoing manager. The new FM must try to get as much information from the outgoing manager as possible by asking direct questions. It is easy for someone to avoid answering indirect questions.

If the outgoing manager was fired outright, the FM will need to move quickly to get the facility under control. A dismissal situation whereby the prior manager committed sexual harassment against an employee, for example, will require sensitivity and quick action on the part of the incoming manager to quell any hard feelings or lingering mistrust by facility personnel. The new FM must do this to prevent the prior manager's misdeed from affecting the new manager's leadership.

A hot start is an uncommon situation. A more likely scenario is that upper management will be interacting with the manager who resigned.

Legacy Start   A legacy start is the situation when the facility manager is retiring and will be spending a few days with the new manager to help in the transition. This is different than a warm start because the retiring manager will frequently try to ensure that his or her legacy continues. The new FM should be sensitive to this need and shouldn't, all at once, discuss the many changes the FM plans to undertake. It may make the retiring manager feel that the valued legacy of years of operational success will be tarnished when proposed new plans become a reality. The retiring manager is an extremely valuable source of facility information and can also provide valuable input about the new innovative ideas. The retiring manager usually has a database of the names of helpful vendors, insights into personnel issues, and great knowledge of the culture of the organization, and might even provide the new FM with some outright wisdom gained from years of experience.

Checklist for New Managers

The following are issues new FMs, in their first days as manager, should discuss with their superiors, employees who report directly to them, and other support staff. The three most important areas in which to seek information are:

- Personnel
- Emergency procedures
- Operations, including access and security, fire alarm systems and the building automation system (BAS), the elevator system, the emergency generator, building equipment operation, work management, lease requirements, budget and purchasing, and contracts

Personnel: Understand the "Who's Who"   In meetings with upper management during the hiring process, new facility managers are frequently given organizational charts and a description of their duties and responsibilities. The FM must understand the reporting relationship with their immediate supervisor and upper management, and between members of the support staff. The FM shouldn't begin a new job by stepping on someone's toes or wasting time doing a job that isn't his or her responsibility. In addition, the FM needs to find out the staff members' abilities and capabilities for performing work under normal and emergency situations.

Emergency Procedures   Disaster can strike the facility at any time. For the FM at an office building, the best place to gain information in preparation for a disaster is the fire plan book or emergency action plan. The fire plan book is an overview of the building and its firefighting systems. It has single-page-size floor plans and piping layouts of sprinkler and fire protection systems. It shows the emergency exit routes for the entire building. The fire plan book is designed to provide first responders with information they may need about the property. An emergency action plan is a document listing the actions to be taken for various emergencies that might occur at the facility. For example, if chemicals are used and stored, an emergency plan will list the actions to be taken and the numbers to call should a spill occur.

After reviewing these documents, the FM should review the actual construction blueprints and tour all emergency equipment locations. The facility manager must also learn how the building's emergency equipment operates, how it is tested, and how it is best secured when not in use.

Emergency Equipment and Operational Procedures   An FM taking over a facility needs to have complete knowledge of the following information and procedures:

- The location of fire and sprinkler equipment and alarm panels, including shutoffs
- The procedure for silencing and resetting alarm panels
- The on- or off-site location of the emergency operations center, which houses communication and safety equipment and acts as a command center in times of emergency
- The availability of emergency communication equipment and the person responsible for its maintenance and distribution during an emergency
- The emergency phone numbers that need to be updated, which include repair, recovery, and restoration services

Access and Security   Keys and other building access devices are usually the first items a new FM receives after a welcome handshake. Frequently, the person handing out the keys might know where only some of them are used! Meeting with a maintenance supervisor to sort all the keys can be very helpful. It is also important to understand the key systems or the card access system employed.

An important question about building access is, How are the building access systems bypassed in order to gain entry during an emergency?

Fire Alarm Systems and Building Automation System (BAS)   Learn how the fire alarm system generates alarms and how the building automation system (BAS) responds to these alarms and handles smoke management. Does the fan system shut down or does it continue to operate to provide positive pressure for smoke-free exit stairwells? Does the ventilation system go into a smoke-management mode to remove smoke from certain areas of the building? This is important information in an emergency.

The new FM will need a password and should have a supervisory level of access to any computerized equipment at the facility in order to make system changes as needed. The FM needs to delete access to any previous employees who may still be able log into the BAS. The FM must also be certain that all current employees have an appropriate access level that meets their job requirements and computer skill level.

Elevator System   A fire call occurs when an elevator system gets a fire signal and the elevator is sent to a location, generally the lobby, where it stays until the alarm is cleared by alarm panel reset. Some elevators go to a secondary location if the alarm came from the lobby area. Generally keys are used to put the elevator into manual or fire service and a key of sorts to unlock hoistway doors. The new manager should also know where any elevator keys or door-opening devices are kept or should carry these keys with other access keys.

Any situation in which an elevator becomes stuck between floors is extremely dangerous. The elevator has the potential of moving while an attempt is made to remove trapped persons, especially if power has been left on. The elevator shouldn't be able to move when its doors are open because of the braking mechanism. But it is not wise (or safe) to rely on a mechanical function when a different mechanical function has already failed. The elevator should not have stopped between floors! Only trained elevator technicians or trained firefighters should attempt getting passengers out of a stuck elevator. For this reason, elevator maintenance contracts frequently specify a period of time that may pass before factory-trained elevator mechanics must respond to an emergency call at the facility.

Unlike the movies, elevator passengers are rarely pulled out through the hatch at the top of the elevator. Passengers are more often removed from an elevator stuck between floors by securing power to the elevator in a safe and verifiable manner, inserting a door interlock-release tool into the hoistway door, and sliding the doors apart. Passengers are pulled out of the elevator to the upper floor. Pulling passengers out using the lower floor would be dangerous because the elevator shaft below the car is exposed on that floor.

When establishing emergency policy and procedures for elevators, a new manager should set up a meeting with the facility maintenance staff, building security personnel, fire department safety personnel, and elevator service staff. The goal of this meeting is to clarify existing procedures and write a set of operational procedures to be followed in the event of an emergency. The personnel involved and the actions to be taken during an emergency should be defined. Existing operational procedures in need of modification should be reviewed. The specific operation, requirements, and troubleshooting procedures of the building's elevator equipment should be addressed at this meeting. The location and access to equipment can also be reviewed.

Another point to discuss is how and when an elevator should be put back in service after a problem, however minor, has been corrected. Should the elevator be taken out of service until a thorough evaluation can be made of the reason for failure, or can normal operations be resumed? All of these issues should be discussed with the elevator maintenance service provider prior to an actual emergency.

Emergency Generator   The new manager needs to know which equipment runs off of the power produced by the emergency generator (see Figure 1.10), which in turn lets him or her know which electrical equipment should be available during a power failure. The new manager also needs to know how to start the emergency generator (see Figure 1.11) and how to secure it. The manager should be aware of how power transitions from emergency power to the incoming power lines (see Figure 1.12) once the emergency has passed. The manager should find out the length of time the emergency generator will operate on its fuel source and how additional fuel is supplied. Additional storage tanks may exist onsite, or perhaps additional fuel must be purchased and brought to the site. The length of time it takes to acquire additional fuel is important for a successful disaster recovery operation.

Building Equipment Operation   The FM should understand how all major equipment starts and stops and also know the equipment's operational parameters. A written checklist for the proper starting

FIGURE 1.10 Emergency Generator
Source: Courtesy of Oconomowoc Memorial Hospital.

FIGURE 1.11 Emergency Generator Start Batteries
Source: Courtesy of Oconomowoc Memorial Hospital.

FIGURE 1.12 Emergency Power Transfer Switch
Source: Courtesy of Oconomowoc Memorial Hospital.

and operation should be on file for each piece of equipment. Chillers, cooling towers, boilers, pumps, and air-handling units should have this information readily available. The new FM should ask which equipment, if any, is not operable at the computer terminal for the BAS and therefore must be started or secured locally by control equipment mounted on the machine or at a remote motor control center.

Work Management   The new FM should obtain the work schedules for the facility workers that includes their responsibilities. How are weekends or after-hours work handled? The FM should learn how to manipulate the software used in the computerized maintenance management system (CMMS). Work-order generation and status lookup is important to keep the flow of work going and to investigate the reason(s) for outstanding work orders.

Lease Requirements   Leases are negotiated contract documents that become legally binding on the parties who sign them. The requirements placed on the landlord and the tenant can vary from lease to lease, even at the same property. The FM should ask if there is a document that describes the special requirements of leases that are in place and ensure that both parties meet their obligations.

Budget and Purchasing   Knowing the amount of money available for each account and how to access the budget information is important. This information is closely associated with understanding the organization's preferred methods of purchasing goods and services. Utilizing an accounting system directly may require some training. It is helpful to learn the system and software as soon as possible by scheduling a meeting with accounting personnel.

Contracts   A new FM should ask if there is a list of contractors currently performing work at the facility and review all service contracts. From the moment a new manager walks in the door, he or she is responsible for all previously initiated contract work. The FM must verify that all the required paperwork is correct (for instance, contracts are signed and insurance certificates are on file) and that the work is being performed as it was specified. A new manager may sometimes find it is necessary to stop the ongoing work and take the project in a different direction before the situation becomes intractable. The FM should review the contract to determine the most appropriate method provided by the contract agreement for terminating the contract. It may also be necessary to seek legal advice before terminating the contract entirely, adding work to the contract, or stopping work.

REVIEW QUESTIONS

1. Identify each of the following structures as a property or a facility:

   Apartment building   _____

   Commercial office building   _____

   Bank building   _____

   Courthouse complex   _____

   City hall   _____

   Army base   _____

   Residential shopping center   _____

   School building   _____

2. The quality of office buildings is defined by _____.

3. The style of management whereby the manager actively participates in the ongoing work at a facility is _____.

4. The style of management that requires the manager to constantly update his or her knowledge of the workings of the facility by visiting the work areas and visually observing is _____.

5. The four LEED levels a building can achieve are _____, _____, _____, and _____.

6. _____ of an organization affects the way a property or facility is managed.

7. The British refer to buildings and the interior spaces they provide as the _____.

8. The Japanese refer to the continuous improvement process as _____.

9. Nonrentable space that is accessible by tenants is referred to as a building's _____.

10. A person who performs lease negotiations and has a state-issued real estate license is known as a(n) _____.

11. Knowing the provisions of the various leases that apply to the facility management operation is important because _____.

12. List the sustainability job titles at large commercial properties and university facilities. _____.

13. Before using the MWBA method an FM needs to _____ to be effective.

Essay Questions

14. You are a manager at a high-rise office building of twenty-five stories and with over 200 tenants (companies and organizations). All leases state that the building must provide an interior tempera-ture of 72 degrees, plus or minus 2 degrees. How can the FM use the four elements of TQM to meet the requirements of the lease and to go beyond lease requirements? Is it is always appropriate to provide a 72 degree space temperature?

15. For an FM at a school district, the cost of maintenance service was continually increasing. The FM wants to improve the maintenance staff's efficiency. The staff has been operating in the *janitorial mode* (performing mostly low-skill-level work such as cleaning, emptying trash, setting up rooms, changing lightbulbs, etc.) for many years. Discuss what the FM can do and how the FM should apply TQM to this situation. What problems may arise when the FM tries to implement change?

16. Why do some organizations have a manager whose job is dedicated to ensuring sustainability rather than passing that responsibility on to their FMs?

SUSTAINABLE MAINTENANCE OPERATIONS

INSIDE THIS CHAPTER

Can Operations Management Provide Greater Building Sustainability While Meeting the Needs of the Stakeholders?

Operations management is the art of keeping the facility running while meeting the requirements of tenants, customers, and staff. Operations management strives to provide a comfortable, healthy, safe, and productive environment for all occupants of the built environment through the efforts of a skilled workforce. A major portion of the job responsibility for most facility managers is successful maintenance and repair management. Facilities employ work-order systems to issue the work, record the results, and provide data for future planning and budgeting.

This chapter discusses the different types of work orders and how the computerized maintenance management system (CMMS) has greatly increased operational effectiveness. Different maintenance strategies are discussed, and how maintenance contributes to sustainability is explained. The chapter ends with information on property conditions surveys and how these surveys contribute to effective maintenance management.

2.1 WORK-ORDER SYSTEM

Work orders are requests to rectify a problem or to have work done. Work orders are generated internally after property condition assessments made by the FM or designated staff, or from tenants, customers or other stakeholders in the property. The work-order system is the method employed to track work orders and their completion status, provide work-order records, and evaluate worker productivity.

A work order that is open has not yet been completed; a work order that is closed has been completed. The act of informing the work-order system that a work order has been completed is known as closing out a work order. Work-order systems may interface with inventory systems to make corrections to the number of parts on hand if parts have been used to complete the work order. The system also interfaces with a preventive maintenance database that can automatically print out work orders according to maintenance requirements as specified by the equipment manufacturer.

The following steps in a work order system begin when a work order is submitted:

1. Work requests are received by telephone, e-mail, direct entry into an electronic work-order system, or paper request forms.
2. Determine a plan for the work to be done. This is especially necessary for a substantial job that causes a large expenditure of funds, uses several maintenance personnel, or causes a disruption to facility operations.
3. Schedule the work and dispatch the appropriate personnel.
4. Monitor work through completion.
5. Deduct material used from inventory.
6. Close out the work order and, if appropriate, submit a bill to the tenant, department, or cost center for whom the work was provided.
7. Evaluate job quality and personnel performance.

The goal of a work-order system is to meet maintenance and tenant needs in an organized and timely manner. Ultimately, a good work-order system aids in providing sustainable maintenance practices for the facility.

Computerized Maintenance Management Systems

Automated work-order systems are often referred to as computerized maintenance management systems (CMMSs). These systems can be very basic: They may handle work orders only. Some systems can be expanded to include other aspects of facility operation by purchasing and programming various modules that can handle additional tasks.

Proper facility maintenance is extremely important. The capabilities of a CMMS greatly aid in the operation of the facility in many aspects, for example:

- Assigning work orders
- Tracking work orders
- Retrieving information from building automation systems through the use of the data acquisition function
- Assigning personnel in responding to automated equipment alarm situations
- Monitoring worker productivity
- Tracking parts and equipment use (see Figure 2.1 for an example of a physical inventory worksheet)

Physical Inventory Worksheet   10/19/2012

Next Physical Inv.:

Stock Location   MAIN

Part No.	Part Name	Location	Shelf/Bin	Qty On Hand
GMBTEST	GMB			_____
MISC. PARTS	MISC. PARTS			_____
OIL1040	VEHICLE OIL			_____
PBL001	PUSH BUTTON, STANDARD	STORES	S4/B4	_____
T400	COOLANT			_____
UREA1	UREA1			_____

Stock Location   TEST

Part No.	Part Name	Location	Shelf/Bin	Qty On Hand
GMBTEST	GMB			_____
T400	COOLANT			_____

Next Physical Inv.:   3/31/2013

Stock Location   CENTRAL

Part No.	Part Name	Location	Shelf/Bin	Qty On Hand
HSE001	HOSE PACKAGE	RACK STORAGE	50/6	_____

Next Physical Inv.:   10/17/2013

Stock Location   MAIN

Part No.	Part Name	Location	Shelf/Bin	Qty On Hand
OIL00I	HYDRAULIC OIL	TOTE STORAGE	T700	_____

FIGURE 2.1 Sample CMMS Physical Inventory Worksheet

Source: Eagle Technologies.

- Issuing chargebacks of the costs to various parties who requested the work, such as other departments within the organization or a tenant in the commercial office building
- Controlling inventory and reducing waste
- Gathering budget data
- Scheduling preventive maintenance

2.2 ASSIGNING WORK ORDERS

Corrective maintenance work orders are easily generated by a CMMS program after keying in the work-order request information. A maintenance tradesperson can then be assigned according to the type of work.

A CMMS can also store and input maintenance requirements to generate preventive maintenance work orders for all facility equipment and assign work orders to meet these requirements. But assigning work orders to the appropriate worker requires some additional thought. If a facility's personnel includes a variety of trade-specific workers (for instance, electricians, boiler operators, carpenters, locksmiths, auto mechanics, plumbers, electricians, landscapers, heating, ventilation, and air-conditioning (HVAC) technicians, refrigeration technicians, and painters), then the task on the work order is matched to the tradesperson. In many such arrangements where workers are unionized, the union contract requirements dictate who should or shouldn't be issued a particular work order. The work-order system can automatically issue work orders per contract and trade requirements.

Some CMMS programs can be tied into the building automation system. When this is done, preventive maintenance work orders can be based on actual equipment runtime data rather than on the frequency of maintenance on the calendar (weekly, monthly, quarterly, yearly). The calendar maintenance schedule is frequently based on the assumption that a piece of equipment will be operated an approximate number of hours every day (for instance, air handler operating twelve hours per day). If the piece of equipment doesn't run as long as the estimate, you will be performing preventive maintenance ahead of schedule. Basing preventive maintenance on actual runtime can greatly reduce costs.

Assigning Work Orders Utilizing Worker Trade Specialization

Most nonunion commercial buildings generally do not have a rigid structure of worker titles based on trade specialization. A facilities maintenance technician (FMT) in this situation is non-specialized and is expected to be able to handle a wide variety of tasks. The FMT must be able to perform several different tasks, each of which could be assigned to a specialized tradesperson. The facilities maintenance technician is truly a jack of all trades. However, FMTs naturally specialize in an area they are interested in or can do well. Managers seeking to get the most work completed at the highest quality level issue the majority of the work orders in a particular trade area to the person who has specialized in that area: Joe likes to do carpentry and Mary likes plumbing, so Joe gets the bulk of the carpentry work orders and Mary gets most of the plumbing work. Although effective in the short term, in the long term it is best to keep FMTs' skill level up in all trade areas and not pigeonhole them into a few areas of expertise. Performing a variety of work is a great way to maintain and increase proficiency. All these factors can be input into a CMMS so that the work order assignment fits worker specialization and facility needs, and maintains worker skills in all trade areas.

2.3 TYPES OF WORK ORDERS

The five basic types of work orders employed at a property or facility are as follows:

- **Corrective maintenance**—fixing what is broken
- **Project work**—work that may take several days and is done to improve the facility
- **Preventive maintenance**—work done to keep a piece of equipment in good operating condition and to prevent failures
- **Predictive maintenance**—directing maintenance personnel to utilize diagnostic tools to evaluate equipment condition (The information obtained is used to adjust the frequency of preventive maintenance service work.)
- **Lean maintenance**—striving to reduce the frequency, and amount, of consumables employed in preventive maintenance service (For example, work orders following lean maintenance principles direct maintenance personnel to evaluate the existing conditions before replacing parts or changing lubricant.)

Corrective maintenance involves troubleshooting equipment malfunctions or making necessary adjustments that require the use of tools, equipment, and supplies. Project work is generally done to improve a condition rather than correct a failure. Project work might be the installation of racks of shelving for more efficient parts storage, or painting the floor of the mechanical room. Anything that takes significant effort to make an improvement is project work.

Preventive maintenance is achieved by issuing work orders that meet the maintenance requirement for a piece of equipment based on the equipment manufacturer's maintenance schedule. Cost effective preventive maintenance has been achieved through the additional use of predictive maintenance. Predictive maintenance is achieved by using monitoring and testing equipment to adjust the preventive maintenance schedule to meet the requirements of the equipment more accurately. Simply stated, lean maintenance is a further improvement on predictive maintenance, achieved by reducing material waste in the maintenance effort.

More detailed descriptions of corrective maintenance, preventive maintenance, project work, predictive maintenance, and lean maintenance are provided in the following sections.

Corrective Maintenance

When something breaks, the efforts to repair it are called corrective maintenance. There are various priority levels of corrective maintenance. The highest priority level is emergency maintenance, which must be performed immediately or further damage will result. A flood due to a broken pipe would be considered emergency maintenance. A dripping water leak on a rusty pipe might also be considered emergency maintenance if the pipe is badly rusted and if a break in the pipe would cause serious damage. The leak itself might not be significant, but it should be treated as an emergency if the potential for damage is high.

Complaints from high-profile individuals or people operating equipment in important areas are not considered emergency maintenance and instead fall to a slightly lesser category of do it now (DIN). Items in this category rise to the top of the list to promote goodwill or to prevent problems that might affect the bottom line. DIN maintenance requests are most likely phoned in to the maintenance coordinator, who then dispatches facility maintenance personnel. More often than not, DIN requests do not come with an official work order and must be entered into the work-order system at a later time.

Other categories of work order priority are high, medium, and low. These levels of importance inform maintenance personnel which order should be performed first.

Project Work

It may seem impossible, but there can be slow times at a facility when maintenance technicians have caught up with most of the work orders. During these periods, improvement projects can be undertaken by facility personnel. Project work is its own work-order category. Some examples of project work including the following:

- Replacing light fixtures with new energy-efficient ones
- Installing new ceiling tile and cleaning the grid
- Painting equipment, piping, or the mechanical room's deck
- Inventorying parts and supplies
- Building storage shelving
- White-boxing vacant areas (getting them ready for brokers who have prospective tenants in tow)

The project work category should not be part of any comparative evaluation of a worker's performance based on time to complete a corrective or preventive work order. Project work generally has a duration time of several days and would negatively affect a worker's productivity evaluation, which is frequently based on the average time it takes to complete one work order.

Preventive Maintenance

Preventive maintenance (PM) is performed per the recommendations of the equipment manufacturer to keep the equipment operating properly. A PM work order is used for this purpose. PM is scheduled work; thus, you can accurately budget and plan for it. A simple example of PM work is replacing lamps in a ceiling light fixture prior to their burning out. By doing so, we prevent the failure of the fixture to provide light. Therefore, the goal of PM is to decrease the amount of unscheduled corrective maintenance. A failure to be proactive by performing PM can put the operation in a reactive state better known as putting out fires.

The General Services Administration (GSA) has produced maintenance guides for the thousands of buildings they manage. Some guides are accessible on the Internet and provide specific procedures for completing maintenance. Guides such as the GSA's Public Building Service, a preventive maintenance guide for the Internal Revenue Service, cover a multitude of maintenance subjects. It also calls attention to many maintenance items that are frequently missed and therefore fail to receive the proper preventive maintenance.

Predictive Maintenance

Predictive maintenance utilizes diagnostic tools to estimate when maintenance should take place based on the actual status of the equipment, rather than simply maintaining equipment to a manufacturer's schedule of maintenance, as is the case in preventive maintenance. Predictive maintenance is most effective when machinery condition trends can be established. To establish a useful trend, predictive maintenance must be utilized with a frequency that is most reasonable for the equipment being tested. If test results indicate no change in the condition of a piece of equipment, then the amount of time between preventive maintenance events can be increased. Conversely, a negative change in equipment condition warrants a more frequent testing schedule and a shorter preventive maintenance interval.

The actual maintenance work order for predictive maintenance falls under the category of preventive maintenance because its job is to vary the schedule of maintenance to meet the actual needs of the equipment.

Some of the methods used to obtain predictive maintenance data are lube oil analysis, thermographic scans or thermal imaging, vibration analysis, eddy current testing, ultrasonic testing, megger readings, and magnehelix readings. Each of these methods will be discussed in the following subsections.

Lube Oil Analysis　Lube oil analysis requires that a sample of oil from a piece of rotating machinery is sent out to a laboratory for analysis. The analysis will define the condition of the oil so a decision can be made about whether the oil should be changed as scheduled or the time interval between oil changes can be extended. The oil analysis lists the types of metal that are present in the oil due to wear or corrosion. If there is iron in the oil, then metal parts are wearing, and metal oxides in the sample indicate rust. If there is lead, antimony, copper, or tin present in the oil, then journal bearings are most likely wearing. Large pieces of metal indicate a part failure, and rubber or plastics are an indication of seal failure. Water in the oil may also be an indication of seal failure.

The analysis will also reveal the oil's viscosity (oil's resistance to flow), which must be correct to provide the proper lubrication for the machinery. Viscosity may change due to a breakdown of the oil as a result of excessive engine heat. This breakdown can lead to oil oxidation; water contamination due to seal leaks; or acid contamination, which is caused by combustion products in the oil.

Thermographic Scans (Thermal Imaging)　Infrared light emitted from a material when it heats up indicates its temperature. An infrared camera is used to scan a piece of equipment to record its temperature. Frequently, these scans are used on electrical switchgear and electric power panels. Loose electrical connections create resistance when current flows through the connection, and resistance creates heat. Hot electrical connections can be identified, cleaned, and tightened to eliminate heat buildup, which could lead to electric insulation deterioration, arcing, and component failure.

It is a good idea to scan electrical equipment once a year, or at least once every two years. This can be done at night when the equipment is normally shut down and when the building engineer can gain total control over equipment operation without having to worry about a disruption to tenants or others. Electrical equipment needs to be in operation (drawing current), so that heat will be generated. Checking a fused disconnect without power flowing through the fuses does not provide any information.

Thermographic scans are generally done by outside contractors working with building staff. But as the cost of scanning equipment continues to decline, more facilities are purchasing their own scanning equipment. Facilities can use this equipment not only to evaluate electrical equipment, but also to scan their roofs for wet insulation due to leaks and heat energy loss from the building's exterior walls. Thermal imaging can even find termite nests by detecting an increase in wall temperature that is due to heat generated by bacteria as it breaks down the wood in the termites' intestines. The maintenance aspect of this process is to target chemical treatments only to the nest area. Also, verification of the destruction of a termite nest can reduce or eliminate the need for subsequent pesticide application.

Vibration Analysis

Vibration Analysis   A piece of equipment frequently vibrates differently when its bearings deteriorate. The equipment may start making a lot of noise. Vibration or acoustical analysis can provide condition information. Vibration meters can detect bearing condition, misalignment, loose parts, and unbalanced rotating elements.

Eddy Current Testing   Eddy current tests are frequently used to inform the manager of the condition and thickness of condenser tubes in an electric centrifugal chiller. This work is often performed prior to a building being sold to ensure a purchaser that the equipment is in satisfactory condition. This testing may be requested by an insurance company that wants to verify equipment condition prior to writing a policy on the equipment.

If tube thickness falls below manufacturer's specification, the tube should be replaced. Eddy current tests also reveal the presence of zipper cracks in the tubes. Circumferential cracks are more difficult for eddy current test equipment to detect because the circumferential nature of the magnetic field created by the test probe masks the problem.

An FM can use these readings when contemplating equipment replacement. Poor eddy current test readings indicate that costly tube replacement is necessary for the equipment to run without a major breakdown. This information can factor heavily in a cost-benefit analysis. It is important information when presenting a case for equipment replacement to upper management. Hard data provided by an eddy current test showing extensive tube problems provide a powerful argument in favor of replacing the equipment.

Ultrasonic Testing   The thickness of steel can be measured ultrasonically using sound waves. A reduction in steel thickness is an indication of corrosion or wear. Ultrasonic testing equipment used in a different manner can detect the high-frequency sound that is made when a pressurized system leaks. An ultrasonic test identifies the location of the leak by detecting the location of the sound emission. Once the leak's location is determined, the correct part can be repaired.

Megger Readings   Megger readings indicate the level of insulation resistance of motor windings. Megger readings are good indicators of a motor's electrical condition. Motors should be tested yearly, and a record of the readings should be kept on file.

When performing an insulation test, the megger impresses a voltage of at least 500 volts across the motor windings. Temperature and humidity levels can affect the results, so the test should be done under approximately the same conditions every year. Readings should be taken with the motor at normal temperatures and low humidity levels. A good three-phase AC (alternating current) motor should have a megger reading in the range of several hundred megohms.

More important than the results of any one set of megger readings is to discover if insulation resistance to ground readings has changed significantly over time. The conditions (high temperature or humidity) under which the motor operates might cause readings to be under 100 megohm, but the motor is operating fine. The key here is that the readings are not decreasing. Decreasing readings alert the manager to a motor whose insulation is starting to fail.

The FM reviewing megger readings needs to consider the environmental and operational factors that have a measurable impact on the validity of the data and ultimately on the subsequent repair/replace decision.

Magnehelix Readings   These low air pressure readings are used to predict when air filters need to be changed in an air-handling unit. The reading reflects the change in air pressure across the filter measured in tenths of an inch of water. This is done with pitot tubes that are connected to a magnehelix gauge and mounted on opposite sides of the filter. As the filter accumulates dirt, the pressure differential across the filter element goes up. Filter manufacturers provide the differential pressure information to tell the user when the filter should be changed. Rather than changing filters on a scheduled basis or when they just look dirty, the filters are changed on a predictive basis based on a pressure differential measurement.

Many building automation systems (BASs) can monitor filter conditions and indicate when a filter is dirty. When the BAS program is set up, the programmer should verify that the appropriate pressure drop was used to activate the filter alarm and that the sensors are properly calibrated.

Lean Maintenance

Lean maintenance strives to eliminate or minimize any process, operation, or material usage that does not contribute to an efficient maintenance operation. It is a natural offshoot of lean manufacturing, the goal of which is to reduce waste; as part of the total quality management (TQM) process, it strives for perfection.

Preventive maintenance has been effective for many years in preventing unscheduled downtime and avoiding major equipment breakdowns. Lean maintenance looks at preventive maintenance and attempts to define exactly how much is necessary to achieve a desired result. Performing too much preventive maintenance wastes parts and materials; causes unnecessary downtime; and takes labor and resources away from doing other, more productive work. Too little preventive maintenance results in excessive downtime due to failure. Lean maintenance provides the balance between too much and too little preventive maintenance.

Example: Oil Change Frequency

This practical example shows how lean maintenance can serve a facility. The goal is to develop an oil change frequency regimen for a fleet of automobiles of yesteryear. Many auto manufacturers in the past recommended changing the engine oil every 3,000 miles. When the condition of the drain oil was investigated in newer automobiles, however, it was found that the oil coming out of the drain plug was as good as new oil poured out of a can. This held true for the first 30,000 miles, and laboratory testing of the oil from several vehicles further confirmed that the drain oil was actually in very good condition.

Following the manufacturer's recommendations for the next 120,000 vehicle miles resulted in a total of fifty oil changes for each vehicle during the vehicle's useful life of 150,000 miles. At 90,000 miles, the oil coming out of the drain plug was dark, with an odor of combustion products, and testing of the oil indicated deterioration. Examination of some of the engine parts showed the engine had normal wear for an engine whose odometer read 90,000 miles. When the odometer hit 150,000 miles, the vehicle was replaced with a new vehicle to ensure reliability of travel. The manufacturer's recommendations created a rigid, inflexible oil change frequency that did not match the average condition of engine parts or the oil condition.

Automobile engines deteriorate with use, and altering the oil change frequency to more closely resemble what an engine really needs is a more sustainable solution. The oil change interval was increased to 6,000 miles for the first 30,000 miles driven. Then from 30,000 to 90,000 miles, the oil is changed every 5,000 miles. At 90,000 to 150,000 miles, the vehicle's oil is changed every 3,000 miles. See Table 2.1.

This new oil change regimen reduces the number of oil changes from fifty to thirty-seven, which is a 26 percent reduction in the total number of oil changes. If an oil change costs $25 in direct fees, and $100 in time and personnel costs, each vehicle will save $325 directly and an additional $1,300 in time and personnel costs.

Note: A 4,000-mile oil change between the 5,000-mile oil change and the 3,000-mile oil change wasn't utilized, but doing so would have resulted in greater savings. This oil change frequency modification would be appropriate only if you have the additional engine wear data to justify the 4,000-mile oil change modification.

TABLE 2.1 Old and New Regimens for Oil Changes

Old Oil Change Regimen		
Mileage	Oil Change Frequency (Miles Driven/Change)	Number of Oil Changes
0 to 150,000 miles	3,000 miles/change	50

New Oil Change Regimen		
Mileage	Oil Change Frequency (Miles Driven/Change)	Number of Oil Changes
0 to 30,000 miles	6,000 miles/change	5
30,000 to 90,000 miles	5,000 miles/change	12
90,000 to 150,000 miles	3,000 miles/change	20
Total number of oil changes		37

The key to being successful when decreasing the frequency of maintenance is not to make drastic changes. Make conservative changes and monitor the results closely. For instance, in our example about oil change frequency, don't increase the mileage in the oil change interval from 3,000 miles to 15,000 miles unless significant amounts of reputable data exist to support this drastic change. An experiment like this could end with disastrous results, such as a major engine mechanical failure. Utilizing research that engineers have gathered in their published reports and benchmarking what other organizations do is a way to begin a program to reduce scheduled maintenance frequency.

More accurate ways to adjust oil change frequency are to install oil-condition monitoring sensors and equipment and to send oil samples periodically to a lab for analysis. In our example about oil change frequency, doing so would be impractical because the cost of monitoring equipment and sample analysis exceeds any savings. The amount of time, record keeping, and effort is also too great to generate savings. On the other hand, if we were dealing with a generator engine with an oil sump of 200 gallons rather than a single, five-quart automobile engine, then monitoring and analysis would be quite appropriate, and significant savings could be achieved.

2.4   PERFORMING PREVENTIVE MAINTENANCE

To get maintenance accomplished, a facility may employ different tactics. The facility can issue individual work orders to facility maintenance technicians. The facility may further define the duties of the FMT through zone maintenance. If a facility has a substantial number of work orders that require more than one person to complete, it may implement the maintenance team approach.

Zone Maintenance

Zone maintenance is a type of maintenance structure used by many hospitals, higher education campuses, and other large facilities with several facility maintenance technicians on their staff. Facility maintenance technicians are assigned specific areas (zones) in a building or buildings that are their personal responsibility. The theory of zone maintenance is that maintenance technicians take ownership of the assigned area. This should make it easy for a manager to rate a technician's performance. By walking through a technician's zone, the FM can see the quality of the maintenance and if any obvious exceptions are present. The FM can also ask the people who live or work in the zone what their perception is of the technician's performance and if there are any problems.

The manager who uses the maintenance quality of the assigned zone to rate a technician's performance must be mindful of the fact that all zones are not alike. A technician who gets a high-traffic cafeteria zone at a university has to work much harder than a technician who takes care of a lower-traffic and lighter-use conference or lecture hall at that same university. Fairness of zone assignments is a sticking point for many maintenance technicians.

Zone maintenance has many good points. It gives workers the responsibility of maintaining a specific area and a sense of pride as a reward. It can give workers focus and makes them experts in the maintenance needs of a particular area as well as the needs of the individual employees and customers who do business in the zone. It shortens the distance that maintenance technicians must travel to answer alarms, and get parts, tools, and supplies, thus saving both time and money.

Maintenance Team Approach

The maintenance team approach is an effective, cost-saving measure used by facility or property owners who have multiple properties. Rather than duplicating the maintenance staff at individual facilities, the maintenance team approach employs a single staff of skilled maintenance technicians who handle the maintenance needs at all the facilities of an organization.

This approach has been used successfully by electric power utility companies and property management firms. In commercial real estate, use of the team approach may be a result of several small properties being located geographically apart. The teams may even be contracted to serve the needs of other property owners and thus serve as a profit center for the organization.

2.5   SUSTAINABLE MAINTENANCE

The adage *Less is more* is attributed to one of the great architects, Ludwig Mies Van Der Rohe. Mies (pronounced "meese") is considered to be the father of the modern skyscraper. Mies eliminated the costly and unnecessary embellishments on buildings (gargoyles, statues, and Grecian columns, for example) and instead found inherent beauty in a building's structure of steel beams and glass exterior skin.

Less is more could also be applied to equipment maintenance. Maintenance that is performed on a simple time schedule goes against this adage. Time-schedule maintenance is wasteful because it always assumes a worst-case scenario about when maintenance should be performed, thereby shortening the interval between preventive maintenance activities. As we've already discussed, a more efficient approach is maintenance based on lean maintenance principles using predictive tools that identify exactly when maintenance should be performed and the precise quantities of parts and supplies that should be utilized. Maintenance is then undertaken at the most cost-effective time, expending the minimum quantities of supplies that are necessary to meet the maintenance requirement.

Sustainable equipment maintenance goes further by striving to meet the following goals:

1. Longer runtimes between equipment overhauls.

2. Less time spent performing the actual overhaul due to less time spent cleaning equipment prior to overhaul; fewer unexpected repairs on equipment in the process of being overhauled; and easier parts disassembly during repairs and overhaul.

3. Reduced energy cost as a result of well-maintained equipment that operates more efficiently.

4. A smaller carbon footprint; less energy input means less carbon dioxide from burning carbon-based fuels.

5. More efficient delivery of maintenance based on accurate estimates of the time and level of maintenance required.

6. Elimination of unexpected equipment failures.

7. Personnel safety and risk reduction as a benefit of using equipment that operates as designed rather than equipment that is out of calibration.

8. Improved indoor air quality (IAQ), which provides workers with a healthy and therefore productive indoor environment.

9. Cost reductions resulting from the prevention of waste, of both time and materials.

10. Supply parts management on a scheduled basis rather than on an emergency basis, thereby eliminating panic purchasing.

11. Avoiding waste by maintaining lower inventories of parts, which leads to fewer part deteriorations due to excessive shelf time.

12. Customer satisfaction; maintenance must address the needs of the user.

2.6   CMMS AND PROCUREMENT MANAGEMENT

CMMS provides an automated tool that allows the facility to keep accurate records of all items in the parts inventory. (See Figure 2.2, which shows a sample CMMS inventory cost report.) Depending on the sophistication of the system and the amount of information input into the software, CMMS programs can also alert the manager when parts are low and then generate the purchase request to replenish the supply.

The FM needs an accurate order history of frequently ordered supplies in order to program the system's reorder points correctly. If historical order records are not immediately available, the CMMS can provide the necessary ordering data over time as the FM places new orders.

Example: Over ordering Supplies

A trades supervisor at a university actually used to put in an order for all the supplies needed for the trade for the entire year. This practice put an extreme burden on storage facilities. It required a major effort to unpack and store all the supplies and materials. The supervisor felt that he was saving money by making this massive bulk purchase. The result was anything but saving money. New supplies were stored on top of older, unused supplies from a decade earlier. Seals and gaskets dried out and the parts, although never used, would require an overhaul before they could be installed. Having so many supplies lying unused provided a great temptation for theft.

Inventory Cost Report				12/21/2013
Part Number	**BLT001**			**BELT, DRIVE**
Stockroom	**Quantity On Hand**	**Unit Cost**	**Carrying Cost**	**Total Cost**
ES100	8	5.60	0.00	44.80
Part Number	**BLT00101**			**DRIVE BELT**
Stockroom	**Quantity On Hand**	**Unit Cost**	**Carrying Cost**	**Total Cost**
EMAIN	3	125.00	0.00	375.00
Part Number	**FLT25161**			**AIR FILTER**
Stockroom	**Quantity On Hand**	**Unit Cost**	**Carrying Cost**	**Total Cost**
EMAIN	60	12.00	0.00	720.00
Part Number	**IMP3**			**Impeller for Pond Pump**
Stockroom	**Quantity On Hand**	**Unit Cost**	**Carrying Cost**	**Total Cost**
EMAIN	1	22.63	0.00	22.63
Part Number	**LUB002**			**LUBE, GEAR**
Stockroom	**Quantity On Hand**	**Unit Cost**	**Carrying Cost**	**Total Cost**
ES100	2	800.00	0.00	1,600.00

FIGURE 2.2 Sample CMMS Inventory Cost Report

Source: Eagle Technologies.

Sometimes parts and supplies go unused for a variety of reasons. Therefore, it is a good idea to mark boxes in storage with an easily readable date or even a simple color band that denotes the month received. This is especially useful for consumable products that cannot be used once they are past their expiration dates. Preventing waste of any kind is an important goal toward sustainability.

2.7 CMMS AND THE PREVENTIVE MAINTENANCE FUNCTION

One of the great values of having CMMS at a property is the automated generation of preventive maintenance work orders. In addition to maintaining daily operations, CMMS information can be very helpful when performing a property condition survey at a later date. Development of a PM work-order system starts with an equipment inventory that accurately describes each piece of equipment at a property. This is done by a site survey, the goal of which is to assign every piece of equipment a unique alphanumeric code that is logged into the CMMS. The maintenance requirements for each piece of coded equipment is then stored. The system can then automatically generate PM work orders (see Figure 2.3). The PM work that is completed by the technician on the equipment is logged into the system.

Gathering the initial equipment data for the equipment inventory, which is the backbone of a CMMS program, is a difficult task. Construction documents can be extremely helpful in pointing out equipment type and location, but all equipment should be physically verified to ensure that it actually exists or that substitutions have not been made. Barcode tags are often attached to the equipment for identification. It may take several weeks to complete an equipment inventory and enter the required preventive maintenance for each piece of equipment. But once completed, the system can issue work orders that accurately reference the equipment and generate equipment maintenance histories. *Equipment maintenance histories* provide a summary of all the work done, equipment runtimes between maintenance events, the person(s) who performed the maintenance, and the parts used.

PM Master Work-Order Form   10/19/2013

Job No.	AHU-002
Procedure Description	INSPECTION OF AIR-HANDLING UNIT

Priority	HIGH	**Maintenance Code**	HVAC
Cost Center	20000	**Work Type**	PM
Shift	SECOND		

Asset No.	Asset Name	Location
AHU002	AIR HANDLING UNIT	MAIN PLANT
AHU002-COM	COMPRESSOR	AHU002
AHU002-FF	FINAL FILTER	AHU002

Cycle Type	Last Scheduled Date	Active Date	Next Scheduled Date
Daily	08/24/2011	08/25/2011	08/24/2011
Annual	09/01/2011	09/01/2012	09/01/2012

Runtime Cycle	Runtime Advance	First PM	Last PM
0.00	0.00	0.00	0.00

Cycle Type	Task No.	Description	Skill Code	Est Hours
All	LOC001	USE PROPER LOCKOUT/TAGOUT PROCEDURES	MECH-E	0.30
All	INS003	INSPECT EQUIPMENT AND AUDIT OPERATING CONDITION. REPORT CONDTION OF EQUIPMENT ON WORK ORDER.	MECH-E	0.00
All	CHK001	CHECK AIR LINE FILTERS FOR SIGNS OF MOISTURE	HVAC02	0.30
All	CHK002	CHECK CONDITION OF GASKETS ON DOORS AND DAMPERS	HVAC02	0.30
A	CLN001	CLEAN COILS AS NEEDED	HVAC02	0.40
A	RPL001	REPLACE FILTER SYSTEM FOLLOW PROPER FILTER DISPOSAL PROCEDURES	MECH-E	1.50
All	SAF001	USE PERSONAL PROTECTIVE EQUIPMENT	INHOUSE	0.30
All	29SAHI	CONTRACTOR TO PERFORM ALL TASK PER CONTRACT	OC01	0.00
All	28SAHI	INSPECT RACK FLUE SPACE	SAFETYCOM	0.00

Cycle Type	Part No.	Part Name	Qty Allocated	Stock Location
A	FLT001	AIR FILTER	1	MAIN

Cycle Type	Employee No.	Employee Name	Skill Code	Est. Hours	Task No.
All	10012003	DONALD BARTON	HVAC02	0.30	CHK001
All	10012003	DONALD BARTON	HVAC02	0.30	CHK002
A	10012003	DONALD BARTON	HVAC02	0.40	CLN001
All	10012002	CARL NELSON	MECH-E	0.50	INS003
All	10012002	CARL NELSON	MECH-E	0.30	LOC001
A	10012002	CARL NELSON	MECH-E	1.50	RPL001
All	10012003	DONALD BARTON	HVAC02	0.30	SAF001
All	10012002	CARL NELSON	MECH-E	0.30	SAF001

Cycle Type	Tool Description
All	SAFETY LOCK
All	RAGS

WO Desc.

09/27/2004:
NOTIFY SUPERVISOR WHEN UNIT IS RETURNED TO SERVICE

FIGURE 2.3 Sample PM Master Work-order Form

Source: Eagle Technologies.

2.8 PROPERTY CONDITION SURVEYS

Properties and their equipment need to be evaluated periodically to verify their condition, find developing problems that may be hidden, and design a plan for correction. A property condition survey (PCS) report, performed by engineering consultants and specialists, is the best way to get a firm basis for making capital improvement, major maintenance, and preventive maintenance decisions.

A PCS report provides the FM with building, equipment, and site condition information necessary to make these important decisions. For instance, the data contained in a PCS report are essential ingredients in developing the five-year master plan for colleges and universities. This report is also used in the property acquisition process to better define a property's value and to alert the potential new owner to existing problems and the cost to correct them.

The PCS report contains a table that defines the expected maintenance costs as specified by the customer generally for the next five or ten years. For instance, condominium associations must put capital reserves in a bank account to pay for anticipated future repairs to common areas inside the condominium and for landscaping and projected site work. The PSC tables define the amount of money that needs to be put in reserve.

Who Performs Property Condition Surveys?

PCS reports may be obtained by engineers trained in PCS investigative methods and report standards. When a difficult problem or question arises with property conditions, engineers may recommend that specialists be employed to further define the problem and provide a solution with accurate cost information. The types of engineers who develop PCS reports include the following:

- Architects
- Civil engineers
- Plumbing/mechanical/HVAC engineers
- Electrical engineers
- Structural engineers
- Environmental engineers
- Facility consultants

In addition to these engineers, the following specialists can help to develop sections of the PCS report and handle specific issues:

- Lighting consultants
- Pavement and parking deck consultants
- Landscaping consultants
- Roof consultants
- Water-use specialists
- Building-code inspectors
- Lead-paint risk assessors and asbestos risk assessors
- Fire prevention specialists
- Security specialists
- Insulation and energy audit specialists
- Construction estimators

Some facilities ask their in-house maintenance staff to perform PCS reports. An in-house PCS report has the following advantages:

- **Cost control**—The facility manager has total control of costs that are incurred during the investigative period and the report-generation period. A situation may arise whereby the FM requires help from outside the organization because of the discovery of problems beyond the capability of staff members. In this case, prices can be obtained for the professional investigative work needed if the FM can define the problem satisfactorily.
- **Familiarity with the facility and the goals of the organization**—The most knowledgeable persons are those who work for the facility organization and understand its needs.

- **Firsthand historical knowledge of facility physical and operational changes**—Many in-house staff members have decades of experience and historical knowledge of the facility.
- **Quick report turnaround**—If well planned, an in-house PCS report can be generated quickly and with minimal disruption to the facility.

Having a consulting firm perform a PCS report for your facility has the following advantages:

- **Depth of knowledge and experience**—Professionals performing a PCS report have a variety of specialized skills. This can provide for a more complete evaluation and the basis for a solution to correct any problems that the report uncovers.
- **Detachment from the facility**—Third-party professionals can produce a report that is fair and impartial. A third-party professional will not consider company politics or its financial position when performing a PCS report.
- **Professional and/or legal record**—Reports generated by professional consultants may be used in litigation. If necessary, a PCS report can provide information that weighs in favor of (or against) a facility, depending on the subsequent actions taken for correction of deficiencies cited in the report. Organizations performing PCS reports generally carry very costly professional liability insurance to protect them against successful lawsuits. This is not to say that a PCS report generated by in-house staff cannot be used in litigation. It can, but without review by a licensed registered engineer, it might only work against a facility because positive report data carry no professional weight in a court of law, and negative information reveals that the owner was aware of a problem.

Property Condition Survey Considerations

Some negative information contained in the report may also be considered a ***material adverse fact*** (legal terminology that is used when there is a transfer of ownership of a property) during property disposition. This occurs when there is information that is known by a seller of the property that would diminish the value or affect the use of the property by the purchaser. These facts must be made known per the state legal requirements and/or the report be made available to the purchaser of the property.

When a third party points out a possible health or safety issue, upper management responds by finding the money to effect a solution. This is not to say that facility personnel are not valuable in performing a PCS report—far from it. The first step in performing any good PCS report is when the PCS professional interviews key members of the maintenance staff. In the process, the professional gains good firsthand knowledge of the problems facing the facility. The completeness of answers provided by staff members gives the professional an understanding of the knowledge and abilities of the maintenance staff. This alerts the professional to likely problem areas that could require further investigation.

REVIEW QUESTIONS

1. Lean maintenance strives to eliminate _____.
2. How do predictive work orders differ from preventive work orders?
3. Discuss how zone maintenance can save a facility money.
4. What is CMMS and what does the FM need to have in place in order to institute CMMS at a property?
5. Describe how a PCS report can aid the facility manager in developing a capital budget and a maintenance budget.
6. What are the advantages in having third-party consultants perform a PCS report?
7. What are the advantages of performing a PCS report in-house?
8. Describe the facility team for a large commercial property. Name the members and describe each in a few sentences.
9. How does CMMS monitor worker productivity?
10. List the ways a facility can receive corrective maintenance work orders.

TABLE 2.2 Inventory and Ordering Spreadsheet

Date	Item Description	Prior Inventory	Reorder Level	Used	Running Total	Reorder Amount
1/2/2013	Pipe nipple $\frac{1}{2}''$ 8″	50	25	21	29	−4
1/2/2013	Pipe nipple $\frac{1}{2}''$—5″	19	5	5	14	−9
1/2/2013	Pipe nipple $\frac{1}{2}''$—4″	20	10	12	8	2
1/2/2013	Pipe nipple $\frac{1}{2}''$—3″	10	10	1	9	1
1/2/2013	Pipe nipple $\frac{1}{2}''$—2″	18	2	5	13	−11
1/2/2013	Pipe nipple $\frac{1}{2}''$—1″	14	12	0	14	−2
1/2/2013	Pipe elbow $\frac{1}{2}''$—90	15	5	0	15	−10
1/2/2013	Pipe elbow $\frac{3}{4}''$—90	14	5	5	8	−3
1/2/2013	Pipe elbow $\frac{3}{4}''$—45	15	10	13	2	8
1/2/2013	Pipe elbow 1″—90	21	10	17	4	6
1/2/2013	Pipe elbow $1\frac{1}{4}''$—90	17	10	17	0	10
1/2/2013	Pipe elbow $1\frac{1}{2}''$—90	16	10	14	2	8

Password: pipe

Essay Questions

11. Discuss the various situations in which zone maintenance, the maintenance team approach, trade shops, and non-specialized FMTs would be most effective.

12. How can CMMS be used to develop a plan to improve worker productivity?

13. When comparing the energy usage data from the time when a building was new to current energy usage data, which is obtained decades later, what considerations should be made to compensate for global warming and the current use of the building?

Project Work

Use Microsoft Excel to create an inventory sheet that automatically deducts the number of parts used from the original inventory total. Use color in a cell that indicates when the number of parts in the running total are below the reorder level. Lock the important columns so that a technician who is e-mailed the inventory can change only the number of parts in the Used column. A password will be necessary to make changes to the other columns. See Table 2.2 for an example of such a sheet.

MANAGING OUTSOURCED SERVICES

INSIDE THIS CHAPTER

Can Facility Managers Increase Service Quality and Operational Efficiency While Controlling Rising Costs?

Outsourcing is the term applied to situations when a facility manager hires an outside contractor to provide maintenance or repair services. Some government facilities want to be completely stand-alone so they can meet all emergencies; however, all well-managed facilities utilize some level of outsourcing in order to meet the facility's needs. In doing so, facilities achieve the highest quality of service at the lowest cost. This chapter discusses the process of outsourcing, including the need for contractual agreements, and how to determine the facility's true needs when outsourcing. The chapter closes with a discussion of how the facility management function itself is sometimes outsourced and the differences between outsourcing in the United States and the United Kingdom.

3.1 REASONS FOR OUTSOURCING SERVICE

Why outsource services that might be rendered by in-house staff? The following list offers some of the reasons a facility manager (FM) or other management personnel might make this choice:

1. Gets the job done in the most cost effective manner.
2. Shifts legal liability inherent in a particular service from the facility to a contracted service provider. For example, services such as security have a potential to generate lawsuits when force or restraint is used.
3. Eliminates the supervisory requirements associated with labor intensive work such as housekeeping.
4. Affords a greater opportunity to have undesirable personnel dismissed or removed.
5. Provides more consistent and higher quality work.
6. Fills skill gaps in your in-house staff members' skill levels, or simply provides increased labor when necessary.
7. Facilitates compliance with government regulations that require specialized licensing (i.e., Boiler Operators License, Environmental Protection Agency [EPA] Refrigerant Handler's License, Pesticide License, Lead Risk Assessor License, etc.)
8. Negates the need to purchase costly specialized equipment that may be used too infrequently to justify the expense or provide a return on investment.
9. Provides additional career opportunities for in-house staff members by having large maintenance/management contractors hire the staff, and then provide their service at the facility.
10. Avoids workman's compensation insurance claims due to injuries sustained on the job (liability is transferred to the service provider).

Service contractors who provide outsourced services such as landscaping; housekeeping; heating, ventilation, and air conditioning (HVAC); and mechanical maintenance and security, to name a few, are specialists in their fields. They frequently complete assignments at a higher quality level than an in-house staff can. One reason for this is that service providers perform the task much more frequently. They also

FIGURE 3.1 High-rise Window Washing is a Service that is Frequently Outsourced.
Source: © Igor Mojzes/Fotolia

have access to specialized tools and devices to do the job more efficiently (see Figure 3.1). Some of the equipment may actually be built by the service provider and refined after years of experience.

In some cases, contractors can also do the job more cheaply than in-house staff because of their expertise, tools, bulk purchasing, and access to the latest techniques. Service contractors can spread out the costs for training and licensing their personnel by having their staff work for several customers, thereby reducing overhead costs to their customers. For example, it may be costly to train in-house staff and equipped to handle blood-borne pathogen issues or to have the personal protective equipment (PPE) available to handle the repair situations for confined spaces or electric arc flash.

Safeguarding Outsourcing Success

Hiring outside contractors is not as trouble-free as it sounds. The FM must find honest, dependable, hard-working, quality contractors who provide their services in the safest and the most professional and cost-effective manner without disruption to the facility operation. This is where an FM's business contacts and professional society membership connections come in handy. By tapping into this readily available, local knowledge base, the FM can find out which contractors other properties are using and why. Are the contractors being used for cost, efficiency, reliability, quality, and/or speed of project completion? Ask the manager you network with about his or her overall opinion of the provider and about the service provider's strengths and weaknesses. Check the Better Business Bureau rating to see if there are complaints filed against the contractor. (The Better Business Bureau has mostly residential complaints, but sometimes businesses have a residential division.)

When the services of a provider are essential to the facility operation or require the outlay of significant funds, it is necessary to research the contractor's financial condition. Credit reports are one source of contractor financial information. For publicly traded companies, the auditor's opinion of financial condition, found in the audited year-end financial statement, is valuable. The FM should also be aware that a contractor's financial condition can change quickly if a contractor is on the losing end of a lawsuit. This is why a contractor needs to have insurance at a coverage level appropriate to the risk the contractor may face. The FM should also find out if there is any pending litigation against the company that could result in a large dollar settlement. Information is sometimes available from the county court website. The FM should also ask the clerk of the courts about pending litigation or the outcome of past litigation.

3.2   THE OUTSOURCING PROCESS

For small, uncomplicated, and well-defined services, the steps in outsourcing may be as simple as getting three quotes for the work, evaluating them, and attaching the best quote to the respective agreement for service. For large, more complicated work that has a significant financial or operational

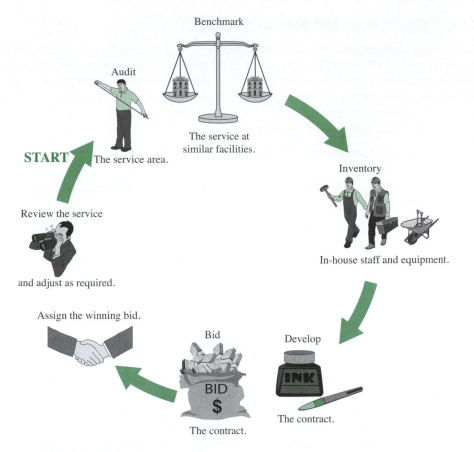

FIGURE 3.2 The Outsourcing Process

consequence for the facility, the steps for outsourcing are shown in Figure 3.2. These steps, some of which can be combined, include the following:

- Audit the service area.
- Benchmark the facility with regard to the service to be provided.
- Review facility staffing and equipment.
- Decide on the type of contract.
- Decide on the contract duration.
- Develop bid specifications and prepare a bid document.
- Hold a contractor site visit.
- Hold a pre-bid conference.
- Bid the contract.
- Evaluate the bids.
- Hold a post-bid conference.
- Award the contract.
- Periodically review the service provided.
- Modify the contract as needed for service improvement.

3.3 PREPARING FOR THE OUTSOURCING PROCESS

Complete working knowledge of the facility's condition, service needs, capabilities of in-house staff, and the equipment that will be necessary to perform the task is extremely valuable before entering into negotiations with service providers. Managers then need to compare the industry's best practice for the service with the level of service that will be provided through outsourcing. This comparison ensures that the service will both meet the needs of the facility and be competitive for the market segment served. The road to successful outsourcing is accomplished by a service area audit, staff review, and benchmarking.

Audit the Service Area and Review Staffing

First, the FM must audit the area(s) under consideration for outsourcing. In other words, the FM needs to know the current condition of the equipment and the quality of the service currently being provided. Then the FM decides on a new service level to be achieved by outsourcing. Corrective work may need to be performed to get the facility ready for the outsourced service activities. For example, flooring may need to be patched or replaced prior to refinishing.

Next, the FM should perform the following steps:

- Review the equipment required to perform the service contract against the equipment the facility already owns.
- Decide what should be done with the equipment (i.e., let the contractor use it, dispose of the equipment, sell it).
- Review staffing requirements for performing the service.
- Decide if the facility's in-house staff should participate in providing part of the service to achieve some savings (and determine the additional liability that may be incurred when in house staff is involved).

Note: The facility manager may discover that facility staff has the ability to perform work that was never required of them in the past.

Example: Work Shared Between Facility Staff Members and the Service Provider

A contractor has been hired to clean the tubes in a condenser that is part of an electric centrifugal chiller. The facility staff drains the cooling tower water feeding the condenser, removes the condenser heads, and later replaces them after the contractor completes the tube cleaning. The reason the facility staff becomes involved in the tube cleaning process is that the facility avoids the cost of having the contractor perform this work. Although avoiding costs, it incurs additional responsibility for the facility. Should the condenser head gasket leak when refilling the tower, it is now the facility's responsibility to make the necessary repairs. The FM must perform a risk-versus-benefit analysis when sharing any job responsibilities with a service provider.

Benchmark Services

The FM should benchmark the facility service against the service at comparable properties in the same location, industry group, and quality level by visiting these properties and asking questions if possible. Membership in professional organizations can be a very useful source of vendor recommendations and benchmarking information. It is important to understand the quality level service that needs to be achieved so that the property is well positioned in the market to compete with similar properties. Decisions on service level should be based on the quality level and overall class of other properties in the area and on the type of property. Paying a premium for a high-quality service may not result in higher rents to offset the extra cost. A housekeeping contract is a good example. An FM seeking to improve the condition of his warehouse property located in an industrial area by increasing the quality of the outsourced housekeeping service . If the level of cleaning was brought up to what would be considered appropriate for a Class "A" office property, this high quality of cleaning would surpass the norm for the property's location and type.

3.4   TRADITIONAL CONTRACTS FOR OUTSOURCING

An agreement is a formal written contract between two parties: the owner and the service provider. *Owner* is a relative term because a property management firm may be permitted to act on the owner's behalf when making decisions for the property. These decisions include signing contracts and negotiating agreements for the benefit of the owner.

A traditional contract consists of terms and conditions for the performance of the contract and detailed specifications. Traditional contract terms and conditions are provided through contract language in contractor-provided contracts, or **agreements for service**.

Contractor-Provided Contracts

Contractors generally have a preprinted contract agreement for the owner or the owner's legal representative to sign. Contract agreements provided by a contractor can have many provisions that require evaluation by a legal professional. When using a contract from the contractor, the owner/representative must perform a contract review process, which can be time-consuming and costly. The owner must also review the contract every time a contract expires and a new one is provided. The owner/representative has no control over the provisions in the contract, which may have changed from the previous contract without notice. Additional review is essential, or the owner's representative risks signing a document that has unknown legal or financial ramifications.

The owner/representative may have little choice in signing a contract provided by engineering and environmental contractors because consulting work in these fields may expose the consultant to possible litigation initiated by the owner or a third party. Litigation may occur years after the service was provided. To be insured for professional liability, consultants, in addition to paying large sums for this insurance coverage, may have to work under a contract approved by their insurer that, among other required provisions, stipulates specific levels of liability the contractor is willing to assume.

Consultants frequently use a basic contract template whose provisions were written by a professional engineering society or other professional organization. The society is comprised of members who provide the same type of consulting services. These contracts are frequently used throughout a particular industry. These professional services contracts provide for a somewhat easier review by legal professionals familiar with the chosen contract format.

Agreement for Service

An agreement for service is a contract form that has been developed by the property owner's legal representation. It can be used multiple times for most outsourcing situations. In the past, it was a preprinted form that set the requirement for contractors working at the property. Today, it is in an electronic format for easy editing and storage. The agreement for service is given to the contractor under consideration by the facility manager for the contractor's legal review. If the agreement is acceptable to the contractor, both the facility manager and the contractor sign, give the agreement back to the owner, and then the owner signs. The agreement does not guarantee that the contractor will be given work, or that the contractor will be chosen over another contractor. The agreement for service provides the contractor with the potential opportunity to work at the facility. Indeed, the facility may have agreements for service with several service providers who provide the identical service. By meeting these requirements, contractors who are parties to the agreement have an advantage in being chosen to provide services over other contractors who are not.

The agreement for service utilizes the contractor's quote for the work, which has the description of the service to be provided and the cost to perform it. The contractor's quote is used as an attachment to the agreement for service, and it is affixed to the back of the agreement. The agreement for service streamlines the process of having work done at a facility because the owner does not need to review a new service provider's contract every time outsourced service work is required. Provisions in the agreement for service also ensure that a contractor has the proper insurance, meets regulatory requirements, and provides the proper amount of coverage for unfortunate events where the contractor has liability. The agreement for service may have a stipulated term or it may be automatically renewed indefinitely.

The agreement for service may also require that the contractor's insurance company provide copies of the contractor's insurance certificate directly to the owner. Another insurance requirement may be that the property owner is listed as **_additionally insured under general liability_** on the service provider's insurance certificate. This requirement provides the owner with additional protection from claims against the work done or any negligence on the part of the contractor. Seek expert legal advice and professional insurance consultation when developing and utilizing an agreement for service or when drafting _additionally insured_ language to be added to an insurance certificate.

Traditional Contract Specifications

Specifications must be accurate in order to provide a roadmap for successful performance. The contract specifications must define the following:

- What is to be done
- When it is to be done and for how long
- The number of persons and their qualifications
- The equipment to be used
- The specified products to be used and the specific amounts of each that will be consumed
- The quality level to be achieved

Specifications help the contractor decide if her or his company is capable of performing the service required and are essential in estimating costs to provide a bid price. Specifications act as a guide when the FM evaluates the service provided. The first question to ask is, Has the contractor met or exceeded specifications?

3.5 SERVICE-LEVEL AGREEMENTS

An alternative to the traditional contract for service is to write service-level specifications for a ***service-level agreement (SLA)***. Service-level agreements concentrate more on the final outcome of the work performed rather than on the methods and materials employed in achieving that outcome. Contract specifications are then very different from the average agreement.

Example: Floor Maintenance According to Service-Level Specifications

Instead of reading, "Finish the floor twice a month using a minimum of 5 gallons of floor finish during each application," the agreement might read, "Maintain the floor at the agreed finish level, per ASTM [ASTM International, formerly known as the American Society of Testing and Materials] specification and mutually agreed quality, on a continuous basis." The contractor would then have the option to employ methods that are not specified. The contractor might spray-buff the floor twice and finish once a month, rather than stripping and finishing the floor twice a month without regard to the floor's actual condition. Traditional contract specifications, that is, specifications not based on service-level outcome, tend to waste both time and materials.

Many managers are reluctant to employ SLAs because they fear that they cannot ensure that they are receiving what was paid for per the contract. In the floor maintenance example, the FM would need to ask, Do we really want 10 gallons of floor finish per month on the floor, or do we really need a clean and attractive floor at all times? However, SLAs give the contractor the ability to use knowledge and expertise to achieve an agreed outcome, rather than just meeting original contract specifications that might not satisfy the current actual needs and conditions.

To the benefit of the FM and facility, some SLAs include specifications that act as a default position should poor performance create a need to utilize the traditional contract format. Such SLAs directly specify what a contractor is required to do and when. Contract termination language that motivates the service provider to perform as agreed may also be included.

It is generally easier to implement an SLA contract if the endeavor is undertaken with a contractor who is familiar with the building(s), the service level it needs to provide, and the expectations of the facility organization. The facility organization should have a high confidence level in the contractor before an SLA is negotiated with the contractor.

Contract Duration

The specified duration of a service contract (housekeeping is a good example) for a facility is typically from one to three years. Contracts may also have provisions for automatic renewal, or automatic renewal with a provision for scheduled contract price increases to be paid to the contractor. Contract duration is especially important when there is an automatic renewal feature. If the facility is unhappy with the quality of work the contractor has provided or the terms agreed to, the FM needs to be aware of the date the contract ends. The termination date is important so that proper written notification can be given to the contractor. Timely notification prevents any automatic contract renewal provisions from forcing the facility to keep the contract for another term. An FM may be perfectly happy with the services that a vendor provides but may wish to negotiate a lower price upon contract renewal. This is another reason for closely watching the contract's end date.

The FM should think about the most convenient time for the facility to start a new contract for a particular service or about the most valuable time to begin a service. The contract should end during a period when the provided services are of lower importance or are not normally provided. A contract to plow snow is a good example. A snow-plowing contract shouldn't end in the middle of blizzard season because the FM's negotiating position might be rather weak at that time.

3.6 WRITING CONTRACT SPECIFICATIONS

Contract specifications can be obtained in the following ways:

- Reviewing specifications in contracts that have been previously written. The service may have been provided in the past or provided from a different facility that is part of the FM's organization.
- Interviewing a number of contractors in a particular field and asking for a bid with specifications. The FM can take the best specifications and have all the contractors then bid on these revised specifications.

- Obtaining written specifications from trade organizations that represent the contract service. Trade organization magazines publish articles in their publications that deal with the services they provide. They are generally intended to be read by contractors who provide the service. Trade publications may include information and articles such as "How to Sell Higher Profit Services to Your Customers." This can be valuable information when you are trying to evaluate a contractor's motives and performance.

- Maintaining professional society memberships can provide opportunities to acquire contract specifications from fellow members. It might even be possible to request a copy of their current contract for a particular service, depending on their company policies about divulging such information.

- Hiring a consultant may be the best answer to an FM's contract needs. Consultants can provide specifications that follow best practices in the industry. Consultants can also aid the FM during the contract bidding process, evaluate the bids, and make recommendations about awarding the contract.

Approved Equals

Approved equal is a construction term that, when inserted in bid documents, allows the contractor to offer the owner a substitute product for one specified in the bid documents. The bidding contractor should be required to submit a price using the specified product and another price using the proposed approved equal. This language ensures that the products under consideration can be compared with regard to price and quality. In a standard service contract, it is up to the person evaluating the bid to decide if an approved equal is really equal to the product specified. In a service-level agreement, the contractor might be allowed to make substitutions as long as the quality of the end result is the same.

Contract Terms and Conditions

Terms and conditions set the basic ground rules for the performance of the contract. This contract section begins with the legal names of the parties involved. It discusses all the legal and contractual requirements, in addition to the attached contract specifications that the contractor must meet in order to be paid for the service. It also has the legal protection language for the facility resulting from the contractor's actions during the performance of the contract. It is sometimes referred to as the boiler plate language of the contract.

3.7 BIDDING

The bid process begins when the owner invites contractors to bid on a service contract. The owner supplies the contractors with a bid document that lists specifications for the work that is to be done. Ultimately the owner or FM collects the bids; holds post-bid conferences with the acceptable lowest cost bidders; and then issues an acceptance letter to the winning contractor.

Some government or municipal contracts require **open bidding**. This is important if you are an FM at a state institution such as a state university. The government agency that wants to contract for a service must first post an ad in a newspaper or other appropriate media providing information about the contract. Public notification means that any qualified contractor can bid on the contract.

Open bidding during slow economic times, when many contractors are looking for work, can be quite difficult. To limit the number of bidding contractors, commercial organizations frequently hold **closed bids** that are open only to a few select contractors that have been prequalified. Closed bidding provides a time-saving advantage for the FM because he or she considers only a few qualified contractor bids. It is true that closed bidding might not provide the facility with the lowest bid that could be had through open bidding, but it allows the FM to quickly, and it ensures that qualified contractors will perform the service for the facility.

Bidding is a very important and telling step in the contract process. Sections of the bid documents that are quite clear to the FM and most of the bidding contractors might be misinterpreted by one or two contractors. Most often the bid numbers tell the story. Too high or too low a bid amount is the first indication that a misunderstanding on the part of a contractor exists.

The bid numbers tell a more complete story if a **breakdown** of prices that comprise the total bid price is submitted. Rather than just asking for a total price to complete the work, a good bid document requires a cost breakdown that provides prices for the individual tasks and items required to complete the work in the contract. If a bidding contractor provides acceptable numbers for most of the sections of the contract but has an unusually high or low number for just one or two sections, a post-bid conference or clarification of the off-price section should be requested. Maybe that contractor thought more (or less) work was needed than what the bid required. Common mistakes a contractor makes are over-estimating the work area, assuming a greater frequency of service than required, or including a higher-priced product than necessary.

Note: Sometimes a large unexpected variance might actually turn out to be correct. It is necessary to discuss this difference with the bidding contractor. It might be discovered that the FM missed something, for example, disposal costs, overtime, increased material costs, energy, protection of surrounding areas, and so on. In this case, the high or low number is the right number.

Pre-Bid Conference and Site Visit

The pre-bid conference covers the particulars of the contract and the overall expectations of the facility. It attempts to answer any questions that the contractors may have before they bid on the contract. Depending on the service to be contracted, items such as scale drawings of floor plans, scale drawings of site plans, and equipment data may be presented at the meeting.

A contractor site visit with the bidding contractors, also known as a walk-through, takes place in conjunction with the pre-bid conference. A *walk-through* is simply a visit to the area of the facility in which the contractor would provide service. This type of visit gives the contractors a chance to see what it is they will be bidding on and any existing conditions that may affect how the contractors do their work. It helps to prevent misconceptions in terms of the area and scope of the project.

Bid Evaluation

The best way to evaluate bids among bidding contractors that are equal in size, reputation, and quality is to have a bid form. A *bid form* is basically a checklist of all the items that will be performed, and how often, as required by the contract. The bid form is frequently developed from the first good bid received, or from specification information obtained from a prior contract. The bidding contractor is required to fill out this separate sheet, listing all the contract items in the same order. This information is then put on a comparison spreadsheet containing the pricing data of all the contractors, thus making comparisons and evaluations easy.

Example: Bid for Landscaping

Let's look at the simple bid evaluation tally sheet for some basic landscaping work in Table 3.1. Assume that you are managing a building with a front area that is visible to customers and tenants and a back area that is less visible and not as important to maintain. You have four landscape contractors bidding on maintaining the lawns and sweeping the sidewalks, drive lanes, and parking areas. Contract specifications specify the products to be used and the amounts. There are no approved equals for this contract. Each bidding contractor is given the first two columns of the sheet; from the information obtained, we put together the bid evaluation tally sheet.

Contractor A is clearly the low bid, and all factors weighed equally, should be the contractor awarded the contract. Because the bids of contractors A and C are fairly close, you would want to discover why there is a difference. The explanation may affect your final decision.

TABLE 3.1 Bid Evaluation Tally Sheet

Item	Bid Price			
	Contractor A	Contractor B	Contractor C	Contractor D
Mow front lawn 28×/season	$ 5,600.00	$ 6,200.00	$ 5,450.00	$ 6,500.00
Mow back lawn 18×/season	6,500.00	7,000.00	6,350.00	6,800.00
Fertilize front lawn 8×/season	1,200.00	1,500.00	1,250.00	1,400.00
Fertilize back lawn 4×/season	600.00	800.00	650.00	900.00
Weed-kill front lawn 8×/season	300.00	400.00	350.00	450.00
Weed back lawn 8×/season	250.00	300.00	275.00	375.00
Blow off sidewalks 30×/year	1,500.00	1,750.00	1,800.00	2,000.00
Mobile-sweep parking and drive	3,000.00	3,200.00	3,100.00	3,350.00
Total contract costs	$18,950.00	$21,150.00	$19,225.00	$21,775.00

Without exposing bid amounts to the bidding contractors, the FM should ask questions such as the following: Exactly who is the manufacturer of materials being used? What are the types of equipment they will employ? What is the skill level of the workers? When will the service be provided (or not)? The FM might discover that the contractor with the highest bid is providing a greater service level or using more trusted materials than the contractors with lower bids. If this is the case, consideration should be given to the contractor with highest bid because it may be the better value.

Once a bid has been accepted, a post-bid conference is held to make sure that both parties fully understand the expectations of contract fulfillment. It is a way to discover and then eliminate any minor issues that may arise in the performance of the contract.

3.8 CONTRACT NEGOTIATIONS

Negotiations for a new contract commence when the term of a service contract is about to expire. In addition to price, three other elements of a service can be negotiated:

- **Quantity**—Obtaining more service for the money.
- **Quality**—Obtaining a higher-quality service.
- **Time**—Obtaining the service performed at a time or date that is more convenient for the facility, extending the length of the contract, or having the service performed in a shorter time frame.

Successful negotiation starts with the FM knowing the best alternative to a negotiated agreement (BATNA) of the opposing side. BATNA is the contractor's cost to provide the service and an additional amount added to it for reasonable profit. This price is the lowest contract amount the service provider will agree to take for its service. Knowing the BATNA of the service contractor gives the FM an advantage in negotiations. The FM can then begin negotiations at a number below this figure. Conversely, the FM also needs to know the highest amount that his or her facility organization is willing to pay for the service. A contractor who exceeds the limit forces the FM to bid for the service.

The FM must also understand other factors in negotiation. Market conditions are a big factor. If there is plenty of work available for the contractor, the FM in this situation does not have as good a bargaining position as when the contractor is looking for work. Bargaining with the contractor when economic times are good would then be based on future work the facility organization might provide when economic times aren't so rosy. To keep this work coming, the contractor needs to provide fair pricing that keeps the facility organization from looking elsewhere. The contractor should also know that it can save money by staying with a particular facility organization because, over time, it will develop service efficiencies that translate into increased profits. Also, avoiding the expense and effort required for bidding benefits both sides.

In addition to market conditions, it is valuable for the FM to know if the contractor wants to expand its business. Even in good times, an expanding business cannot afford to lose its current customers, so this situation gives the FM an advantage in negotiations

Example: Contract Renewal

The contractor makes an offer to extend the service for another year at a 15 percent increase over the prior year. This initial offer from the contractor should be regarded as a point from which negotiations can begin. Not being a state or federal government agency, the owner has no legal obligation to hold open bidding for the new contract. The contractor has performed well for the facility organization, so there isn't any reason not to renegotiate the contract for another term.

However, the FM provides a counteroffer. The counteroffer may be based on what other, similar facilities are paying for the same service. (This comparison work should have already been done for the existing contract.) The FM might counter with an offer based on inflation. The contractor might then counter with an offer based on inflation and on additional factors that are unknown to the FM. For example, for labor-intensive services, a wage increase for the workforce could raise the contractor's BATNA. The cost of a particular product used by the contractor may also have risen much higher than the overall inflation rate. These factors ca drive up a contractor's cost.

If the contractor still holds to a 15 percent increase, the FM might offer to change the length of the agreement but demand a price fix for subsequent years. Conversely, the FM might end the negotiation for now and state that he or she will seek bids from other contractors. Additional pricing has long been an effective tool in contract negotiations. The threat of going out for bid might be enough to force a negotiation. The FM needs to make sure that the contractor has time to think about the possibility of the FM taking this action. The contractor is immediately faced with the possibility that the 15 percent increase makes his or her service uncompetitive in a bid situation.

In any event, the FM needs to distance him- or herself from the negotiation. The FM should remember that some negotiations will be losing events. The FM needs to move quickly away from the loss, not dwell on the event, and begin thinking about the future.

3.9 OUTSOURCING FACILITY MANAGEMENT

Top-level corporate managers (the chief executive officer [CEO], chief operating officer [COO], and chief financial officer [CFO]) are ultimately responsible for the overall financial and operational success of their organization. A COO of a manufacturing company may decide to hire a facility maintenance/ management contractor and get out of the facility management/maintenance function that is ancillary to their core business of producing manufactured goods. The corporation can then concentrate its efforts on the core function, which is product manufacturing. The added costs of outsourcing this service may be offset by greater concentration on the main profit center components of the business (production, marketing, and sales). A school district is another example. Its core function is education; by using a facility management/maintenance contractor to handle the day-to-day operations, district superintendents and principals can focus on educating their students.

The contracted FM shows his or her value through the work he or she does on a daily basis and through effective planning for the facility's future needs. A value of having contracted maintenance/ management with a large firm is that a management company has the resources to back up the manager should a difficult situation or an emergency arise. The management contracting company may also offer a plethora of other services that they can provide to the customer: landscape maintenance, energy management, worker training, pest removal, housekeeping, disaster cleanup, construction management, and human resources management. The contracted FM is often asked by his or her company to market these services to the client. In this situation, the ability to sell services to the client is an important skill because the management company can profit from providing these additional services as well as the management fees for the manager's service.

Facility managers may benefit by becoming managers who are under contract to the facility. Facility management/maintenance contracting firms can provide a career path for facility management and maintenance personnel because the contractor manages many facilities of various sizes and complexity. As the manager's skills increase, she or he can take on increased responsibilities. A company that can provide assignments at increasingly demanding facilities can provide the FM with a path to advancement.

Another issue with both positive and negative aspects is that the FM is not a direct part of the client organization. The contracted FM is an outsider. Being an outsider can be advantageous because it places the FM apart from the client's internal politics, and the FM is frequently viewed as an outside expert. Conversely, it can be difficult for the contracted FM when the facility profits greatly from the FM's efforts, but the FM doesn't gain directly from these efforts. Ultimately the FM will gain through contract renewals and the selling of services that provide profits for the management contracting company.

Next, we will discuss the following three types of management contractors:

- Institutional facility management contractors
- Large property management contractors
- Small property management firms

Institutional Facility Management Contractors

Hospitals, universities, school districts, stadiums, and military bases are some of the organizations that institutional facility management contractors target to provide their services. They usually seek contracts that are three years or more in duration. Sometimes contracts have a trial year whereby the facility is under no obligation to continue the contract if the services rendered or costs aren't what they expected. Contracts might also have a contract cancellation clause if the contractor doesn't perform satisfactorily. The facility that has a cancellation clause in the contract may be able to cancel the contract with no more than one month's notice to the contractor.

Institutional facility management contractors may take on the entire facility operation either sending in their managers and employees to take over the current management and office staff, sending only manager(s) and hiring only the best members of the current facility staff, or bringing in an entirely new contracted staff. In addition to facility management, the service most often provides maintenance and housekeeping. Sometimes the facility management contractor provides all the housekeeping equipment, staff training, and chemicals. The service might also include a computerized maintenance management system (CMMS), discussed in greater detail in Chapter 2, to provide work orders, personnel management, and worker productivity tracking. In this case, temporary staff is provided to catalog the facility's building mechanical equipment in order to enter the data into the CMMS.

A facility management/maintenance contractor can be an extremely useful resource. Such a contractor can be a lifesaver for facilities that want to regain control over their maintenance personnel and/or management function. A facility may then employ a management contractor for a few years to get their facility back on track. Later, when the operation is running efficiently, the facility can decide once again to handle all or part of the maintenance/management operation in-house.

Large Property Management Contractors

Large property management contractors are similar in many ways to institutional facility management/maintenance contractors, except that these large companies differ by providing management for commercial properties only. Property management firms also have additional services to sell, and these services differ from one company to the next. Property management firms may provide services such as real estate brokerage, financial asset management, accounting, tax services, property acquisition and disposition services, property development, construction, and project management services. They may be small operations managing a few small apartment buildings, or they may be huge international corporations managing residential, industrial, and commercial office properties worth tens of billions of dollars.

A new service that some management contractors offer their clients is a path to achieve Leadership in Energy and Environmental Design® (LEED®) certification for their properties. The management company can provide the personnel with the experience of having completed several certification projects. With LEED Existing Building Operations and Maintenance (EBOM) criteria, existing buildings, some decades old, can achieve certification.

Small Property Management Firms

Small property management firms provide maintenance/management services directly to small property owners. They target their services to apartment owners, shop owners, and small businesses. They can either provide a variety of services directly or they subcontract to get them. A large portion of their business may consist of performing small maintenance or light construction tasks. They can change light bulbs, paint, provide minor remodeling services, put up signs, collect rent, answer tenant work orders, sweep parking lots, pick up trash, and basically do whatever the client needs. These firms generally do not provide many professional growth opportunities for FMs, but many FMs have started their own small property management businesses and done very well financially.

3.10 OUTSOURCING IN THE UNITED KINGDOM

Outsourcing in the United States can be very different from outsourcing in a foreign country. For example, in the United Kingdom a 2006 law called Transfer of Undertakings Protection of Employment (TUPE) is written so that the United Kingdom meets the requirements of the Acquired Rights Directive of the European Union. This law protects employees from losing their job when another company or contractor is hired to do their work.

Example: Enforcement of TUPE in the United Kingdom

A new cleaning company wins a contract to perform housekeeping services at an office property. The new company must hire the employees of the company who lost the bid at their previous wage and benefit level. TUPE prevents workers in the United Kingdom from losing their job, and it prevents companies from driving down the wages of the particular service industry as a whole by the company seeking the service provider who offers the lowest priced contract and therefore frequently pays the lowest wages.

In the United States, union rules in employee contracts may sometimes prohibit or limit outsourcing. The restriction is generally based on situations where work would be taken away from a union employee(s) by a nonunion contractor. An example is a situation in which a company finds that it can outsource a particular function or process to an outside contractor to reap higher profits. Another example is when a nonunion contractor is hired to work with a union contractor, which could happened on a construction project that has mostly union companies doing the work. If a nonunion contractor is hired to do the electrical work, the union contractors on the job may institute a job action to slow down or disrupt the project.

Some owners may require that a percentage of the contractors hired for a construction project be union contractors or minority contractors. A municipal transit authority is one example of an organization that might require that a percentage of the outsourced contractors are union. It may also require that a percentage of the contractors are minority owned. A facility management contractor might be charged with the same hiring responsibility when employing contractors for an owner.

REVIEW QUESTIONS

1. How is an SLA different from a standard service contract?
2. Name five reasons why a facility operation would want to outsource services that it could provide in-house.
3. Explain the different reasons why the duration of a contract for services is important.
4. What is an approved equal?
5. When should the FM ask for a cost breakdown of the work items in a bid?
6. When should a facility or property utilize the services of a maintenance management contractor?
7. Describe how a walk-through is used in the construction process.
8. Explain the difference between open bidding and closed bidding.
9. Why is it important to have a bid form?
10. In what ways can outsourcing be limited or restricted in the United States?

Essay Question

11. Describe a situation in your own life where you negotiated an agreement. The agreement may have been successful or unsuccessful, but explain in about one and a half pages why you think it was so. Be prepared to discuss your thoughts in class.

FINANCIAL MANAGEMENT AND CONTROL

INSIDE THIS CHAPTER

Financial Management: Is It a Friend or Foe of Green Initiatives?

Financial management is the process of managing budgets and resources for current and future needs so funds are available when needed. The chapter begins with the basic financial tools managers use to pay for goods and services received. We then discuss planning for future expenses through various budgeting methodologies, and the two major budget divisions of capital and operating budgets. The chapter ends with a discussion on financial control and the use of budget data to find problems with the facility operation.

4.1   UTILIZING FUNDS

Determining the organization's purchase procedures is a good first question for a new facility manager (FM) to ask the accountant or office manager. Facilities often have their preferred tools methods for payment of the day-to-day items and another set of tools and methods for payment of items provided under contract. Proper utilization of the accounting system is predicated on an understanding of the organization's accounting system as a whole, and the individual parts (budget types, account types, account numbering system, and methods of utilizing funds and providing appropriate backup for proof of proper fund utilization). Knowledge of the accounting software also enhances use of the organization's purchase procedures.

Using the proper methods for obtaining the money to pay for goods and services for the facility is extremely important. Accidental or intentional misuse of facility funds has been the downfall of some FMs.

Purchase Orders

The *purchase order (PO)* is an agreement between the FM's organization and the vendor company providing the product or service. When a facility manager purchases supplies or requires service for the facility, under normal nonemergency situations, the company contacted for these supplies and services has an account with the facility organization. The important financial information about the organization contained in the account enables the company to be paid for their work or product supplied. Once an account is established, the FM asks the company providing a service or a product for the total amount the item or service will cost. A purchase requisition is then filled out, and a purchase order number is issued by the accountant. When the vendor provides the product or service that is stated in the purchase order, the vendor is paid the amount requested in its bill.

When the FM's organization receives many products or services from a particular company, the organization might issue an *open purchase order*, which is a standing purchase order assigned one account number for all future uses. An open PO is a convenient arrangement with suppliers of hardware; plumbing; electrical parts; or heating, ventilation, and air conditioning (HVAC) equipment, for instance. Often, these supplies are needed on short notice, and the open PO allows for quick purchase. Open PO amounts are generally based on an estimated amount that covers the amount the organization will owe the vendor over a specific time period. This period may be as long as one year. In this way, a new PO does not have to be generated every time the facility calls on the vendor for a product or service. The supplier simply invoices the facility for the items supplied. As the amount of available funds dwindles in the open PO, more funds can be added to the account without changing the PO number.

A list of persons who are permitted to purchase items from the account is given to the supplier. It is a good idea to issue different PO numbers to each person who is on the open supplies account with a particular vendor so that the accountant knows who is spending the company's money and what items they purchased. If there is only one open PO number for everyone purchasing a variety of items at a supplier, tracking individual expenses becomes difficult, and preventing one person from using all the funds is extremely difficult.

Corporate Credit Cards

Some facilities issue credit cards to their managers and supervisory personnel for minor purchases from vendors where no account currently exists. Credit cards are easy to use and extremely valuable when emergency repairs are needed (e.g., a critical tool or part breaks and a replacement is needed). But the use of credit cards must adhere to your company's guidelines and procedures. Failure to follow the rules for using credit cards can result in severe disciplinary action or even dismissal.

Petty Cash

An organization may wish to keep a small amount of actual cash and coins on hand for purchases that require cash only, or other instances where giving someone cash is appropriate. Some organizations keep petty cash so they can tip someone making a delivery to the office, pay highway tolls, or purchase postage stamps. But most of this is no longer necessary because of credit cards and online services; thus, keeping petty cash is becoming less and less important.

4.2   THE BUDGETING PROCESS

A *budget* is an item-by-item, projected cost breakdown of all the expenses a facility is projected to incur during the coming year. Obtaining these individual budget amounts can be done in several ways, and the method used defines the budget produced. Budgeting methods are as follows:

- Vendor data budget
- Formula-based budget
- Zero-based budget
- Percentage budget
- Combination budget

Vendor Data Budget

Vendors who do business with the facility are asked for a budget number for the services they normally provide the facility for the following year. Depending on when the vendor invoices for their services, the FM may ask vendors to provide their costs on a monthly, quarterly, or biannual basis.

Supplies also should be budgeted with a vendor data budget. A *vendor data budget* is a budget that is constructed from cost data provided by a vendor for goods or a services that the vendor provides to the facility. For example, if parts for a piece of equipment are purchased on a consistent basis, the distributor of the part should be asked about scheduled price increases or decreases during the coming year. If the cost of a part will go up 15 percent in June of next year, the budget for those parts should be increased by 15 percent starting in June.

Formula-Based Budget

We can use one or more of several formulas to calculate a variety of costs in a *formula-based budget*. The most popular one is dollars/square foot, but any formula that equates cost to amount can be used. A formula-based budget is especially helpful in situations where prior numbers for performing the work are unavailable. For example, if the FM knows that prepping and painting an interior wall costs $0.50/square foot, and the facility will be painting 1,000 square feet in the coming year, the cost can be calculated as follows:

$$(\$0.50/1 \text{ square foot}) \times 1,000 \text{ square feet} = \$500$$

Zero-Based Budget

A *zero-based budget* is a negotiated budget between the people with the authority to provide the budgeted funds and the person requesting the funds. Every item on the budget is challenged, and the FM is required to give justification in the form of facts and figures for all budgeted items. Zero-based budgeting forces the FM to research costs thoroughly for the budgeted item. It also requires the FM to analyze the associated risk to the facility if the budget item is eliminated or reduced in scope. By presenting an analysis of facility needs in relation to the budgeted item and the risk and cost, the FM can be successful using the zero-based budget format.

Percentage Budget

A *percentage budget* works only for items for which budgeted amounts can be obtained from the prior year's budget. A percentage is added to the amounts of the previous year. The common situation is a percentage increase in the budgeted amount, rather than a percentage decrease. For example, if the service charge for security personnel will go up 5 percent in the current year, the FM can multiply the previous year's budgeted amount for security by 1.05 to get the total amount that must be added to the current budget to account for the 5 percent increase.

Combination Budget

Most budgets employ a combination of vendor data budgets, formula-based budgets, zero-based budgets, and percentage budgets to achieve budget numbers that are as accurate as possible in what is called a *combination budget*. For example, if the FM is confident about the yearly increase for certain account items, budget numbers can easily be obtained using the percentage budget method. For large budget amounts, upper management may require the FM to provide a great deal of justification, and these numbers can be provided through a process of zero-based budgeting.

4.3 BUDGET MANAGEMENT

There are two major budget categories:

- Operating budget
- Capital budget

The *operating budget* provides the fund of money for all the maintenance work; the wages paid to the FM and the facility staff; and everything necessary for the maintenance, operation, and repair of the facility. When making cost estimates for the operating budget, the FM breaks down all the types of work and necessary supplies that will be required for the following year into topics such as plumbing, electrical, HVAC, landscape, wages, and so on.

Funds for major maintenance, new construction, and large building improvements are provided in the *capital budget*. The size and financial situation of the facility organization determines the dollar amount that justifies budgeting for an item or a project in the capital budget. *Major maintenance* is a maintenance item that represents the expense of a significant amount of funds. Major maintenance consists of items such as reroofing a building for $400,000, or overhauling an electric centrifugal air-conditioning chiller for $100,000. Capital improvements can also consist of large construction projects that cost millions of dollars. Some examples of capital improvements are new building construction, lobby renovations, and building rehabilitation or conversion to a new use.

Sustainability is a very important factor when budgeting for items in the capital budget. Major construction and building improvements are frequently part of an overall plan for organizational success that may span several years. Higher educational facilities may have a five-year plan to guide capital expenses, and these items are part of the capital budget.

Budgeting Using Account Types

Budget topics, whether items in the capital or operating budget, have an account code number assigned to them. Let's look at the plumbing account, which is part of the overall operating budget (see Table 4.1, which does not include dollar amounts). The plumbing account is further broken down into subaccounts, such as plumbing supplies, plumbing repairs, plumbing fixtures, and

TABLE 4.1 Breakdown of Plumbing Budget

14560 Plumbing	January	February	March
14561 Plumbing materials			
14562 Plumbing tools			
14563 Plumbing fixtures			
14564 Drain cleaning			
Plumbing total			

plumbing maintenance. Breaking the plumbing account into various subaccounts for plumbing functions forces the estimator to be more accurate because he or she is no longer providing an overall plumbing estimate for everything that has to do with plumbing; instead, he or she must budget topic by topic. The manager who has multiple sites under his or her control may budget location by location. Topic-by-topic and/or location-by-location budgeting provides for greater budget control because it is easier to track when amounts paid out for the work have gone over or under the budgeted amount.

The total facility budget for maintenance is coded, for example, as 14500. Plumbing is a subaccount of the maintenance budget 14500 and is denoted as 14560. Adding the "60" tells the accountant that it is plumbing. Other accounts could be coded as 50 for electrical maintenance, 40 for HVAC; any unique number for each type of maintenance will do. A further breakdown of the plumbing account to provide more accuracy and control might be similar to the line items in Table 4.1: 14561 (plumbing materials), 14562 (plumbing tools), 14563 (plumbing fixtures), and 14564 (drain cleaning).

The budget is set up in rows and columns. Budgets may be set up in many different formats to meet the accounting needs of the organization. But all spreadsheets are set up the same: Columns run across the page; rows are listed from the top to bottom. The first cell in the first row defines the expense item account code, then the name of the item; the dollar amounts, which are the costs estimated for the item, are placed after the item in the month column when the work will be paid. After the December column, there is a column for the total amount budgeted, a column for year-to-date amounts spent, and another for amount of variation from the budgeted amount. Variations are either positive dollar amounts, indicating that the FM is within budget, or the amount is surrounded by parentheses ( ), which means the FM is over budget by that amount. The amounts in parentheses are added as a negative number to all the other numbers in the budget variation column. If there isn't an expense for a particular month, there is no need to put a number in the column for that month, even if budgeting is done on a month-by-month basis.

The FM could also budget quarterly, semiannually or even yearly. The FM puts the total cost for the period. Organizations often start their budget year on January 1, which is called a calendar year. Or organizations may also use a fiscal year, which frequently starts July 1, but it could actually start at any time they choose as long as it runs for a twelve-month period. The U.S. government uses October 1 of the year prior as the beginning of its fiscal year, and its budget ends on September 30 of the budgeted year.

If the organization is using the calendar year for budgeting, an item budgeted quarterly has the first dollar amount go under January, the second quarter budget item under April, the third quarter under July, and the fourth quarter under October. Semiannual budget figures are placed January and July.

Whichever time periods your company uses for budgeting, the FM must budget the estimated dollar amount in the month the facility will pay the bill.

Financial Control

Facilities use budgets to gain financial control. It is the same reason why households budget money: to know that sufficient funds are available and when it may be necessary to borrow funds. It gives financial professionals a snapshot of how a facility is doing with regard to revenues earned and expenses. Many facility managers are concerned only with the facility operating budget, which consists of all the major and minor work expenses, parts supplies and services, and facility department project costs that will be expended over the course of the next year. Traditionally, many facilities put forth the most effort on the budget process in August for the following year's budget. But budgeting is something that should be done continually because it is too easy to forget about a necessary future corrective action once the incident or equipment failure has passed by several months.

Example: Budgeting for Water Line Replacement

In April, an FM's maintenance team made an emergency repair to a leaky pipe fitting on a water line. The leak ruined the carpet and some ceiling tiles, but it was caught before any real damage occurred. At the time of the leak, the FM noted that the entire line was in poor condition. If nothing else is done, the water line could fail completely in two or three years and cause a flood, and the FM will be viewed as being negligent. At the time of the lea, the FM should have immediately budgeted for a water line replacement. It is all too easy to forget about a budget item such as a water line replacement because, after several months and many other maintenance emergencies, a water line that is not actively leaking is easily forgotten. The FM should have a file, either written or electronic, to record all necessary future work.

Example: Budgeting for Landscaping

To gain more financial control, always budget known amounts separate from unknown amounts. For example, a landscape maintenance contract is for $1,500 per month, to be paid in the months of March, April, May, June, July, August, September, and October. This timeline results in a contract total of $12,000 for the year. The FM is also planning on having the landscape contractor perform about $4,000 worth of new plantings, but the FM hasn't decided exactly what he wants and when. This brings the total amount the FM expects to pay the contractor to $16,000.

Poor budgeting practice would be to budget $2,000 each month during the term of the contract, which runs for eight months. The problem is that, although the amount budgeted totals the $16,000 that the manager expects to pay; the extra amount can hide contract overcharges by the contractor. If the contractor mistakenly charged $1,600 per month for March through October instead of $1,500, the FM might not pick up the mistake right away because the line item total would still be under budget. The budget should be set up to act as a second check on what the FM agreed to pay in the contract.

An acceptable budget that accurately covers the amount to be paid is shown in Table 4.2.

A better budget is one that alerts the FM to possible overcharging by the vendor. The budget report shown in Table 4.3 separates the costs for the new plantings from the basic landscape contract.

A better and more efficient budget, like the one shown in Table 4.4, results from budgeting the variable new planting amounts during the month that you would actually be paying for the work completed. This budget is better and more efficient because the facility will not be over budget for the work done in March and paid for in April. In July, an accountant discovers that the FM did not do any new plantings yet this year. With this budget, it is easy to delete the $2,000 budgeted in March from the budget total, or, if it is still needed, move the amount to November.

TABLE 4.2 Acceptable Landscaping Budget

	Jan	Feb	Mar	Apr	May	Jun	July	Aug	Sept	Oct	Nov	Dec	Total
Landscape contract			2,000	2,000	2,000	2,000	2,000	2,000	2,000	2,000			16,000

TABLE 4.3 Better Landscaping Budget

	Jan	Feb	Mar	Apr	May	Jun	Jul	Aug	Sep	Oct	Nov	Dec	Total
Landscape contract			1,500	1,500	1,500	1,500	1,500	1,500	1,500	1,500			12,000
New plantings			500	500	500	500	500	500	500	500			4,000

TABLE 4.4 Better and More Efficient Landscaping Budget

	Jan	Feb	Mar	Apr	May	Jun	Jul	Aug	Sep	Oct	Nov	Dec	Total
Landscape contract			1,500	1,500	1,500	1,500	1,500	1,500	1,500	1,500			12,000
New plantings				2,000							2,000		4,000

TABLE 4.5 Budget Variance Report

Supplies	Budget March	Actual March	Variance March	Percentage of Budget Year to Date (YTD)	Budget for the Year
Plumbing	1,000	1,200	−200	26.7	12,000

The efficiency of the budget becomes more evident by examination of a budget variance report. The report can be generated from the computer at any time, but it is typically generated quarterly. A **budget variance report** compares **actuals**, which is a term for the amount(s) of money actually spent as compared to the amount budgeted.

Let's assume that a facility has budgeted $1,000 per month for plumbing supplies. In March, a budget variance report like the one shown in Table 4.5 is run.

The report indicates that the facility is over the $1,000 amount budgeted for March by $200 [negative balances such as the −200 are frequently shown as (200)]. If the manager had spent the exact amount budgeted for the first three months and not gone over budget, the total spent would be $3,000. This three-month total would equal a percentage of the budget year to date (YTD) of 25 percent. Because the facility went over budget in March, it is 26.7 percent of the total budget for the year. This isn't a serious situation because three-quarters of the year is left to make up the variance. Budget numbers are estimates, so the facility is actually doing fairly well. It is 20 percent over budget for the month of March, but only about 7 percent over budget for the quarter.

4.4   BAD BUDGETING PRACTICES

Bad budgeting practices (not including outright fraud) are as follows:

- Padding the budget
- Wasting funds to keep the future budgeted amounts in the account higher than they should be (also known as *eating up the budget*)
- Miscoding accounts
- Failure to report amounts that will not be spent
- Failure to accrue monies that will be needed at the budget's year-end and having them taken away just when they are needed
- Lumping together fixed contract payment amounts with variable payment amounts

Adding more funds into a budget item than what is required is known as padding the budget. It is always a good idea to consider what accurate price increases and to add an appropriate amount to cover costs that are likely to occur during the budget year. Adding large sums of money to cover work just to be on the safe side is bad practice for the organization as a whole, for the following reasons.

Financial managers (also known as accountants and comptrollers) must be able to report your organization's financial condition accurately. The negative effect of unnecessarily adding to a budget is compounded when many FMs in the same organization practice it. The result could be a bleak profit projection for the quarter when, in reality, results should have been positive. Estimating low profits can effect investment activity and end up reflecting negatively on a property investment management firm.

Facility managers point out that they aren't criticized when they are under budget, but they are often treated harshly when they are over budget. But when the FM is dealing with large sums, a bad budget estimate can be just as damaging to an organization, whether it results in the account being significantly over or under budget. The bleak profit projection for the quarter and the inability of the organization to use the excess monies that the manager budgeted may also result in necessary improvements being postpones or not being done at all. Unless the FM can show that the performance of the budgeted work resulted in unexpected savings, putting an account substantially under budget should result in the same reprimand as an account that is over budget.

Some managers who want to look good hide an under-budget condition by adding extras to a project or by purchasing items that aren't needed in order to make up the difference, Their justification is that, if the funds aren't used, the funds will be lost, and the budget category will get less funding next year. An ethical manager knows that wasting funds is wrong. When such actions are discovered by upper management, the FM's position will be indefensible. It is better to admit that the budget was more than what was needed and allow the excess funds to be removed from the budget. FMs should sit down with the accountant on a regular basis and review the budget variance report. The FM should discuss what funds could be removed from the budget, or which money should be moved to a later month because the work was not completed as scheduled.

Example: Budgeting for a Painting Contract

During April, the weather was unusually wet. The painter contracted to paint the pylon sign for a shopping center couldn't do the work as planned because of the weather. So the FM decided to put the work off until the fall when the weather might be better, and the painter wouldn't disturb the flowers planted around the pylon sign. In addition, the sign would be freshly painted for the winter holiday season. The money budgeted for painting the sign should therefore be moved from April to November (or whatever month that the painting was rescheduled). The FM no longer has to worry about his April money "disappearing" in say, July, when the accountant at the home office would have noticed that the money for the painting work had not been used for over two months beyond the budget month and would have deleted the money for the painting from the budget.

A few months prior to the end of the budget year, the accountant may ask the FM to point out which monies in the budget will definitely be spent in that final quarter of the year. The monies that are identified by the FM as still being needed will be reserved. Accountants use the term *accrued* when they reserve funds for later use. The term *accrued* is just a way of saying that the monies you intend to spend will be put on the books as an expense prior their actually being spent in order to prevent the money from being deleted in the rush to close out the accounts at the end of the year.

Another poor practice is miscoding expenses to take money out of accounts that have excesses and to hide shortfalls in other accounts. The justification is that it is better to use the money where it is needed. But in miscoding expenses, the FM loses the ability to track expenses accurately. Account monies are jumbled together, and previous budget mistakes are easily repeated. The previous year's budget is the best starting point for developing the budget for the coming year. If budget items are miscoded, the data for the new budget will be flawed.

Finally, a common error is to mix fixed-price recurring contract costs with variable costs. For example, a landscape account may have a monthly fixed charge for mowing the grass and a variable cost for fertilizing depending on how much fertilizer is applied. Budgeting one amount to cover both fertilizer and mowing makes it difficult to track the cost of supplies used in each separate task.

4.5   THE BUDGET AS A TOOL TO FIND PROBLEMS

An accurate budget alerts the FM to problems because the actuals are less than or greater than the money that was budgeted. A significant variance means that a problem exists, for example:

- A water bill that is too high could be a sign that the lawn sprinkler system may be faulty or that the timer to operate the sprinkler is set for too long a period.
- Being over budget for roof repairs may signal that the roof is deteriorating at a rate faster than predicted in a roof survey.
- Too much overtime pay could indicate a failure in supervision or improper servicing for equipment.
- If more funds are constantly needed to be budgeted for a service, it might be time to find another service provider.

A budget can indicate more than just the finances of an organization. A budget is a window through which to view the success or failure of an FM's operational and management abilities.

4.6   FINANCIAL MANAGEMENT
FOR GREEN INITIATIVES

Green initiatives summarize the hard work, planning, team building, investigating, and benchmarking that have led the FM to the point of green project execution. In the past, green initiatives were avoided. They were thought of as being at best a costly way of showing environmental stewardship, or as an expensive marketing tool to attract tenants with green sensibilities. The justification for this conclusion was a simple return on investment (ROI) calculation. For example, a property costs $10,000,000 and generates $1,000,000/year in rental income. This property has a return of 10 percent. The same property, after spending $2,000,000 for green initiatives, still receives $1,000,000/year in rent, but now has an ROI of $1,000,000/$12,000,000, or 8.3 percent. A project such as a lobby renovation was thought of as being a better way to attract new tenants.

Today, the value of a well-planned green initiative is well known. Green initiatives comprise projects such as lighting retrofits, xeriscaping, green and cool roofs, solar energy, geothermal energy, building HVAC system energy-efficient equipment upgrades, lowering the carbon footprint, heat island effect reduction, permeable paving with water collection, and water-saving devices. Green initiatives have additional value because they frequently provide a building with a lower energy cost per square foot, decreased maintenance requirements, and/or reduced water consumption. Green initiatives not only provide a direct cost savings, but they make the tenant spaces within the property easier to lease because of reduced costs for the tenant. It makes the property stand out among properties of the same type or class. Higher leasing rates (percentage of space leased as compared to total rentable space) provide a greater income stream, which can increase the return on investment to levels higher than that of initiatives that provide no dollar returns.

REVIEW QUESTIONS

1. *Eating up the budget* is a term used when managers _____.
2. Facilities budget in order to attain _____.
3. The type of budget that provides the funds for the day-to-day maintenance operation is the _____ budget.
4. Major maintenance and new facilities are paid for from funds taken from the _____ budget.
5. A report that compares actuals to the amount budgeted is the _____.
6. We can formulate a percentage budget by _____ to the numbers budgeted the previous year.
7. The process whereby all numbers in a budget are challenged is known as _____.
8. Formula-based budgeting can be especially useful when _____.
9. When budgeted money is moved so that it is not lost and so that it can be used later in the year, we say that the money is _____.
10. Allocating more funds to a budgeted item than what is actually required is known as _____ the budget.

Projects

11. Use a spreadsheet program such as Microsoft Excel to complete this project. Write a budget for plumbing supplies for a one-year period.
12. Use the following information and the information in Table 4.6 to evaluate a project involving remodeling a brick apartment building. The project includes repairing and replacing equipment as necessary. The apartment building is twelve stories tall, has 125 apartment units, and is twenty-five years old. The goal is to keep the building viable in the marketplace, expend the $350,000, and get the most benefit for the money from a maintenance and marketing stand-point. Inspection of the building exterior has indicated areas of loose mortar between some brick joints. The apartments are cooled by window air-conditioning units, and each apartment has two air-conditioning units. Half of the air-conditioning units were replaced twelve years ago. Bathrooms are showing their age. The bathrooms consist of a combination tub and shower, vanity, mirror cabinet, and toilet. Heating is hydronic and is provided by a pair of fire tube boilers. Elevators are serviceable, but they are running slow and could benefit from a system upgrade. Each unit has six windows and they are serviceable, but they show signs of their age as well. There are twenty vacant apartments in the building. Other than replacement of one air conditioner in each, none has had any work done since the apartment was built. Carpeting is changed only when an apartment is vacant.

TABLE 4.6 Replacement, Upgrade, and Repair Costs versus Equipment Useful Life for an Apartment Renovation Project

Item	Replacement Cost (Including Installation)	Cost to Upgrade or Repair*	Useful Life
Window air conditioner	$ 325	NA	15 years
Hydronic boiler	$135,000	$50,000 (replace burners with energy-efficient units)	30 years
Windows	$ 350	NA	15 years
Toilet	$ 300	NA	28 years
Vanity/lavatory	$ 700	NA	20 years
Mirror	$ 200	NA	20 years
Tub/shower	$ 1,200	$500 (recoat interior of tub; life of recoat = 10 years)	25 years
Bathroom floor and interior finish	$ 800	NA	25 years
Carpeting	$ 1,250	NA	10 years
Building exterior	Tuck point mortar joints $150,000 (total building tuck pointing)	$45,000 (repair)	35 years

*NA = not applicable.

Develop a plan to utilize the budgeted funds efficiently. Provide a rationale for your choices and present the information in the form of a spreadsheet. Note that it is important to consider the useful life of the equipment discussed. Assume that any apartment/building items not mentioned are in very good condition.

What items would you complete? Explain. How much money should be budgeted for the following year to complete the work?

Essay Questions

13. You are a facility manager at a new facility without any prior budget records. How would you develop a budget for the coming year?

14. Why is it important to a national real estate investment company that FMs report when they know they won't be using the funds budgeted for major maintenance or capital improvement projects?

CONSTRUCTION MANAGEMENT AND SUSTAINABLE DESIGN

INSIDE THIS CHAPTER

What Makes Construction Sustainable?

Construction project elements are described from conception to completion. The chapter discusses the important contribution facilities managers (FMs) make to construction projects by acting as owners' representatives. The chapter shows how the construction contract greatly influences how the project will be performed and the relationship among construction personnel. It discusses the different types of bonding and other legal instruments used in construction. Sustainable design and sustainable construction methods are highlighted in the chapter, along with newer construction document production systems such as building information modeling (BIM) and laser scanning. The chapter also gives an example of a sustainable construction scheduling method known as the critical path method (CPM), which was developed during the U.S. Navy's Trident submarine project in the mid-1950s.

5.1 THE FACILITY MANAGER'S ROLE IN BUILDING CONSTRUCTION AND REMODELING

The facility manager needs to be part of the design team to ensure that the building will be economical to maintain as well as pleasing to the eye. The FM needs to insist on this point prior to concept drawings being finalized. Once construction plans are completed, it is too late to redesign for maintainability because of the time and cost required. A building that is efficient to operate and economical to maintain is well on its way to being a sustainable building.

Facilities are constantly changing. To meet the changing physical building requirements, facility managers frequently need to put on another hat—*a hard hat*—and perform as a type of construction manager (CM). For small projects, where the FM is very knowledgeable in the steps and trades necessary to complete the work, the FM can hire an architect to draw up plans, and obtain the building permits necessary. The FM can then directly hire the different building trades needed. The FM is in fact acting as a general contractor (GC). Other projects that are larger in scope require the services of a general contractor. General contractors are construction firms or individuals that hire the various trades necessary to complete the project. GCs take total control of the construction project from start to finish. The term *general* is appropriate because a GC acts in many ways like a military general, marshaling the various construction trade groups to produce a finished built product.

When a section of an office building's interior requires remodeling to meet the needs of a new tenant, a GC is hired by and works in concert with the facility manager, who has plans drawn up by the architect. This team arrangement is normally sufficient to complete the task.

When construction projects are much larger, such as when universities and hospitals undertake extensive, multimillion-dollar, new-construction projects, upper management may choose a different solution. Upper management may hire a firm who specializes in construction management for the project's duration, use the project architect, or hire a separate construction manager/facility manager to become part of the facility organization directly. In either case, the construction manager supervises the project and works with the existing FM. The CM/FM position may turn out to be permanent position, or it may be completely eliminated once construction levels at the facility return to normal.

The difficult nature of these projects, created by adding on to an existing facility and integrating all the building systems while keeping the original facility in operation, can result in a project lasting for several years. In this case, the direct hiring of a CM/FM is a viable cost-saving alternative to outsourcing this function. The CM/FM has all of the facility's construction and construction planning responsibilities. The CM/FM might also take over the facility operation for the FM when he or she needs additional help, is offsite, or is on vacation. Frequently, the position of FM and CM/FM are of equal rank in the facility hierarchy. In the following section, we will describe the various roles that are discussed throughout this chapter.

5.2 PROJECT PERSONNEL

It is most satisfying to create a lasting physical structure where there was originally nothing or nothing of value. For instance, a rundown old building can be made into a beautiful and valuable asset to the community. Being part of a team performing new construction or building renovation gives this satisfaction. Everyone associated with a construction project has a level of pride. At all participation levels of a construction project, members of the construction team care about the project's execution and the quality of the end result.

There are three sides to a construction project: the owner's side, the contractor's side, and the side of the general public. When these three sides work in a spirit of cooperation, with each side seeking to build or maintain long-term relationships, construction projects can move along quickly and with fair compensation or pricing for all parties.

Owner's Side Personnel

The following personnel work in the best interests of the owner or owners. In traditional construction, these construction professionals must work in the best interest of the owner.

Owner's Representative   On commercial construction projects, construction contractors rarely meet the person who actually owns the property. As stated in Chapter 1, an individual rarely owns a large property completely. More often than not, large corporations, real estate investment trusts, and pension funds are the actual owners of large commercial properties. When an FM takes on the additional responsibility for oversight of construction, the FM becomes what is known as the owner's representative or owner's rep in the vernacular of construction. This is the person who interacts with construction contractors. The job of owner's rep is to work in the best interest of the owner. The owner's rep makes sure that construction is performed on time per the agreed upon price and quality level and that construction operations do not cause unplanned disruptions to other tenants and customers. Owner's reps work with the top managers of the construction project. Owner's reps are the final decision makers about whether the project was a success, and they decide if more work needs to be done. They have the ultimate power, which is the "power of the purse."

Owner's reps decide whether to release progress payments to the construction firm as work is completed during the course of the project. They can also hold back retained funds at the end of a project. *Progress payments* are payments made to the construction contractor or general contractor as a project is completed. Progress payments are generally required for large projects that extend over several months. The intervals for progress payments are based on the stages of project completion, and they are decided prior to the start of the project. A progress payment schedule is defined in the contract for construction. *Retainage* is an amount of payment, usually 5 to 10 percent of the total project cost, that is withheld from the general contractor for a period of thirty to forty-five days after the project is completed to ensure owner satisfaction with the work. These funds provide motivation to the general contractor to make any necessary corrections in a timely fashion.

The owner's rep does not have to be an employee of the facility organization. The facility's upper management may decide to contract a professional construction management firm to assign an owner's rep to the project, or they can pay the architectural firm that drew up the plans to act as owner's rep.

Architect   Architects have the big picture ideas and designs for building construction. They provide the owner with building designs for the beauty and/or functionality and for sustainability. They can refine their designs with architectural details that show the builders how the custom elements are to be built by tradespersons. Architects provide estimates of project costs and may act as the owner's rep, or they may work with the facility manager who is acting as owner's rep to supervise the project.

Engineer   Engineers are also frequently referred to as consultants. They have numerous areas of expertise: site survey; heating, ventilation, and air conditioning (HVAC); fire protection; mechanical systems; structural design; and electrical design. They receive licensing as a registered professional engineer in their area of expertise, and licenses must be obtained by the individual on a state-by-state basis. Engineering consultants provide detailed drawings, perform the necessary calculations, and specify the methods and materials needed to construct. This information is added to the plans as various plan sheets and are denoted by letter designations at the bottom of plans (e.g., P for plumbing, FP for fire protection, A for architectural, M for mechanical, H for heating, etc.).

Contractor's Side Personnel

Contractor's personnel can vary with the type of contract signed, and the composition of personnel found in the traditional construction format may vary. The following personnel work on the side of the contractor and/or construction company.

General Contractor   As discussed previously, the general contractor is the key player for the actual construction. The general contractor hires and coordinates all the trades, sets the construction schedule, and is responsible for the success or failure of the project from a construction viewpoint. The GC is a construction company or an individual. The GC may perform some or all of the work required for the construction. Even when the GC hires subcontractors to perform most of the work, the GC frequently performs work such as carpentry with its in-house staff. Such work continues for most or all of the construction project. This is done for practical reasons: The GC has construction supervision responsibility and must frequently have someone on site. The GC can take advantage of the situation and perform work that will keep a staff member on site and at the same time supervise both the project and the GC's construction workers. This is one way a GC can earn additional revenue. The person who represents the GC on the construction site might be a construction project manager or a construction superintendent.

Construction Estimators   Estimators are extremely valuable players on the contractor's side because they can make or break a project. Estimators provide the dollar amounts used by the GC in bidding contracts. The GC will lose money if the estimate is too low, but the GC might not get the contract if the estimate is too high. Estimators frequently work with the construction manager who develops the construction schedule. Time is money, so the time it takes to complete a project is very important. Estimators also work with subcontractors in obtaining bid pricing for subcontractor work.

Construction Project Manager   The construction project manager (or simply construction manager) represents the general contractor on several construction projects. Construction managers go from construction site to site checking on construction progress, handling a multitude of issues, and making sure the suppliers and subcontractors are paid. They are also responsible for making sure that the construction company receives payment(s) from the owner when they are due and that the construction budget is balanced.

Sustainability Project Manager (SPM)   A sustainability project manager is utilized by construction firms for sustainable construction projects that will be seeking Leadership in Energy and Environmental Design (LEED) or other green certification. The SPM is generally certified by the organization that will issue certification.

Construction Superintendent   The construction superintendent stays at the site and represents the GC. The superintendent makes sure that the job stays on schedule and that the subcontractors do the proper work.

Safety and Compliance Manager   The safety and compliance manager makes sure that all Occupational Safety and Health Administration (OSHA) and environmental regulations are followed and that the job site is safe. The person in this role ensures that proper procedures are followed and that everything regarding safety and compliance is properly documented. This provides some protection from claims levied against the contractor and helps prevent fines resulting from lack of documentation from government agencies.

Construction Staff with LEED-AP Credentials

LEED Accredited Professional (LEED-AP) is a designation awarded by the U.S. Green Building Council that denotes a professional knowledge of sustainability issues and the point system employed to certify buildings under the LEED rating system. Many owners and government agencies performing new construction or system upgrades award contracts to companies who have a high percentage of key people with this credential because they want to ensure that personnel working on their project are skilled in green building. LEED-APs in construction may have several different job titles (estimator, construction manager, scheduler, safety manager, construction superintendent, architect, materials manager, interior designer, mechanical designer, etc.). A third-party LEED-AP may be assigned to the project so that the work can obtain points toward LEED certification. The individual in this case would be working with all the professionals to ensure that the LEED certification level sought is achieved.

Subcontractors   Subcontractors are contractors in many different specialized trade areas who work for a general contractor. Subcontractors provide bid estimates to the general contractor for the work they hope to provide toward completion of the project. They frequently base their bid on a section of the construction documents that deals with their trade. For example, electrical contractors are given the electrical plan sheets out of a full set of construction documents on which to base their bid. Subcontractors are responsible for their work, pulling their own building permit, and being present for building inspection. Here is a partial listing of subcontractor specialties:

- Mechanical and plumbing
- HVAC
- Building automation
- Electrical and/or lighting
- Concrete and/or masonry
- Drywall and/or carpentry
- Painting, staining, and/or wall coverings
- Carpeting and floor coverings
- Paving
- Insulation and spray-on fire protection
- Structural steel and/or welding
- Landscaping
- Grading and/or excavation
- Environmental remediation
- Recycling
- Acoustical (suspended ceilings)
- Sprinkler and fire protection
- Roofing
- Demolition

Personnel Working in the Interest of the General Public

The following government personnel ensure the safety of buildings for the general public. They work directly for federal agencies, such as the Department of Commerce, National Institute of Standards, and Technology NIST, that investigate fire, building collapse, and other major safety issues, or state and local government agencies that perform inspection and permitting work.

Plan Examiner   Before any commercial project can be constructed, or rather before any work can begin, construction documents (blueprints) must be drawn up by an architect or engineer. The plans must then be reviewed by a state-licensed architect or professional engineer, and they must be signed by them and sealed with their stamp. This process ensures that the plan examiner for the municipality or state agency having jurisdiction knows that the design has been thoroughly evaluated for correctness of design and calculations. It also establishes a point of responsibility should problems arise due to faulty design. The examiner reviews the plans to see if they are compliant with the building codes in force by the jurisdiction. The code used may be a ***model code*** developed by a standards organization (codes such as the International Building Code or the Uniform Plumbing Code are model codes) or a state building code. If the plans are in compliance with all codes and zoning restrictions, the examiner issues a building permit.

Building Inspectors   Building inspectors ensure safety surrounding issues of construction and use. When they enter a building, inspectors review the structural components of the construction, plumbing systems, and electrical work. Departments that employ inspectors include the fire, health, and zoning departments. Fire and health inspectors prevent hazardous conditions from creating major disasters.

Inspectors must be contacted just prior to their being needed to inspect the various phases of construction. Contact is made by the general contractor or a subcontractor who is responsible for the work being inspected. In this way the inspector can investigate the construction at the point when the item to be inspected is not yet hidden by subsequent construction.

Zoning personnel make sure that the use of a building, or tenant's business within an existing building, conforms to zoning regulations for the area. These regulations ensure that a building is used in a way that does not conflict with the surrounding community. For example, zoning regulations may prevent an industrial building from being constructed in a residential area, or a retail business from occupying office space in an existing building.

5.3   STEPS IN BUILDING CONSTRUCTION, REMODELING, AND BUILDING IMPROVEMENTS

Building construction and remodeling can be broken down into a series of well-known steps. It is more common for the FM to be involved with renovations and building improvements rather than the construction of a totally new building. But major renovations that involve mechanical systems improvements and tie-ins to existing systems can actually be more challenging than new construction. Breaking down the events in the total construction of a building allows the FM to eliminate steps that do not apply to his or her project. This process is useful when performing a remodeling project where most of the construction already exists or when performing a building improvement such as the replacement of an existing roof. A broad list of steps included in the construction process is as follows:

- **Conceptual phase**—A project concept is an idea brought forth by development professionals within the organization. These professionals are frequently the decision makers in a facility, property, or industrial organization. The concept may be a few notes and a pen drawing on the back of an envelope. Concept drawings appear to be very simple, but they are based on the designer's years of experience and numerous strategy sessions, marketing data, and massive amounts of financial and demographic information from multiple sources.

- **Presentation drawings**—Professional-scale drawings consisting of a series of elevations and site plan drawings are produced by an architect or engineer. These drawings are used to help convince other decision makers about the merit of the project.

- **Preliminary drawings**—The conceptual ideas are refined and presented in a professional format by architects. Basic building parameters, for instance, type of building, size, and overall quality, are identified. These architectural drawings are generally shown as three possible choices or options: A, B, or C.

- **Budget numbers**—Budget numbers are estimates of construction costs based on similar projects and data derived from cost estimation tables and software. Budget prices are given for the three options: A, B, or C.

- **Site selection**—A site is chosen for the new construction. Site selection can influence the type of building chosen because different types may work better under different environmental scenarios (suburban, rural, urban, high ground, low ground, wet and dry areas, etc.). Sustainability of the chosen site is extremely important.

- **Option selected**—One of the proposed options is chosen based on cost estimates and economic analysis of financial outcomes resulting from successful completion. The project with the best overall value is chosen.

- **Plans rendered**—Construction documents, also known as blueprints or plans, are drawn up by the architect and are approved by the owner's rep.

- **Bids sought**—General contractors are selected to bid on the project, or the project is open for bidding to any contractor that wants to bid. Bidding contractors are given a set of plans to bid on.

- **Costs estimated**—General contractors estimate construction costs, get pricing from various subcontractors, and submit their bids.

- **General contractor selected**—In general, the GC is selected based on price, completeness of the bid documents, and known capabilities of the GC. It may also be based on the quality and reputation of the subcontractors.

- **Post-bid conference held**—This meeting is held if there are questions or items are unclear in the bid submitted. The GC meets with the architect and owner's rep in order to clarify various items in the bid. If the questions are answered to the owner's satisfaction, the contract is awarded. If the GC's answers are unsatisfactory, another post-bid conference is held with the second lowest bid, and so forth.

- **Permits obtained**—The general contractor (or the project architect) and the various subcontractors submit construction documents to the building inspection department in order to obtain construction permits.

- **Construction started**—Once building permits are secured, materials and equipment are brought to the site.

- **Building inspection**—Inspectors interact with the construction superintendent and various construction trades during the various construction phases.

- **Construction is completed**—The actual construction work and installation of equipment, fixtures, and finishes are complete, which means that *substantial completion* has been achieved. At this point, the owner can use the building as intended, and only commission work and punch list items remain. Contract language can alter when substantial completion occurs.

- **Mechanical and HVAC systems commissioned**—This process tests equipment to see that all building systems that were installed function as designed. The process compares design parameters with equipment test results. It is an important step that ensures that the building will achieve the energy efficiency, environmental comfort, and indoor air quality levels specified by the design.

- **Punch list**—Outstanding construction items are identified by the development of a punch list. A *punch list* includes items that have been missed, are found to be not in compliance with the specifications, or have been damaged during construction. Corrections are made so that payment will be issued.

- **Payments issued**—Final payment is made to the general contractor and closeout of the project is completed.

- **Beneficial occupancy**—The owner takes beneficial occupancy and moves into the building or tenanted space, and can use the building for its intended purpose. This date may be important for the equipment warranty starting date(s) and may take place prior to issuing of payment or completing the punch list.

5.4 IMPORTANT CONSTRUCTION PROJECT STEPS IN DETAIL

This section highlights the important phases of construction. It is written from the viewpoint of the owner's representative because this is the job most FMs will be responsible for from project conception to closeout.

Bidding

Once the plans have been drawn up, the next step is to find a general contractor willing to do the work for the best price. Typically plans are provided to the bidders and then they are invited individually, or all at once, to come to the location where the work will be done. This site visit allows the contractors to verify the actual *field conditions* at the site. Verifying field conditions helps to eliminate possible extras tacked on to a project.

Example: Demolition

If the project calls for the demolition of all existing flooring materials, the contractor should verify what is on the floor and whether layers of other types of flooring exist underneath it (i.e., glue-down wood flooring under carpet). If only one room or a couple of rooms are carpeted and the rest are vinyl tile, then the demolition contractor should check under the carpet in each room, especially if there is a visible height difference. With many rooms, of course, the bidding contractor couldn't be expected to check under the carpet of each room, and if only one room has the hidden wood flooring, the contractor would be entitled to extra money to remove it. A wise demolition contractor would state in its bid that *only carpet will be removed.* Then it doesn't matter what is under the carpet; it would be an extra charge to remove it.

On some government-financed projects, the bid must be open to allow all contractors a chance to bid. This process can be very expensive because a rolled set of plans can be several inches thick. Contractors may be charged a fee in order to cover the costs to provide them with plans. Another constraint is that the FM may be forced to pick the contractor who bid the lowest, even if its reputation as a contractor is poor. Unfortunately, the lowest bidder might be a contractor who made a mistake in the bid price. This can result in the contractor trying to cut corners to save money, provide shoddy workmanship, and may even result in a project failure. As a result, some municipalities allow the FM to pick the second lowest bidder, which is somewhat of an imperfect way to increase the chances of a successful project.

On many commercial projects, the bid is opened to only a select group of general contractors. These contractors have a reputation for quality work at a reasonable price. In this case, the cost to provide construction documents is borne by the owner. The architect can often make suggestions about contractors they have worked with, or the FM can find contractors by networking with other facility managers. In this case, facility managers are not bound by law to pick the lowest bidding contractor.

Why wouldn't the FM pick the lowest bidding contractor if the general contractors are all truly equal? The problem may be that the general contractors are equal, but the subcontractors they are using might be of varying quality. This might be especially true if they are trying to submit the low bid. In addition to a quality issue, one GC might be using union subcontractors, and another might be using nonunion. It should be verified that the subcontractors were listed on the GC's bid will actually be the ones performing the work.

Contract language may be necessary to prevent the GC from utilizing a process known as bid shopping. In bid shopping, the GC who has been awarded a fixed-price construction contract searches for lower-priced subcontractors as a way to increase profit. This may or may not be legal and depends on state law. This raises the question of the quality of the subcontractors who will now be performing the work. Ultimately, it is the GC's responsibility to meet project specifications and ensure the quality of the workmanship. FMs acting as owner's reps must make sure that the project work results in a quality project that can be turned over to the owner, whether or not bid shopping has occurred. The FM must have a thorough understanding of contract specification, contract provisions, and the requirements to fulfill the work as stated in the construction documents.

If a project becomes rushed (as new tenant work frequently does), the property manager (PM) might review the bids and tell the FM to give the work to the GC who has performed exceptionally in past projects regardless of price. The property manager knows that the tried-and-true GC will get the job done quickly, and if the property organization is funding the work for the tenant, the PM can pick whichever contractor he or she wants. But eliminating a general construction contractor based solely on good past experiences with another GC can have negative consequences. It can damage a possible new relationship with the GC who was eliminated because it provided a good bid in good faith. The new GC might also be reluctant to provide bids on projects in the future. Bidding projects can be a significant cost to the GC. The FM should explain to the GC the reason why the bid was not accepted and possibly offer a chance to bid on future projects.

Construction

After the contracts are signed, construction may proceed as shown in the detailed plans and specifications. The owner's rep is responsible for seeing that work progresses, construction project milestones are met, and the building's rules are followed by the general contractor and subcontractors. *Project milestones* are specific events in the construction process that must be achieved before other construction work can continue. Project milestones are important because they serve as excellent indications about the project's schedule.

Building rules address construction issues such as freight elevator use, noise, carpet protection, storage of construction materials, areas that are off limits to contractors, use of smoking areas, where trash and recycled materials need to be contained for removal, and so on. A construction schedule should be finalized, and project milestones identified and made available to all parties. An important item in the schedule is when and where construction progress meetings will be held. Usually these meetings are held at or very close to the job site.

Once the project is underway, project meetings are held. The construction manager who represents the GC discusses issues, possible change orders, and the project schedule. Sometimes subcontractor representatives are invited to the meetings to discuss how their individual work is proceeding and any issues relating to it. As the work proceeds, the percentage of job completion is evaluated by the architect, and progress payments are made to the GC on a schedule that was agreed to in the contract. If new building mechanical systems were constructed or old systems upgraded, then near project completion, equipment commissioning is necessary to evaluate the actual performance of the new system. Third-party consultants skilled in equipment commissioning should evaluate the system.

As-Built Drawings

Unforeseen issues in the field result in subcontractors having to vary from what is shown on the construction plans. In many cases, this variance is communicated to and must be approved by all affected parties prior to the change in the actual construction being made. These changes happen frequently in construction. As long as the spirit and intent of what the architect or engineer designed is achieved, in many cases there is no problem with deviating from the plan to avoid conflicts with other trade work. Sometimes a variance from the plan also makes good economic sense, and a modification can result in savings in both time and money.

When a variance from the plans is completed, the contractor doing the work notes the change by hand, drawing the modification on the plans with a red pen or pencil. These red marks are known as *red-line drawings*. Mechanical contractors frequently need to make reasonable changes to the plans. A contractor running an underground water line may encounter a huge boulder and opt to go around it, or obtain water from another location. An electrical contractor might need to move ceiling light fixtures over a bit because a supporting I-beam above the ceiling doesn't allow enough clearance.

At the end of the project the red-line drawings are given to the owner; the collection of these plans make up the *as-built drawings*. In today's digital age, it is important to have the architect collect all the red-line drawings and produce a digital copy of the as-built drawings, to have one point of responsibility for collecting the digital as-built drawings from the various subcontractors and drawing in the changes to create an as-built plan set. The architect can store a digital copy at her or his office for future use and send the facility a digital file. FMs need to make sure that they have the proper software to view the plans. Often computer-aided design (CAD) software companies provide free plan-viewing software for this purpose.

Punch List

When construction is completed, a walk-through is conducted by the supervising architect with the GC's construction manager and the owner's rep present. Everyone is looking for items that are incomplete, forgotten, or damaged. A list of discrepancies is made, and the GC is responsible for making the corrections. The work is most often completed by the subcontractors. The most common punch list item on projects whose interior walls are painted is touch-up painting. On projects where jute-backed, roll-type carpet is used, carpet installers frequently scuff the paint when the rolls are unrolled, or they try to navigate a large roll around a corner. Carpet installers are not the only ones who might scratch the paint. So it is rarely known which subcontractor did the damage. GCs get around the problem by having the painter add a best guess amount to their bid to cover touch-up painting.

Common punch list items include the following:

- Missing electric outlet covers
- Scraped, missing, or damaged ceiling tile
- Equipment manuals not turned over to the owner
- Red-line drawings not turned over to the owner or architect
- Doors that were cut down to proper size by planning were not sealed along the exposed wood edges
- Locksets not installed, or locks not rekeyed

- Transfer ducts blocked or missing
- Plastic covers left on light fixtures
- Filter elements missing or dust covers not removed from air return grills
- Missing thermostat covers
- Diffuser airflow not balanced

Once the punch list items are completed, a final walk-through is done. If everything is satisfactory, the retained funds are released, and the contract is satisfied. Another successful project is completed.

5.5 CONSTRUCTION CONTRACTING AGREEMENTS AND PROCESSES

A contract is the roadmap for how construction will be performed. Choosing the correct contract forms for construction is essential for project success. A source of prewritten contracts for the various construction types can be obtained from the American Institute of Architects (AIA), who issues forms known generally as AIA forms. These contracts provide protection for the owner's interest, but the contracts are not specific to an organization or building's needs. An important but frequently overlooked step is to incorporate the building's rules that contractors must follow. The contractor should be made aware of the rules at the first meeting between the owner's rep and the contractor. There should be no surprises for the contractor when building rules are referenced in the contract documents. For instance, rules for when contractors can use the freight elevator, rules for the protection of the existing building and its mechanical systems, and safety and emergency procedures should all be referenced as an attachment to the contract. Once these provisions are in the contract, they become binding on the contractor. The contractor then must follow the rules as listed. The AIA also has contract formats that cover the contract relationship between the owner and the architect or engineer.

We will discuss the different forms of contracts for construction projects. The chosen format is important because the type of contract can radically alter the organizational structure, loyalties, and responsibilities of project personnel. Different formats also have different benefits, risks, and rewards. The following are the major types of construction contracting formats:

- Fixed-price contract (prices achieved through bidding)
- Owner as general contractor
- Construction management
- Design build
- Guaranteed maximum price (GMP)
- Cost plus fixed fee
- Unit price
- Time and materials (T&M)
- Incentive and penalty

Traditional Construction Process and Fixed-Price Contract

The traditional construction process results in a fixed-fee contract for the construction work to be performed. The process begins when the owner or owner's rep hires the architect to design the project. The architect will be part of the project throughout the construction process. Then construction plans are sent to selected general contracting firms that can do the work at the desired quality and cost level, and they go through the process of open bidding. The general contractor is then selected and the work begins. If everything goes exactly as planned, the owner will pay the fixed-price fee amount when the work is completed. But the "fixed-price" aspect of the contract is somewhat misleading. *Change orders* are requests to perform additional work, and they can add to the fixed price. Change orders are sometimes the result of necessary modifications to the construction documents that were uncovered as the project work commenced. They can also be a result of hidden field conditions that require additional work. (Whether the bidding contractor should have investigated the field conditions more thoroughly or whether the field condition should result in a change order is an issue that is often hotly debated among owners, architects, and general contractors.)

Owner as General Contractor

In this form of construction, the owner or owner's rep acts as a general contractor. The owner's rep hires the architect and all the contractors necessary to complete the project. He or she organizes the building inspections, provides the construction schedule, and pays all the bills. The owner or owner's rep acting as GC tries to save money by taking the GC, and GC markups, out of the picture. The advantage of an owner acting as GC goes beyond any potential dollar savings. It provides the FM (who can act as the owner's rep) with a different relationship between him- or herself and the subcontractors by expanding the FM's importance to them. In the future, the FM may assign small jobs to these subcontractors. With the enhanced income stream provided by larger projects, subcontractors frequently become more responsive and generally provide better pricing. But established GCs provide an even greater income stream, so they also get preferred pricing from the subcontractors who work for them. The projected dollar savings by the FM/GC undertaking the project thus might not materialize.

Acting as GC is often a fun and rewarding experience. It is a great way to develop planning and construction knowledge skills. Note, however, that the time the FM spends being the GC on a construction project is time lost on operations and master planning for the facility. This time lost costs the facility organization money. It eats away at the savings achieved by not having to pay a GC for construction management services.

Construction Management Contract

At the start of the project, the FM hires a construction management firm for the project's duration. This construction manager is different from the construction manager employed essentially as a project manager by a construction firm. In this case, the CM works for a construction management company that generally does none of the actual construction. The CM hires all the subcontractors, creates a project schedule, and may even hire the architect for the building design. The CM is responsible for the success of the project. The CM is an expert in the construction process and may have additional skills in professional disciplines, such as estimating and architectural or civil engineering. Hiring a CM firm frees the FM, who might otherwise be acting as an owner's rep, from day-to-day involvement with the project. Some construction management firms are small one-person operations; others are huge corporations that perform this service around the globe.

Design-Build Contract

Many organizations realize that they are not architects, engineers, or construction managers. They are experts in their business, and the construction knowledge learning curve would be too steep for them to take on a large construction project where mistakes are costly. The solution is to enter into a design-build contract with a design-build contractor. Design-build firms do exactly what their name implies: They design the building and then they construct it for the owner. Design-build firms frequently market themselves to a particular industry and become known as experts. For example, a design-build firm may specialize in buildings used for drugstores, supermarkets, gasoline stations, or medical clinics, to name a few. Their specialization leads to efficiency. They develop efficient building designs for a particular business type. Their construction workers have skills that mirror the requirements of the design. The building designs can be modified to accommodate the specific needs of each new customer, which gives the buildings some individuality. But buildings are always similar enough to cut down on architectural, construction, and engineering costs significantly.

Guaranteed Maximum Price Contract

GMP is very similar to a traditional construction project, with two major exceptions: (1) Competitive bidding among general contractors based on a full set of contract documents is eliminated, and (2) a project can begin without already drawing up a full set of plans.

The construction development process starts with the owner choosing an architect and a general contractor who is willing to negotiate a GMP contract. For this form of construction to be successful, the owner must have a very high level of comfort with the architect and general contractor. The owner should be thoroughly convinced that the general contractor is one of the best for the job. The architect then defines the project requirements and provides an overall estimate of total project cost. The general contractor takes the plans and makes his or her estimate of cost. The GC might call in potential subcontractors who will play a major part in the project and get individual cost estimates for their part of the work directly from them. The owner, architect, and GC then agree on a *maximum price* that the project will *not* exceed. This maximum price is based on the initial plans and specifications provided by the architect and an assumption about what additional work will need to be done to complete the project. If the project goes over the maximum price without the scope of the project changing, the GC must pay any additional costs above the GMP. If the project costs are lower than the maximum price, then the cost savings are shared between the owner and the GC.

GMP contracts can be very useful when an owner wants to fast-track a project or when a project is very complicated and the construction documents are not easy to complete in a reasonable time frame. GMP allows the project to get underway quickly, and a maximum price can be given first for the parts of the project that are well defined. As more plans and specifications are developed, further pricing can be given. In this way the parties to the project don't have to wait until a full set of construction plans are created. The other advantage is that if the GC or one of the subcontractors can find a less expensive way to get the job done, both the owner and GC can share the savings. This approach is called value engineering, and it is all about discovering more efficient ways to get the job done. The theory is that the shared savings provided in a GMP contract will inspire a GC to look for savings and project efficiencies.

A GMP contract that is not used properly and does not include the proper safeguards can put the owner at a disadvantage. The first is an economic disadvantage created by a conflict of interest between the GC and the architect. The architectural firm representing the owner may work with the same GC on many projects and may have had this relationship for decades. This familiarity between architect and GC might not be in the owner's best interest. For example, the owner may be given an overly high estimate of the construction costs by the GC. A high GMP benefits the general contractor, who splits with the owner any cost savings achieved by the final project cost coming in below the GMP. A high estimate should be challenged by the architect for the benefit of the owner. But a good honest relationship between the architect and the GC can benefit the owner because of the greater insight the architectural firm has into the construction capabilities of the GC. The architect can specify work the GC is known to be able to do efficiently, which leads to cost savings during design and construction.

A way to try to prevent high estimates is to hire a third-party construction estimator. In the United Kingdom, these consultants are known as quantity surveyors. Many quantity surveyors work in the United States and around the world. Construction estimators or quantity surveyors accurately estimate construction costs and negotiate the guaranteed maximum price with the general contractor. The construction estimator is also valuable in deciding when progress payments should be made and when funds set aside for *retainage* should be released. Construction estimators can also reestimate the new project costs when project efficiencies are found and a cost reduction is achieved.

Comments on Construction Estimates and Pricing for GMP   Accurate estimates are essential to a successful GMP project. If owners agree to a price for the work, then they should not have to face financial difficulties if no construction efficiencies are found. Owners should not always declare victory when actual construction costs come in lower than the guaranteed price. Lower prices might reflect a high estimate rather than construction efficiencies. The owner might be required split these false savings evenly with the general contractor. An accurate construction cost estimate achieves much greater savings for the owner.

To be fair to general contractors and architects, cost estimates for GMP work should never be "right on the money." There is too much risk with giving an owner a price that is accurate to the *penny*. If anything occurs to drive up cost, the general contractor pays for it. So the GC must build in a reasonable price cushion. But with the risk there should be reward. The reward for a GMP contract is being able to split savings with the owner. These savings must be real savings achieved through the general contractor's knowledge, ability, and hard work.

Cost Plus Fixed-Fee Contract

The general contractor is reimbursed for the actual costs of the project and also receives a fixed fee as profit. The general contractor has incentive to complete the project quickly because the quicker the project is completed, the sooner the GC is paid. Cost plus fixed-fee contracts are frequently bid out, meaning that requests for proposals are sent to several contractors to find the lowest priced contractor for the project.

Unit-Price Contract

The contractor provides a table of all the construction tasks, wages, and materials, and completes the unit cost on various bases.

Unit-price contracts are usually not stand-alone contracts. They are usually sections added to fixed-price contracts. Unit pricing for work helps the owner avoid situations whereby the contractor must offer competitive pricing for additional work. The cost to provide additional work is spelled out clearly in the contract document through unit pricing. If the FM is unsure about the scope of the project, obtaining unit prices for additional work is a smart approach. For example, a painting/interior finish contractor did a superior job at a very reasonable price. The work was so good that the FM decided to ask for a price to perform additional work. The price that the contractor offered was much higher, on a percentage basis, than the original bid price that got the contractor the job. Some contractors like this one base their initial price on the hope of receiving add-on work for which they can receive a much higher profit by not having to provide a competitive price.

Time and Materials Contract

Most contractors refer to time and materials as T&M. *Time and materials (T&M)* is a contract format where the owner pays for the labor time and the materials used. A profit factors is built into the labor rate, and there is a markup on materials. T&M is most often used for small jobs that need to be done quickly and where a high confidence level already exists between the contractor and the owner. The owner must be sure that the contractor won't use more time than is necessary or overcharge for materials. Having the contractor perform T&M might be the only way to address emergency situations quickly. A contractor who provides a fair price under such circumstances is a sure sign that a sustainable relationship has been achieved between owner and contractor.

Incentive and Penalty Contract

Any contract can have incentives and penalties added to the contract language. A fixed-price contract might include a penalty for not completing the work by a specific date, or it may give the contractor a bonus if the work is completed early. No set rule exists for this situation, but it is considered unprofessional and unfair by many general contractors if the contract has a penalty clause but does not also have a bonus or incentive clause. A penalty clause carries risk for the contractor and, as such, the FM can expect higher bid prices to compensate for this penalty risk. The greater the penalty, the greater the risk to the contractor. An incentive clause tempers this somewhat because it is a form of a reward, making performing the work worth the risk. The contractor might then take the chance of offering a lower bid price with the hope of getting the bonus.

5.6 ADDITIONAL CONSTRUCTION CONTRACT AND LEGAL ISSUES

Some additional construction contract issues include bonds, insurance, liens, and warranty issues. We will discuss surety bonds (including bid bonds, performance bonds, and payment bonds) in addition to lien waivers, insurance, and warranty.

Surety Bond

A *surety bond* is an agreement among owner, contractor, and surety company that guarantees that the contractor will perform the contract in accordance with the contract documents. Different types of bonds are used for a construction project. Bonds may be mandatory for federal government construction projects, or bonds may be necessary to secure bank funding for very large commercial projects. The three common types of surety bonds are bid bond, performance bond, and payment bond.

Bid Bond   A bid bond guarantees the bid price the contractor has given to provide the work. Should the contractor who is awarded the bid decide to back out of the project, the bid bond covers the cost difference to hire the next lowest bidder to complete the project.

Instead of a bid bond, bidders can be required to submit certified checks, cashier's checks, or money orders for 20 percent of the bid price. For the successful bidding contractor, this large amount of money would be lost if the bid price were not held to by the contractor. Once all the contract documents are executed, the check or money order is returned to the contractor winning the bid. All unsuccessful bidders have their checks or money orders returned immediately after the bids are opened.

Performance Bond   A performance bond ensures that the contract will be completed for the price stated in the contract documents and that the project will be finished by the stated completion date. Should a contractor go bankrupt during construction, the bond provides the funding to hire another contractor to complete the unfinished work.

Payment Bond   A payment bond ensures that the subcontractors will be paid for their work once it is completed. If a general contractor fails to pay the subcontractors, the payment bond will cover the cost of paying the subcontractors. A payment bond is important because the owner is responsible for paying the subcontractors for their work, even if the owner has already given the money for the subcontractors' work to the general contractor.

Lien Waivers

Owners are responsible to pay for items, materials, or work they received from the general contractor, the GC's subcontractors, and the distributors who supply construction materials. The GC is paid by the owner, and then the GC pays the subcontractors. To protect the owner from paying twice, the owner obtains a signed waiver of lien from the GC and his subcontractors that performed the work before paying the general contractor. The lien waiver should be for the amount of payment requested. Once an owner has a waiver of lien rights from a contractor, an owner cannot be forced through litigation to pay for this portion of the contract work again, even if the GC has not actually paid the subcontractors. A subcontractor should not provide a lien waiver if he or she has not been paid by the general contractor. It is a little tricky because the contractor doesn't want to give a waiver without being paid by the owner, but the owner doesn't want to issue payment without lien waivers. Sometimes this situation is resolved by the owner handing the general contractor a check, and the GC handing the owner the lien waivers at the same moment. If an owner fails to obtain lien waivers and the GC doesn't pay the subcontractors, the subcontractors can file a mechanic's lien against the owner. A mechanic's lien claimant can sue to have the owner's property sold to satisfy his or her claim.

Insurance

Before a construction contract is signed, the FM should make sure that the contractors have the appropriate type, amount and quality ranking of insurance required by the FM's organization. The most frequent requirement is a specific dollar amount of liability coverage (usually $1 to $3 million) to be provided in the event of a claim. The insurance company should also have a good ranking by an independent rating agency, such as A.M. Best Company. The FM may even wish to have the contractor provide a copy of its insurance certificate that lists the facility organization as being additionally insured under general liability. Additional insured language should be discussed with the facility organization's insurance carrier because of its knowledge in these matters. It should also be discussed with the facility's legal counsel to verify the appropriate legal wording on the certificate. The FM should also discuss the amount of insurance necessary for the risk involved for the type of contract and the work being done.

Warranty

When a warranty should begin on a piece of equipment that was installed at a construction site is a question that must be answered before the construction contract is signed. A new building project might take several months or years to complete. Owners don't want to find themselves taking occupancy of a new building where all one-year equipment warranties are already expired, or not enough days are left under warranty to test equipment under different seasonal situations.

Does the warranty start when the equipment is received, when it is installed, when it is commissioned, or when the owner takes occupancy? This question must be settled in the contract documents, and is a point that must be clear to all parties working on the project.

5.7 USING SOFTWARE TO SCHEDULE AND ORGANIZE CONSTRUCTION PROJECTS

Computerization of the construction process enhances its sustainability with better project organization and better coordination of the trades, and having materials available when they are needed. The critical path method (CPM) and building information modeling (BIM) are two of the most important construction scheduling and design tools in use today.

Critical Path Method

CPM utilizes a network diagram for scheduling project activities. The network diagram shows the tasks and the time it takes to complete them. The network also displays the relationship between construction activities. Construction tasks that are related to one another are considered to be a phase of construction, which is called an activity. For instance, in a remodeling project, all the tasks that relate to construction demolition (removing ceilings, walls, and flooring) are part of the demolition phase, and demolition is noted on the diagram as a single activity. The construction manager (CM) defines all the required activities needed to accomplish the project, puts them in proper sequence, and estimates the time it will take to complete them (activity duration). The CM begins by making a Gantt chart listing all construction activities. A Gantt chart (see Figure 5.1) is a type of bar chart that lists the project activities in the order of their start date, with the bar for each activity extending to the finish date for that activity.

The Gantt chart provides information readily understood by even the untrained viewer. Some general contractors are reluctant to show their customers the network diagram of a project because such a diagram is complex and might be misunderstood by customers. GCs are generally happy to give their clients a Gantt chart of a project because it is easy to understand.

A CM is able to point out several project efficiencies directly from a Gantt chart. But a network diagram developed from the Gantt chart provides the CM with a far greater insight into the functionality of the project, which in turn provides even greater project efficiencies. To use time, labor, and resources efficiently, the CM must discover, through the use of a network diagram, any or all of the following:

- Activities that should be completed simultaneously with other activities
- Activities that should be started prior to the completion of the preceding activity
- Activities that could or should be delayed
- Whether another concurrent activity needs extra time for completion
- How to control the overall project duration

Gantt charts and network diagrams both display the activities in the sequence in which they will be completed. A network diagram is better at displaying how an activity is interrelated with other activities. It shows how the scheduled completion of activities has an effect on the other activities along the path. It also shows the resulting consequences when there is a variance in the time scheduled for completion of an activity. Several paths through the network diagram may lead to completion. The path through the project's network diagram that has the longest completion time when all the activities along it are added together is referred to as the critical path (see Figure 5.10). The path is critical to the project because it controls the overall project duration. Shortening the time to complete any activity along the critical path shortens the length of time to complete the entire project. The *critical path method* should be used for construction projects that are complicated, have multiple phases, have activities that are performed concurrently, and/or span a considerable amount of time.

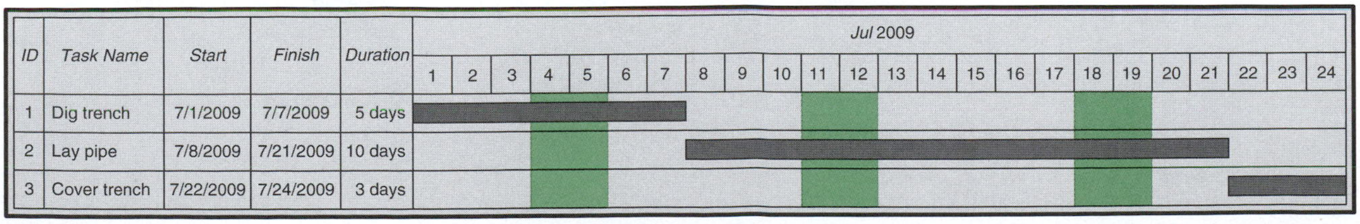

ID	Task Name	Start	Finish	Duration	Jul 2009
1	Dig trench	7/1/2009	7/7/2009	5 days	
2	Lay pipe	7/8/2009	7/21/2009	10 days	
3	Cover trench	7/22/2009	7/24/2009	3 days	

FIGURE 5.1 Gantt chart created with Microsoft® Visio program

TABLE 5.1 Pipe-laying project schedule spanning 18 days

Activity	Duration	Date Complete
Dig trench	5 days	July 7, 2012
Lay pipe	10 days	July 21, 2012
Cover trench	3 days	July 24, 2012
Total time	18 days to complete the project	

A pipe-laying project is a good way to see how CPM scheduling works (see Table 5.1). The table indicates that the project began on July 1, 2012. There is no work done on the weekends for this project. A Gantt chart of the pipe-laying project is shown in Figure 5.1.

The Gantt chart clearly shows the project activities, their start and finish dates, and their duration. Gantt charts are very linear, and they are somewhat difficult to interpret when trying to make connections between activities. Gantt charts can be very practical, however, for displaying the work required to complete projects. On the chart shown on Figure 5.1, Task ID 1, "Dig trench," is a five-day task shown as spanning seven days! The reason is because no work is done on weekends, in keeping with a real-world construction practices. Weekends are shown as a vertical yellow highlighted area on the chart. July 1 is a Wednesday, so work for Task ID 1 is accomplished Wednesday, Thursday, Friday, Monday and Tuesday.

Gantt charts don't readily promote the discovery of project efficiencies. Computer-generated Gantt charts are easily modified to make them a valuable quick check of project activities and their timelines.

The other way to display project information is by drawing one of several types of network diagrams. One of the simplest diagrams is called activity on arrow (AoA). An example is shown in Figure 5.2. Circles, which are called nodes, display the beginning and end of an activity; the letters denote a new task. The numbers inside the circles show the sequence for completing the activities. Frequently the length of the arrow is in relation to the time it takes to complete the task.

A network diagram like the one shown in Figure 5.2 is drawn directly from the Gantt chart in Figure 5.1. This example is drawn without any thought of construction methods and efficiencies. The resulting diagram is very easy to understand, but it doesn't provide any more information than the Gantt chart.

Reducing Project Duration

Project efficiencies can be discovered by analysis of the network diagram shown in Figure 5.3. In the pipe-laying example, the trench is long enough so that the construction workers don't have to wait until the entire trench is completed before laying the sections of pipe. The workers can start laying pipe in the trench after one day of trench digging. All of the pipe must be laid prior to being inspected by the plumbing inspector. Because of these inspection requirements, the pipe can't be covered in sections to save even more time. Node 3 shows the point in the project when all pipe has been laid in the trench. To complete the diagram in a logical manner, a dummy task is drawn from node 3 to the ending node 4. Dummy tasks are not real tasks; they simply help complete the diagram and show its intent more efficiently via a dashed line.

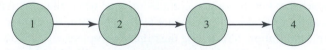

FIGURE 5.2 Activity on arrow network diagram of construction

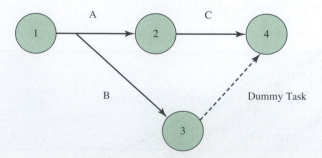

FIGURE 5.3 Activity on arrow network diagram showing reduced construction duration

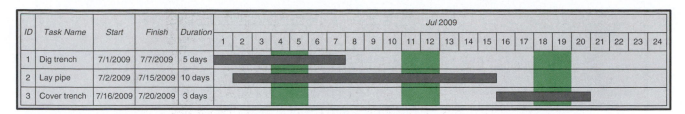

FIGURE 5.4 Revised Gantt chart showing reduced project duration (revised using Microsoft® Visio)

Laying of the pipe is now shown to begin four days sooner, resulting in the job being finished four days earlier. This reduces the total project duration by 22 percent. When labor hours remain unchanged, getting the work completed sooner reduces the cost for construction. Getting the project completed faster may also mean that revenue from the finished project will flow to the owners sooner.

The Gantt chart is then modified to reflect the reduced project duration, as shown in Figure 5.4, so that it continues to be a valuable document. Network diagrams used in combination with Gantt charts provide a major tool in construction project management.

Activity on Node Diagram

The activity on node (AON) network diagram displays more project information than the AOA diagram. An AON diagram provides the data necessary to find the project's critical path. The first step is to create a table showing all the project activities, the duration of the activities, and the activities that are *successor* activities. The importance of this step is to define the project activities and to provide a map for drawing the network.

Activities before an activity node are defined as *predecessor* activities; those after an activity node are *successor* activities. During construction, predecessor activities must be completed before a new activity can begin. A thorough understanding of the construction process and the time it takes to complete each activity for the size of the project is necessary in order provide accurate data for an AON diagram and to know which activities are predecessors or successors to one another.

The beginning of an AON diagram is described in Figure 5.5 and Table 5.2 using a wall remodeling project. The activities in the wall remodeling project are as follows:

- Demolition of existing walls and cleanup of space
- Layout of wall studs and wall stud framing installation
- Electric conduit and outlets, which are run concurrent with the wall installation

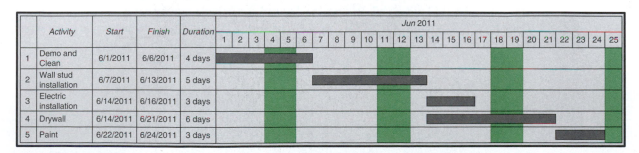

FIGURE 5.5 Gantt chart of a wall construction project

TABLE 5.2 Wall construction project: activity, activity letter designation, predecessor activities, and the duration of each activity

Activity Name	Letter Designation (ID)	Predecessor Activities	Activity Duration (DU)
Demo and cleanup	A		4
Layout and framing	B	A	5
Electric installation	C	B	3
Drywall and taping	D	B	6
Paint	E	C, D	3

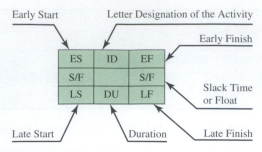

FIGURE 5.6 AON diagram with node designations

- Drywall installation and seam taping
- Wall finishing—painting the walls

Early Start, Early Finish, Late Start, Late Finish, Duration, and Float/Slack Time

A node for an AON network diagram is shown in Figure 5.6.

Obtaining the Data for an AON Node   The AON diagram's node is also a table of information for the activity. The time information is generally in days or weeks for construction projects. The first piece of information needed is the duration (DU) of the activity. Durations are obtained by constructing a table of the construction activities and a simplified AON diagram consisting of arrows and circles for nodes to see the relationship between activities. The next two pieces of information—early start and early finish—are obtained by completing a process called a forward pass (see Figure 5.7). To complete a forward pass, the CM moves through the project from start to finish (left to right). The CM takes the early start (ES) time, adds the duration (DU), and gets the early finish (EF) time; therefore ES + DU = EF. The early finish time is used as the early start time for the succeeding node. If there is more than one predecessor node, the longest early finish time is used.

Early start (ES) is the soonest time an activity can begin. It is the *longest* early finish time of successor activity. By convention, for the start node that begins the project, a *zero* is used for the early start time. Early finish (EF) is found by adding the early start time and the activity duration.

Backward Pass   Late start (LS) is a combination of slack time and early start (ES) time. Late finish (LF) is the early finish (EF) plus the float. Both the first and last activities on the network have zero float time.

Figure 5.8 shows how the backward pass is performed. Beginning with last activity (E), start at the top right corner, which is EF (early finish time). Then moving down, 0 float is added to 18 (early finish

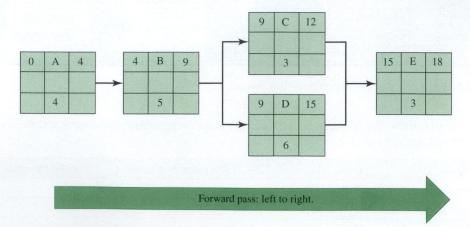

Forward pass: left to right.

FIGURE 5.7 Forward pass

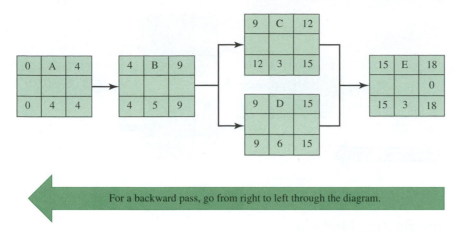

FIGURE 5.8 Backward pass

time) to produce 18 (the late finish time), which is written in the lower right corner. The next step is to move backward by taking the 18 (late finish time) and subtracting the duration (3). This equals 15, which is the late start time in the lower left corner of E. The late start time of E is the late finish time of activities C and D.

Float Float, also referred to as slack (S) time, is extra time to complete (see Figure 5.9) a task because another task must be completed concurrently and takes longer to complete. The shorter task will have float time equal to the amount of extra time the longer task takes to complete.

Float can also result when one of the two simultaneous tasks requires time to dry or cure. An example is when foundation footings are poured, the foundation workers need to wait until the concrete hardens and gains strength before pouring the foundation walls on top of the footing. In the meantime, the setting of the forms for the foundation walls will have some float time due to the cure time of the footings. If it takes two days to set up the foundation wall forms and it takes five days for the footing to cure, then setting the forms for the foundation wall has three days of float. So if the footings start curing on Monday morning, the foundation workers can wait as long as Thursday morning to start setting the forms without causing a delay in the project. The float days are Monday, Tuesday, and Wednesday.

Developing a network diagram using the critical path method (see Figure 5.10) to govern the activities of a construction project requires additional study in the areas of construction management and practices, scheduling, budgeting, and project management. The goal of this section is to introduce you to the theory behind the CPM and give you the ability to read a network diagram and accurately discuss a project's critical path with a CM. The actual creation of a CPM network diagram requires knowledge of a project management software package and is beyond the scope of this text.

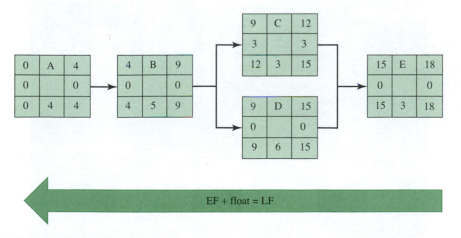

FIGURE 5.9 Project showing backward pass and float

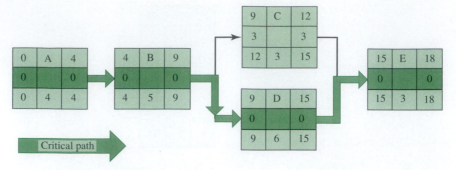

FIGURE 5.10 Project's critical path

Building Information Modeling

Building information modeling (BIM) is a virtual color representation of a building in three dimensions (3D). BIM program software is an object-oriented, model-based, graphics design program. BIM creates a virtual building environment as a 3D picture that can be viewed from any angle and also rotated and flipped. Figure 5.11 shows a three-dimensional image of a mechanical room, and Figure 5.12 shows an as-built photo of the same mechanical room. Anyone viewing a BIM illustration gains a sense of size and depth and the intricacies of the project. This is not possible with a flat two-dimensional (2D) drawing of straight lines and symbols that has been used for centuries to provide a project blueprint.

FIGURE 5.11 3D image of a mechanical room

Source: Image courtesy of Mortenson Construction.

FIGURE 5.12 As-built photograph of the mechanical room shown in Figure 5.11.

Source: Image courtesy of Mortenson Construction.

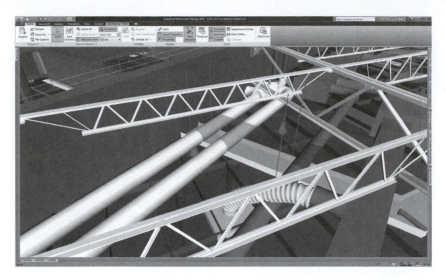

FIGURE 5.13 3D computer image, created using Navisworks™ by Autodesk®, shows a clash in a building information model.

Source: Image courtesy of Mortenson Construction.

BIM doesn't stop with 3D. BIM gives the designer a fourth dimension: time. BIM produces a building model that has the ability to show the various phases of construction in the order they will actually be constructed. This is extremely valuable information for the construction contractor because it shows how and where equipment and materials must be placed so they are as close to the point of use as possible without getting in the way.

Builders work in a 3D world, but they are given 2D construction documents. A 2D view presents everything to the viewer on one horizontal plane. It's as if everything is built on the floor. (In fact, a set of plans, called a reflected ceiling plan, which shows the light fixtures, ceiling grid, and tiles as a reflection of the ceiling on the floor.) Two-dimensional plans are drawn separately by the various construction trades (plumbing, electrical, HVAC, architectural, structural, etc.). Even when the plans drawn by the various trades are combined together on one page, it is not easy to see where space conflicts, known as clashes, might occur. See Figure 5.13, which shows a clash illustrated using a 3D image. The designers are alerted to a problem where three pipes (the blue pipe and two white pipes) are currently running through a cable tray access the area. The clash shows the pipes running through a red transparent access area above the yellow cable tray.

Construction elements are built at various heights within a space. In a 2D drawing, height is not defined, and it is all too easy for two trades to want to build in the same location. And lines drawn on a set of plans do not represent their relative size. Clashes are frequent problems when different construction trades are drawing their plans independently.

Example: Problem with 2D Construction Documents

A line on a plan representing a run of 1-inch pipe looks the same as a line representing a run of 6-inch pipe. Mistakes can be made when pipes are run close to other features, such as walls and floor supports. Let's say that a 1-inch pipe fits nicely when it is installed next to a beam. A 3-inch pipe is shown by a line drawn with the same line thickness or lineweight used for the 1-inch pipe. On the plan, it looks like the 3-inch pipe should clear the beam, but it simply won't fit and must be moved. In cases like this, the construction item that is more difficult or expensive to move remains in place. In this case, the support beam stays where it is, and the 3-inch pipe will be moved. This error has caused additional rework and costs, possible arguments between the trades, and quite frequently a project delay.

BIM avoids this problem because everything is shown to scale, as it exists in space, in a 3D, picture-like format. The designer will see that the 3-inch pipe won't fit, and it will even be highlighted as a clash on the BIM model. Corrections to the plan can be made before the pipefitter is even on the construction site; in fact, the correction is made before the work is bid, saving everyone time and money.

BIM modelers are alerted to clashes via different visual methods. Clashes are called out on the computer's monitor by color change highlighting, as shown in Figure 5.14; darkening the background, as shown in Figure 5.15; and by transparent dimming and highlighting the clash, as shown in Figure 5.16.

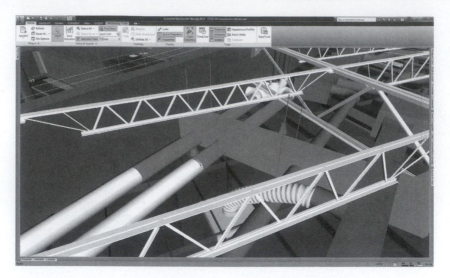

FIGURE 5.14 Screen capture of a 3D image showing the clash highlighted.

Source: Image courtesy of Mortenson Construction.

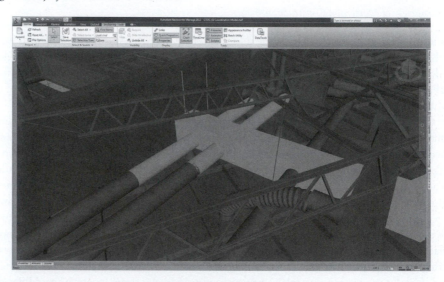

FIGURE 5.15 Highlighted clash with background darkened.

Source: Image courtesy of Mortenson Construction.

BIM also allows the modeler to *fly through* (a BIM term) the project and view the construction work as if the user were a comic hero like Superman. BIM can also give the FM information and purchasing data about important items on the plans. Rather than crawling up into a ceiling on a step ladder to try to get nameplate data from a leaky valve, the FM finds the valve on the BIM construction plans, which have a database of construction information, and get the manufacturer's specifications displayed on the computer screen.

Example: BIM's Information Capability

The owner of a strip mall needs to strengthen an existing roof due to increased snow loading caused by reconstruction of a center section of the mall, which added a much taller new building construction section. A major savings in construction costs for the project could be achieved if the existing strength of a roof supported by steel bar joists is known or if support joist data is discovered, so the roof support strength can be calculated. The owner has a full set of construction plans, but the strength information that could be obtained from the steel bar joist manufacturer's cut sheets has long since disappeared. The alternative solution for joist data collection is to climb up to several joists using a 22-foot extension ladder and search for an identifying steel tag from the joist manufacturer. Once the joist is identified, a structural engineer can review the supports and spacing to determine the roof's strength. If the plans had been a BIM model, a mouse click would provide the joist data once the joist was located on the plans.

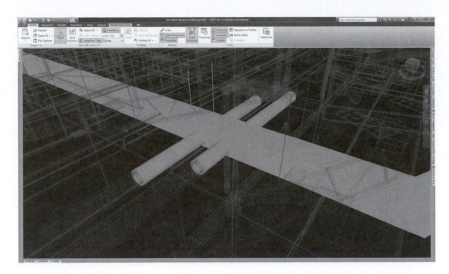

FIGURE 5.16 Clash shown by transparent dimming and highlighting.

Source: Image courtesy of Mortenson Construction.

BIM can act as a database of building information. The Construction Operations Building Information Exchange (COBie) is the standard that allows different equipment manufacturers using a variety of software platforms to provide data about their equipment for insertion into a BIM file. BIM data is then available for review and can be checked for its ability to meet project sustainability needs and future maintenance needs.

BIM has the following additional uses:

- BIM can measure quantities of items that will be used in building construction, thus providing data for construction cost estimating.
- BIM can also take the lumen data from light fixtures, the reflectance of interior finishes, and the amount of glass, and figure out how much ambient light and electric illumination is available. A plan for managing lighting can be developed from this information.

Stanford University Center for Integrated Facilities Engineering performed a study on thirty-two major construction projects that used BIM to generate their construction documents. The center found that there was a drop of up to 40 percent in unforeseen change orders, a cost estimating time decrease of up to 80 percent, a project completion time reduction of 7 percent, and a cost savings of 10 percent for the contract.

BIM Using 3D Laser Scans

BIM is still quite new, and only a very small percentage of existing buildings have their construction plans drawn in a BIM format. The challenge is to convert existing building information to a 3D interactive model, even in situations where, quite possibly, old 2D blueprints do not exist. Laser camera scans that can be integrated into an architectural computer-aided drawing (CAD) program are the answer.

Example: Laser Scanning to Tie New Building Systems to Existing Systems

A hospital building has existing heating, plumbing, electrical, and air-conditioning systems that must to be tied into the systems in a new wing of the hospital that will be completed shortly. In the past, an engineering firm would send a group of technicians to the mechanical space of the hospital to measure and hand-draw all the systems on a set of floor plans, which would take the better part of a week. The drawings that were created might still lack some important spatial detail. Verification of a drawing, for whatever reason, would require a trip back to the facility. Also items of smaller importance would simply be left out (like a ½-inch electrical conduit and supports) due to the vast amount of measurements and drawings needed just to get the basic concept of how the system functions. But in this case, hospital management has embraced the BIM concept and realizes the future advantages and savings by using this technology.

FIGURE 5.17 Raw point cloud created by a laser camera for a chiller plant installation. The center of the bright area shows a dark circle on the floor. This dark circle is where the laser camera was set on a tripod. If you look closely to the right side of the picture, you can see ghost-like images of construction personnel.

Source: Image courtesy of Mortenson Construction.

To begin, a 3D laser scanner (see Figure 5.17) is set on a tripod in the mechanical room of the old hospital described in the example. Targets are set up in the space, and the laser camera rotates and obtains a seemingly infinite amount of dimensional data on the location, size, and configuration of all existing equipment and piping structural supports. The scanner detects millions of points on an object, which is referred to as the point cloud. The data file of points is so large that a 1 TB (terabyte) portable hard drive attached to the laptop is often filled up with data. Software converts the cloud data to a 3D image that can be rotated, panned, and enlarged. The laser camera can operate in total darkness, and the picture resolution is said to be 1,000 times greater than even the best digital camera today. The dimensional information is then inserted into a 3D modeling software program. The resulting photo image is a black-and-white 3D picture. The images can be downloaded and converted into a color 3D architectural design program, resulting in an interactive 3D model. Figure 5.18 shows the beginning of the chiller plant construction by the addition of the chiller piping. Figure 5.19 shows the installation of the chillers. The result is a BIM of a chiller plant nestled between supporting columns of an existing space.

FIGURE 5.18 Adding a 3D mechanical system model to the cloud for the chiller plant

Source: Image courtesy of Mortenson Construction.

FIGURE 5.19 BIM computer-generated model of the chiller plant

Source: Image courtesy of Mortenson Construction.

Laser scanning can also be used to duplicate complicated shapes of a building façade. Laser scanning can re-create building details of the distant past and the artisans who created them are long gone.

5.8   SUSTAINABLE DESIGN AND CONSTRUCTION PARAMETERS

A sustainable project starts several months before the first shovelful of dirt is turned or the first brick is laid. Sustainable design starts with the creation of a team dedicated to taking on sustainability issues and developing a plan of implementation. The design must utilize sustainable construction methods, materials, and environmental design, and always strive toward the goal of sustainability for the building and the surrounding community. A thorough understanding of the design and construction of sustainable facilities requires a discussion of the following additional topics (some of these topics are discussed in this chapter; others are discussed in subsequent chapters):

- Resource conservation and material selection
- Site selection
- Transportation
- Energy efficiency
- Maintainability
- Opportunities for daylight use and natural ventilation
- Carbon footprint of construction materials and operations
- Recycling and reuse of building materials
- Site excavation and water collection
- Life cycle of materials and building systems
- Environmental concerns
- Habitability of built environment
- Space utilization and flexibility
- Project cost

Site Selection

Sustainable construction team members must consider the project location. Team members need to make sure that what they are proposing to build is in the best location for the building itself, its future occupants, the environment, and for the surrounding existing buildings and community. Sustainable site selection requires that a building site provide a synergistic relationship between itself and the buildings around it, providing the goods, services, and employment that people need. Team members must also consider the effect the building will have on the environment around it, including changing wind patterns and increases to the heat island effect that can be created by a new building. Site selection should also consider how a building will be oriented on the site. Building orientation affects opportunities for daylight in interior spaces by taking advantage of exposures that provide more opportunities for light gathering. It is also important for controlling solar gain created when ultraviolet light passes through window glazing and heats the interior space.

Additional factors to be considered for site selection include the following:

- **Zoning**—All land surface areas are zoned by communities according to acceptable use. Zoning must be considered in any site selection; failure to do so could result in not being able to obtain a building permit. But zoning for an area may have been set up decades ago and may no longer relevant to the current needs of the community. Zoning may be challenged by appearing before zoning boards and discussing the merits of the project to obtain a variance.
- **Redevelopment**—Existing sites can be redeveloped in order to meet the organization's needs. Redevelopment saves natural raw land sites from development. Redevelopment also utilizes areas that may be unsuitable for other types of development. A brownfield's site may be unsuitable for a daycare facility or residential housing, but it may be fine for light industrial or retail use.
- **Transportation**—Providing for transportation for customers and staff members is an important consideration for reductions in a property's carbon footprint.
- **Raw land sites**—Raw land sites can be natural sites that have value in their current state. The development plan should consider the natural habitat.
- **Construction materials, equipment, and storage**—The site must provide space for the equipment needed during construction, or provisions need to be made to lease space at adjacent properties.

Zoning

All areas are zoned for a specific use. The general use classifications are business/commercial, agricultural, or residential. Communities may have their own zoning classifications and sub-classifications. Research is necessary to define where a project should be located. Sometimes the best location might not have the proper zoning for the project. In this case, project team members will have to meet with the community's zoning board. Zoning officials might require that a public meeting be held and that notices go out to property owners that surround this potential new addition to the neighborhood. Members of the public are given the opportunity to voice their opinions on the project. This is especially true if the owner wants to build a property that has a nonconforming use. A *nonconforming use* is the use of a building for something that differs from the buildings surrounding it, or the building has the same use but is of a much different scale (larger or smaller) than the buildings surrounding it.

All zoned areas have restrictions on what can be built in a particular zone. If the developer wants to construct a building that is different from what is allowed in the current zoning rules, the developer must apply for a *variance*. If a variance is granted, then construction can proceed. A nonconforming use might occur when a developer wants to buy some houses, tear them down, and build a shopping center in a neighborhood that is zoned for single-family residential buildings. The people who have lived in the area under the current restrictions have a right to voice an objection to giving the developer a variance. A zoning board takes into consideration the wants of the builder and the desires of the current property owners in the area, and then renders a decision on the variance. Presentation drawings shown to a community's zoning board can be very helpful in the decision-making process. Boards don't always side with the homeowners. The board might decide to give a variance to the developer if the project will serve the overall good of the community. Disgruntled homeowners living near the project will always be able to voice their opinion once again—at election time.

Raw Land Site

Important natural features on a raw land site might have to be left undisturbed so that the site will be more livable and have biodiversity. Some of these features include the following:

- Trees (large old-growth trees are not replaceable and are difficult, if not impossible, to move)
- Wetland and natural drainage areas

- Grasslands and sandy areas for animal habitat
- Wooded and brush areas that provide animal cover

Undeveloped sites in suburban or remote rural areas might not be suitable for building occupants and visitors. Consideration should be given to provide the infrastructure for:

- Public or carpool transportation
- Lunch facilities

It would be unfortunate to build a very green building and later discover that much of the energy savings were lost by employees having to travel 5 or 6 miles by car just to get lunch every day.

Previously Developed Sites

Successful urban building construction requires the designers to consider the building's look, size, orientation, configuration, and use when deciding on a location. The building should visually compliment the other buildings around it or in some way compensate for differences. The answer for a new high-rise building in Chicago was the installation of mirrored glass for the façade. The new steel-and-glass building was totally out of character with the old, highly ornate stone buildings adorned with statues and columns that surrounded it. But the mirrored glass on the building's exterior reflected the façades of the old buildings, and when pedestrians gaze at the new, they see the old. The feelings evoked by period architecture were not lost or diluted by the new construction.

Existing properties afford the opportunity to reuse buildings through system upgrades and renovation. Large amounts of building materials can be saved in this way. It is often more difficult to renovate than to demolish and rebuild. The new design tools available to the designer today, such as laser scanning and BIM, provide new, more cost-effective ways to renovate existing buildings.

A large massive masonry building can absorb a lot of solar heat, which will radiate outward to the surrounding area and possibly creating a warm microclimate once the temperature drops. This microclimate could be objectionable. Consideration should be given to the placement, configuration, and positioning of a building to reduce the effect of heat radiation. Airflow around the building and its effect on the airflow to other buildings should also be considered.

Even previously developed sites can sometimes afford an opportunity to improve the natural environment. The property can be brought to a more natural state by planting indigenous plants in a natural arrangement. The recycling and reuse of surface runoff water to create water features or a water garden are some additional ways to improve the natural state of a property.

Transportation

Transportation issues during construction, as well as the transportation issues that will be encountered by tenants and customers once the building is occupied, must be considered for sustainability. If people can't get to the property on a regular basis without extensive delay or effort, the property won't be sustainable in the long term because it decreases worker productivity. The important transportation issues include the following:

During Construction Issues

1. Material and equipment storage on site
2. Bringing materials and workers to and from the site (temporary roads may have to be built)

After Construction

- Parking availability, type, and location
- Access to mass transit (rail and bus)
- Pedestrian issues (sidewalks, bike lanes, pedestrian malls, bus shelters, etc.)
- Building in functionality for tenants arriving to the property, working at the property, and leaving the property. (Examples include a coffee shop for arriving tenants, office supply shop for working tenants, and a supermarket for tenants who take mass transit and need to make supper when they arrive home.)
- Provisions for charging electric vehicles and bicycle storage
- Speedy evacuation during emergencies

Material and Equipment Storage

Discuss with the GC where construction equipment, tools, materials, and waste will be stored on the construction site. An agreement is necessary in advance so both the owner's representative and the contractor are aware where materials will be stored and the impact of storage on site operations. The following considerations will help to avoid misunderstandings:

A GC performing a tenant remodeling project at a strip mall may decide to place a construction Dumpster in the drive lane in front of the space because of the easy double-door access through the store front. This would affect traffic movement at the strip mall and may disrupt the business activities of stores located on either side of the construction. The owner's rep might prefer to locate the Dumpster behind the center where all the regular trash and recycling Dumpsters are located.

The GC might have assumed that an adjoining vacant space can be used for secure storage of tools and construction materials. However, this space is in the process of being leased to another tenant. The GC might then have to provide a trailer for the storage of tools and construction materials, and the location of the trailer will have to be determined. The GC may also have to erect temporary fencing to prevent theft and limit access to the construction site. The owner's rep also should reach an agreement with the GC about storing the materials on site, and for storing only what is needed for the project at that site.

Project Execution

Project team members are part of the construction project from start to finish. All members need to be aware of the project schedule and the progress of the work. Verification of the work completed in relationship to the schedule is important. Construction experience is a valuable asset when evaluating the accuracy of the schedule. For example, it may seem as though the remodeling project won't be completed on time because painting, carpet, locksets, ceiling tile, and furniture have not been installed. Experience tells the team member that the final interior finish work can move very quickly, allowing the project to be completed on time. Conversely, if plumbing or HVAC work hasn't been completed per the schedule, the project will likely be completed late.

Project Completion

Different terms are used to describe when a project is complete. **Beneficial occupancy** is the terminology sometimes used to describe the owner's takeover action (moving into a space) for office buildings. **Mechanical completion** is the term used to describe completion of industrial facilities with lots of machinery that must be built. **Substantial completion** is the term used to describe when the construction project is complete for buildings or new tenant spaces. It might seem obvious when substantial completion occurs, but really it is subjective, depending largely on whose interests the person represents (owner, architect, equipment manufacturer, general contractor, or subcontractor). Every member of the project has different view of this matter. Parties to the project must agree to specific terms in the construction contract about when substantial completion has occurred. They must also agree on who is to decide when completion has occurred.

Substantial completion is frequently decided and certified by the project architect. When the project is substantially complete, construction warranties become active. The date is important because a constructed or installed item generally has a one-year warranty. The completion date is especially important when the construction project takes several months or even years to complete. It becomes even more complicated when warranted mechanical systems must be operating months before the overall project is complete. The owner moving into his or her newly constructed building may find little recourse from the equipment manufacturer if defects are discovered after the warranty period, even if the owner has never actually used the equipment.

5.9 SUSTAINABLE BUILDING CONSTRUCTION METHODS

Sustainable construction, although a fairly new term in "green speak," is not a new concept. It was practiced by architects of ancient cities around the globe. When the bulk of wealth and power of a nation resided in the hands of a king, construction decisions of major importance were based not on internal rates of return on investment, but often on more far-reaching goals that would attempt to provide into perpetuity a sustainable environment for the citizens of a nation. The factors limiting the ancient builder's success in sustainability were the available technology and unforeseen natural and manmade disasters.

Construction Material Recycling

Sustainable construction methods provide for the most efficient construction and deconstruction, with the goals of reducing waste and recycling as many building products as possible. Why is sustainability such an important topic? Builders in the United States have constructed cities, towns, and villages, and the infrastructure to support them, for hundreds of years without consideration for sustainability. The result is mountainous landfills, which are sad testimonies to the wasteful methods of the past.

Today, government, academia, owners, and construction contractors themselves are demanding that sustainable construction practices be followed. Their reasoning is simple mathematics. If one plumber cuts out several feet of old copper tubing and tosses it away, it might not be a big deal. The problem is when thousands of plumbers do the same thing. The wasted material reduces the supply of easily obtainable copper, and the price of copper tubing goes up. If supply is low and demand is high for copper tubing, the price goes up even further. Copper is a finite resource, so recycling provides additional copper that adds to the available supply, which in turn keeps the price from rising sharply. Recycling copper also reduces the amount of energy required to produce copper, and keeps energy costs down. Rising energy costs negatively affect all aspects of construction.

One could also argue that copper would be much higher in price if it were not for the discovery and use of alternative plumbing materials, most notably polyvinyl chloride (PVC) pipe, acrylonitrile butadiene styrene (ABS) pipe, and cross-linked polyethylene (PEX) tubing.

Recycling costs money, and the financial returns for the recycled material are not always certain because commodity prices vary. Sorting, storage, administration, and transportation all add to the cost of recycling. Recycling sometimes necessitates environmental remediation activities. Hazardous materials must be properly contained, removed, and/or treated prior to recycling. Security is another expense and is needed to keep the recycled materials secure.

If builders don't recycle the materials they consume, material costs silently rise as the material supply diminishes. At some point, the cost of construction becomes unsustainable because owners cannot achieve a reasonable rental return. It will simply no longer be economically feasible to complete projects that were once cost effective. New building costs rise and rents at new and existing properties may also rise. High rental costs will then cause marginal businesses to fail. An economic slowdown ensues until a balance is once again achieved between costs and investment return.

State and Local Recycling Ordinances

States and local municipalities have specific construction and demolition waste recycling laws. Construction materials that frequently must be recycled include the following:

- Gypsum board
- Cardboard and paper
- Unpainted and untreated wood
- Metal, including steel studs, pipes, conduit, and plumbing fixtures
- Concrete and brick
- Asphalt pavement and asphalt roofing shingles
- Window glass
- Landscaping materials

To recycle building materials efficiently, the materials must be collected in a manner that does not interfere with construction or require a lot of handling, which would add significantly to the cost of construction. Containment of recycled materials that is convenient for the construction worker and simultaneously provides efficient transport to storage areas is essential. The use of wheeled carts, enclosed chutes mounted on the side of the building, and containers that can be lifted by crane can make the recycling job easier and more cost effective.

The following questions should be considered when developing a plan for recycling:

- Does the construction site have enough space to recycle a variety of building materials generated during demolition and construction?
- Can offsite areas be utilized for recycling?
- Can recycled materials be stored for the duration of the project, or is a coordination of material pickups necessary to keep the site clear for material movement?
- Will the materials need to be monitored, secured, or protected to avoid damage or theft?

If there isn't enough space to handle all the materials that will be generated, then a more creative recycling plan is necessary. The plan will need to include the frequency of recycle material pickup and address other planning and storage issues to provide the most efficient material movement from the site to the recycling center and to avoid site space issues during the course of the project. For example, the demolition phase will generate the majority of gypsum board for recycling; later in the project, the builder will be dealing only with the cut ends of pieces while fitting the new wall board. The cut ends will take up much less room. The construction contractor may need to increase or decrease the availability, pickup frequency, and/or the size of recycling containers at certain points over the project's duration to meet the demands of the flow of recycled materials.

Waste and Pollution Avoidance

Methods that reduce pollution, waste, and energy use contribute to sustainability. Pollution avoidance can provide some unique advantages, such as the following:

- Filtering and reusing rainwater collected in an excavation, rather than pumping it into a storm sewer, can provide a ready source of non-potable water rather than having to truck it to the site.
- Grinding up old concrete pavement to be used as aggregate under roadways saves money and resources, and reduces landfill use.
- Saving cut pieces of steel studs or lumber for possible reuse, rather than immediately tossing them into a recycling Dumpster, helps make construction more sustainable by eliminating reprocessing costs associated with recycling.
- Donating trees and shrubs when clearing a site for construction, or saving them for a second life at the site, reduces waste.

5.10   POSTCONSTRUCTION MAINTENANCE ISSUES

Postconstruction maintenance issues, which are new to modern construction design, have come to the forefront. The reason is cost. Over a building's life span, it turns out that the design and construction costs are a small percentage of maintenance personnel costs, building operation costs, and maintenance/repair costs.

A major issue for maintainability is the building's exterior finish. A beautiful, smooth, white-plastered concrete façade may be bright and clean-looking when newly built. But later it becomes a magnet for dirt, stains, and graffiti, giving the building a poorly maintained, unsafe, and unsavory appearance. Another problem is when an architect designs a façade with too many different materials, or materials that have conflicting maintenance requirements. To maintain one surface, the chemicals applied or the method of application may damage an adjacent surface. To get the job done, extensive surface protection is required. Giving careful thought to the various exterior surface types during the design phase can greatly mitigate future problems.

A way of analyzing exterior building surface types to learn which provide the best overall return for the property is through a strengths, weaknesses, opportunities, threats (SWOT) analysis, which we will discuss in Chapter 13.

Maintenance Logistics

An all-too-frequent issue facing FMs is lack of forethought concerning logistical issues of maintaining a property and its equipment. The maintenance team's ability to get needed supplies and equipment into a building and store them safely in a convenient location is of utmost importance. FMs and their supervisory maintenance personnel should provide input to address this issue at the design stage.

Design topics, operations, and equipment that can help with maintenance logistics include the following:

- Loading docks at the proper size to handle both tenant needs for moving in and moving out and the needs of the building operation
- Dock levelers to make up the difference in height between the truck's bed and the loading dock
- Overhead garage-type doors, or even double doors large enough to allow a pallet jack or forklift to move equipment into the building easily
- Back hallway access from the loading dock to the freight elevator so equipment and supplies won't have to interfere with the movement of tenants and their customers
- Monitoring and directing loading dock use for fair and efficient use of docks by all parties
- Placement of lifting pad eyes in strategic locations or the addition of I-beam trolley tracks welded to building support structure for heavy lifting and movement of equipment
- Safe storage of barrels and buckets of chemicals
- Safe storage of pressurized tanks
- Storage of deicing chemicals

REVIEW QUESTIONS

1. How does BIM differ from traditional paper blueprints?

2. Why is it important to reduce waste in construction?

3. What is the job of the owner's rep in a construction project?

4. _____ protects the owner from having to pay for the failure of a subcontractor who did work on the owner's project but did not pay his or her material suppliers.

   A. A warranty
   B. Liability insurance
   C. A bid bond
   D. A lien waiver

5. Owners can protect themselves from a failure of the low-bid contractor to live up to quoted construction cost in their bid by obtaining a _____.

   A. performance bond
   B. bid bond
   C. payment bond
   D. lien waiver

6. It is a good idea for an owner's representative to get _____ from the contractor when there is a good possibility that additional work will be needed or desirable. It is done to prevent haggling over the costs associated with this extra work.

   A. estimates from other contractors
   B. unit prices
   C. budget data from construction cost data books
   D. a performance bond

7. To prevent future issues with items not shown as they were actually constructed, it is a good idea to collect _____ and using them to have a final set of plans drawn up by the architect.

   A. a last set of construction plans
   B. the first set of construction plans without revisions
   C. a last set of construction plans showing all previous revisions and red-line drawings
   D. a list of plan modifications obtained from building inspection personnel

8. List five common punch list items.

9. List five services that subcontractors provide.

10. The construction superintendent works for the _____.

   A. general contractor
   B. subcontractor

    C. owner's representative

    D. architect

11. Shortening the length of time to complete a construction task that lies along the critical path _____

                             .

    A. has no effect on the project's duration

    B. reduces the project's overall duration

    C. provides more float time, but overall project duration is the same

    D. none of these

12. How does a GMP contract differ from a fixed-price or lump-sum contract?

13. Name five construction or demolition waste materials or debris that can be recycled.

14. Why is it sometimes difficult to recycle construction wastes?

15. How is a laser camera used in the generation of BIM construction documents?

16. What are red-line drawings on a set of construction documents?

17. Describe how an owner uses retainage to his or her advantage.

18. Give some reasons why the lead time of a particular construction material can vary.

Essay Question

19. A company wants to build a 57,000-square-foot grocery store. What type of construction contract would best suit this project? Explain.

FIRE AND SECURITY SYSTEMS AND DISASTER PREVENTION

INSIDE THIS CHAPTER

Can Disaster Prevention Be Integrated into the Normal Activities of the Facility Maintenance Operation?

Many types of disasters can strike a facility. A facility's ability to quickly (relative to the type and size of the disaster) return to full operation after a disaster event is evidence of a sustainable disaster recovery operation. The chapter begins with disaster categories and the risk analysis that initiates the process of designing a disaster plan for the facility to help it overcome the disaster. Fire is the only disaster type that the building itself provides equipment to combat. The chapter discusses the various fixed fire suppression systems, smoke control, fire spread control, and structural fire insulation that either mitigate the effects of fire or attempt to eliminate the fire threat entirely. The greening of America has even encompassed fire suppression chemicals that provide alternatives that are both safer and less toxic to the environment. The chapter then discusses other potential disaster events involving security, access control, alarm systems, and cyberattack.

6.1 FACILITIES MANAGERS FACING A DISASTER EVENT

The solution to an emergency draws heavily upon the strengths of the facility manager (FM). It is also a time of great need for the facility, its workers, and its customers. There is no easy formula to define for the FM the appropriate level of resources and personnel to commit to correcting the effects of a disaster. Facility managers have their own unique capabilities, knowledge, and experience for handling a disaster.

The FM needs to be thoroughly prepared before a disaster strikes. This preparation may include actual disaster drills. Preparation should include providing a suitable replacement for the FM. The FM may become physically or emotionally exhausted, or injured, during a disaster and be unable to help. In large organizations that have several facilities being managed across the state or country, a good manager replacement strategy includes cross-training and familiarizing managers with different facilities and emergency capabilities.

An FM at a facility that has no sister facilities within the organization might engage a group of managers at surrounding facilities to come up with a solution to a manager replacement problem during a disaster. Managers in urban locations frequently participate in security groups that share information on criminal activity that is going on in the area. FMs may even participate in an e-mail or phone list that immediately alerts facilities to the current threat. These types of groups could easily be extended to working together to handle disasters at member facilities. Another solution is to contract with a company that specializes in disaster management.

Facilities have different rankings in importance to the community. Greater government assistance is normally more forthcoming to essential service facilities such as police department facilities, fire department facilities, water purification plants, electric and gas production facilities, sewer treatment plants, and hospital facilities. A facility's ranking in importance needs to be taken into consideration by the FM because it relates to the number of services the FM must obtain for the facility through private contractors.

The FM must mitigate the disaster, even if it means passing the responsibility to another manager or to a company that specializes in disaster recovery. The outcome for missteps taken during and after a disaster event can be great. Financial loss not only happens from the disaster event itself, but also when companies who did lucrative business with the facility organization disassociate themselves from both the event and the organization.

Making leadership decisions in a time of crisis can accelerate a return to profitability. Making such decisions shows enormous value to the top management of the facility organization.

6.2 DISASTER RISK ANALYSIS AND PLAN

A multitude of individual disaster types and combinations of disasters can strike a facility. One disaster can set off a chain reaction leading to subsequent additional disasters. *Risk analysis* is the first step in defining the probable disasters that can occur and the likelihood of their happening. Risk analysis aids the manager in discerning how funds should be spent for disaster preparation. Hiring consultants to perform hazard analysis, an identification of hazards both internal and external to a facility, is an example of an expense incurred in disaster preparation. An external hazard, for example, could be from adjoining properties that store hazardous or flammable gases and liquids.

Disaster Scenarios

The goal of risk analysis is to discover which types of disaster have the greatest likelihood of occurring and to estimate the disaster's impact on the facility. A possible source of disaster information is the company that insures the property. They have performed their own analysis in order to price the insurance policy. A disaster risk analysis must also include scenarios in which multiple disasters occur simultaneously. For example, a flood might result in service and supply interruptions, which could lead to disruptions in delivery of food and heat to a large number of people.

The FM should also provide scenarios for further analysis based on common activities that occur at a facility that could influence the disaster event. Here are some examples:

- A tenant is moving out of the building; large areas of the hallway are congested and the freight elevator is turned off.
- A tenant picnic is underway with a barbecue on the upper elevated parking deck, which blocks a secondary escape route.
- Lobby access is restricted due to remodeling.
- Blocked streets due to paving or excavation work restrict the access of emergency vehicles.
- A fire sprinkler branch line isolation valve is closed to prevent accidental activation during construction demolition.
- Tenants create a traffic jam by driving off the elevated parking deck and causing an accident.
- A water line breaks during an evacuation.
- Large numbers of tenants or visitors do not follow directions or know the safety procedures.

Example: Spill Containment

Facilities with parking should be especially concerned with spill containment from fuel or oil leakage from parked vehicles or vehicles involved in an accident. Facility maintenance personnel are not trained hazmat workers, but their quick action can greatly mitigate a potential disaster. Spill cleanup kits can be very useful if everyone knows where to find them. But creating a simple dam around a spill with sand and/or granulated clay kitty litter can buy precious time to institute more aggressive cleanup measures and make the area safe. A simple containment pan that can be placed under a leak can save a tremendous amount of containment materials and make cleanup much easier.

"What if" scenarios should be discussed with fire inspection personnel. For example, at a loading dock, a truck that is backing up punches a hole in the diesel fuel tank of a truck in a loading bay. After calling the fire department, should the truck that caused the damage remain at the dock, or should the truck move to open up access for the fire trucks and equipment? Should building ventilators and electric equipment (sources of possible ignition) be secured in the spill area? Every building is different and decisions about how to act must be based on the actual location and the specific situation.

Disaster Response Plan

A disaster recovery plan should be created to address the serious issues of each disaster type identified in the disaster risk analysis. The plan provides information that is critical to the facility during the restoration phase. It also provides a well-thought-out checklist so that important items are not forgotten during this difficult time. The plan should be set up to cover a number of specific disaster types. Speed is just as important in the recovery phase as it is during the disaster itself. The FM needs a sequential plan that is tailored to the needs and risks faced by the facility. The Federal Emergency Management Agency (FEMA) has developed a free guide, *Emergency Management Guide for Business and Industry*, which can be obtained at the FEMA website. This step-by-step guide aids businesses in developing a disaster response plan.

A team must be organized in order to implement the necessary procedures during a disaster and to prepare for future disasters. The team should include persons familiar with the building's safety, security, and mechanical systems. The response plan should list the persons, by name and position, who will be active participants in emergency activities in response to a disaster. The plan defines the actions that participants need to take and the procedures they need to follow during and after a disaster event. The plan checklist should also provide the names and contact information of the people who should be informed about the status of the recovery, and all the vendors and government offices that are useful for recovery. Priority is given by the facility to vendors who have an active service agreement on file, which means that their insurance is up to date and the work will be performed according to this agreement. The plan should include methods of communication during disaster and disaster recovery; for instance, updates should be made on the organization's website and mass e-mail notifications should be sent.

Note: As with all items mentioned in this section, refer to the authority having jurisdiction (AHJ), which may be different from the local authorities, prior to making any decisions, new installations, replacements, or modifications to any safety related equipment because code requirements can vary.

6.3 PREVENTIVE MEASURES AND FIRE PROTECTION SYSTEMS

Measures taken prior to a disaster can greatly mitigate the size of the fire event. The two most important items are ensuring an adequate supply of firefighting water and the testing of fire alarm systems.

Water Supply

One of the most important aspects of fire protection is water supply. The water supply must be checked regularly, according to the requirements of building insurers and fire prevention codes, to ensure an adequate supply of water is available to fight a fire. This involves testing the flow capacity of the source that is the fire main water supply (see Figure 6.1) and the ability of the building's fire pumps to boost the pressure and capacity of the water supply (see Figures 6.2 and 6.3).

FIGURE 6.1 Fire flow testing of hydrant

Source: Image courtesy of Hydro Flow Products, Inc.

FIGURE 6.2 Fire pump test

Source: Image courtesy of Hydro Flow Products, Inc.

Fire Alarm Systems

Fire alarm systems consist of an alarm control panel, fire and smoke detectors, pull stations, and both visual and audible alarms. The alarm control panel provides visual indication of system condition and alarm location. The control panel provides the means for silencing the alarm after the condition triggering the alarm has cleared and the alarm device is reset. A keyed access glass and steel door to the alarm panel prevents tampering by unauthorized persons. The alarm panel box houses batteries (for use during power failures) and phone-line hookups (for automatic calling of the alarm-monitoring company or directly to the local fire department).

The equipment is maintained by a vendor who is properly trained and licensed to perform this work. These vendors also provide the required yearly testing. The periodic testing and inspection of this

FIGURE 6.3 Fire pump test in Chicago, Illinois, on April 12, 2012

Source: Image courtesy of Hydro Flow Products, Inc.

equipment generally falls to facilities maintenance personnel. Facility maintenance technicians activate alarms on a weekly basis. First, they contact the alarm-monitoring company to put the system in a test mode so that fire personnel are not called. Then they activate fire pull stations, setting off the alarm on the panel and all the alarms throughout the building. Maintenance personnel make sure that all alarm strobes, horns, and alarm-indicating lights, inside and outside the building, are working. They then reset the panel.

If the building has a wet-pipe fire sprinkler system, maintenance personnel, on a quarterly basis, once again contact the alarm-monitoring company, open the sprinkler system's *inspector test valve* to simulate flow from one sprinkler head. This triggers the fire alarm, requiring a reset. In both cases (testing both a fire alarm pull station and a wet-pipe sprinkler system), after resetting the alarm panel, the alarm-monitoring company is contacted again to reactivate the alarm system. Refer to the National Fire Prevention Association (NFPA) standard NFPA 25 "Standard for the Inspection, Testing and Maintenance of Water Based Fire Protection Systems" and local requirements for information about meeting all testing, maintenance, and inspection requirements.

6.4 FIRE SPRINKLER HEADS

The two basic types of sprinkler heads are open and closed. Open heads are essentially open-spray nozzles attached to systems designed to spray water over the entire protected area rather than just the area where the fire was detected. Open heads have no valves and no temperature designation.

Closed heads are used on pressurized systems containing air, water, or a water and antifreeze mixture. Closed heads have either a glass bulb filled with a low-boiling-point liquid that holds the valve in the head closed, a solder pellet that will melt at a low temperature or a two part metal link soldered together with a low-melting-temperature alloy (melting point between 135 and 225 degrees Fahrenheit) as the activating mechanism. The liquid inside the glass bulb is colored. The color of the liquid denotes the temperature at which the bulb will break and allow the head to open.

Orange	135 degrees Fahrenheit
Red	155 degrees Fahrenheit
Green	200 degrees Fahrenheit
Blue	286 degrees Fahrenheit
Mauve	360 degrees Fahrenheit

6.5 WET-PIPE SPRINKLER SYSTEMS

As the name implies, a wet-pipe sprinkler system (see Figure 6.4) is completely flooded with pressurized water from the city's water main. As soon as a sprinkler head in the system opens due to the heat of a fire, water immediately becomes available and sprays out of the head. The taller the building, the greater the opposing force that needs to be overcome before water will flow out of a sprinkler head.

The parts of a wet-pipe sprinkler system include:

- **Fire main line**—A large water pipe that enters the building from the water main located under the street.

- **Sprinkler riser line(s)**—Pipe takeoffs from the fire main that provide water to the main sections of a sprinkler system. Sprinkler risers are frequently identified as north riser or south riser for the purpose of identifying where in the building a fire has activated a sprinkler head.

- **Branch lines**—A series of lines in a grid pattern coming off the riser where the sprinkler heads are mounted.

- **Branch shutoff valve**—A valve that secures the water to a branch. In high-rises, each floor has a branch shutoff valve. These valves must either be locked open or include a tamper switch that activates a "system in trouble" alarm.

- **Check valve or alarm check valve**—When shut, the check valve, or alarm check valve, keeps water in the system from flowing back into the main supply line. It sets off an alarm when opened due to flow in the branch lines, due in turn to sprinkler head activation.

- **Flow indicator switches**—Risers and branches can have flow indicator switches that initiate a fire alarm when flow is detected. Flow switches also show the location of a fire on the liquid crystal display on the fire alarm (i.e., flow north riser 26th floor).

FIGURE 6.4 Wet-pipe sprinkler system

Source: Photo courtesy of Oconomowoc Memorial Hospital.

- **Main drain**—System drain valve used for draining the system. In some test procedures, the task is performed by a sprinkler technician.
- **Main shutoff valves**—Main shutoff valves are frequently outside stem and yoke (OS&Y) gate valves either chained and locked open or with tamper switches installed. Their purpose is to secure water to the building should a major break occur.
- **Alarm devices**—Alarm panels provide a point where alarms can be viewed and silenced once corrective measures have been successful.
- **Inspector test valve**—Valve used during alarm testing to simulate the flow out of one sprinkler head at the end of a branch line.
- **Inspection and test record**—A record of all inspections and required maintenance that is frequently attached to the sprinkler riser for viewing by a fire inspector.
- **Spare sprinkler heads and sprinkler-head installation/removal wrench**—A steel box with the proper code-required amount and type of sprinkler heads and removal tool mounted on the wall of the fire sprinkler room.

Utilizing city water pressure is only appropriate for some low-rise buildings a few stories tall, depending on fire main pressure and capacity requirements. The actual weight of water in a vertical riser pipe creates a head pressure in opposition to the force of water in the main. The higher it goes, the greater the head pressure. When head pressure and pipe friction losses are equal to the city water main pressure, almost no water will flow out of the sprinkler head. To overcome head pressure, an automatic fire pump is needed. The fire pump is generally a centrifugal pump driven by an electric motor. The pump is activated either by switching over from automatic operation to manual operation, or by a drop in pressure in the riser detected by a pressure switch. To make sure that the drop in pressure is sufficient to activate the fire pump, a *jockey pump*, also known as a *pressure maintenance pump (PMP)*, is used to boost system pressure to a higher level. This boost ensures that the last head located the farthest away from the riser on a branch line will cause a sufficient enough pressure drop to activate the fire pump when the head is opened by the heat of a fire. A jockey pump has its own pressure switch that activates the pump periodically in order to compensate for a minor drop in water pressure. This is normal operation. According to NFPA-20 2007 "Standard for the Installation of Stationary Pumps for Fire Protection," the jockey pump should be sized to make up any normal fire sprinkler line pressure loss in 10 minutes. (NFPA standards can be purchased from NFPA's website.) If the pump starts and stops frequently or operates continuously, then a problem with the pressure switch or a significant leak may be present and should be investigated immediately.

When activated, wet-pipe sprinkler systems put out a fire faster than any other sprinkler system. They are also the most economical to install, test, and maintain. A major problem with wet-pipe systems is that they must be installed in heated areas when freezing outdoor temperatures are possible. Also, due to the black iron pipe used throughout the system, water flows out colored black and smells like sewage. The water flows like black and with an odor until the lines are cleared of the stagnant water that was sitting in the piping. This characteristic is normal, but it can create quite a mess when a sprinkler head is accidentally activated by someone bumping it with, for example, the end of a ladder.

Wet-Pipe Sprinkler Issues

A major problem with wet-pipe systems is that they must be installed in heated areas when freezing outdoor temperatures are possible. Subzero temperatures can cause the water inside the head to freeze. The ice exerts a pressure that will force the valve open, activating the head and dumping a considerable amount of water into the hallway. If an icicle is hanging from a sprinkler head, the head and possibly the piping to the head are frozen. In this condition, the head could become active at any time. Fast action is required to prevent a larger problem. The branch line needs to be secured and sprinkler fitters called in to repair and unthaw the line.

Due to the black iron pipe used throughout the system, water flows out colored black and smells like sewage. The water flows black and with an odor until the lines are cleared of the stagnant water that was sitting in the piping. This characteristic is normal, but it can create quite a mess when a sprinkler head is accidentally activated by someone bumping it with, for example, the end of a ladder. Or perhaps the damage may have occurred due to shock when the attached piping is bumped by material handling equipment, as may be the case in a warehouse. When that happens, the head might not go off immediately; sometimes a very fine mist of water is seen coming from the head. (It is so fine that the water evaporates before it hits the floor.) In this situation, there is no time to waste: Shut the branch line valve, notify the alarm company that there is a problem with the system, and have the sprinkler head replaced by a sprinkler technician.

Deluge Systems

All the sprinkler heads are open in a deluge system. Deluge systems are fast-acting systems that will put down a lot of water on a fire very quickly. They are used in cases when a fire could spread rapidly. To prevent this type of hazard, all the heads discharge water at the same time. Deluge systems are also used to provide an escape path for people by providing a curtain of water that keeps fire from blocking their path. An automatic deluge valve is required in an open head system. The deluge valve prevents the flow of water to the heads until fire is detected. The valve has a latch that, once opened, does not let the valve close until it is manually reset. Once fire is detected, the valve opens, and all the heads become active. Deluge valves are controlled either electrically or pneumatically.

Antifreeze Systems

Antifreeze systems provide an economical choice when installing sprinklers in low-temperature areas. Antifreeze sprinkler systems are wet-pipe systems that have been modified by injecting a factory premixed solution of an NFPA-approved antifreeze solution (either propylene glycol or glycerin) into the sprinkler piping. The antifreeze prevents the pipes from freezing due to cold temperatures. These systems are used in existing residential locations, at loading docks, and under canopies that are exposed to ambient temperatures. Antifreeze systems must have backflow prevention in order to prevent possible contamination of the domestic drinking water line that is attached to the same incoming source of water.

The NFPA has now expressed safety concerns regarding a fire risk associated with the combustibility of the two previously approved antifreeze solutions (propylene glycol or glycerin) The risk occurs during the period when the antifreeze charge is sprayed on a fire. The NFPA has responded to the concern by issuing a Tentative Interim Amendment (TIA) to their regulations in August 2012. NFPA no longer allows the use of these antifreeze solutions for new installations. An exception is allowed in residential situations if the jurisdiction having authority approves the system after receiving documentation from a designer that shows why using antifreeze in the system is acceptable. According to the NFPA, several companies are researching alternatives and hope to find an acceptable antifreeze solution.

6.6 DRY-PIPE SYSTEMS

Dry-pipe sprinkler systems generally use compressed air to fill the riser and branch lines leading to the closed sprinkler heads. In addition to the equipment found in a wet-pipe system, dry-pipe systems have an air compressor and a valve to hold back the water. Dry-pipe systems are used in parking garages, loading docks, canopies, and any location that can be exposed to freezing temperatures. Under normal conditions, the compressor runs for a couple of minutes to replace air lost in the system through leaks.

Activation of a dry-pipe sprinkler head occurs just like a wet-pipe sprinkler head with one exception. Heat causes a link to break apart or a bulb to burst, opening the sprinkler head. The difference is that, once opened, the dry-type head gives off a blast of compressed air. The force of the air pressure is used to hold the main riser water valve closed. Once the air pressure is lost through the sprinkler head, the water valve opens and water flows into the riser and branch lines and out the sprinkler head(s) activated by fire. All other sprinkler heads remain closed.

Dry-type systems require additional maintenance prior to being exposed to freezing temperatures. Systems generally have provisions for draining off the accumulated condensate from the compressed air system. When draining accumulated condensate, care must be taken not to drop the air pressure to the point where the water valve opens. Some systems have two valves and a length of pipe, about a foot or so long, which is attached vertically and increased in size by bell fittings. Branch piping is sloped toward this condensate drain so that water accumulates in the line above. Water is removed by opening the upper drain valve and allowing the pipe chamber to fill with condensate. This upper valve is shut, and the lower valve is cracked open to allow the moisture to drain off. Once the water is drained out of the system, the lower valve is shut.

This process is also a good way to find out if water is in the branch lines for any other reason. Maybe the valve that holds back the firefighting water has leaked or was activated accidentally. Checking this prior to the system being exposed to freezing temperatures prevents sprinkler pipes from freezing and breaking.

Preaction Sprinkler Systems

To protect valuable, electronic, or easily damaged equipment and materials from unintended sprinkler head activation, facilities frequently install preaction sprinkler systems. These systems are used in places such as hospitals, libraries, art museums, data centers, and telecom centers. There are two types of preaction systems: single interlock and double interlock.

Single-Interlock Preaction System   Single-interlock preaction systems also have a deluge-type valve, but these systems use closed sprinkler heads just like those found in a wet-pipe system. When the heat of a fire, smoke, or the infrared light of a fire is detected, the deluge valve opens and fills the riser and branch lines with water. If a fire is present, water flows only out of those heads that were activated by the heat of the fire. Single-interlock preaction systems have a great advantage over wet-pipe systems: They do not activate because someone breaks a sprinkler head or because of a pipe failure. This could also be a disadvantage if construction work was done in the space, and damage occurred without anyone knowing it; the damage would be discovered only when the system is activated by a fire or during a test. Preaction systems should be tested whenever construction or demolition work was done in a space.

Double-Interlock Preaction System   Double-interlock preaction systems have a deluge-type valve and closed sprinkler heads; their riser and branch lines are pressurized with nitrogen gas or air. If air or nitrogen is lost through pipe or head damage, the deluge valve remains closed and no water flows. If fire is falsely detected and the deluge valve opens, no water flows into the space because the heads are closed. Two conditions must be met *at the same time* before the deluge valve opens. The two conditions might be a loss of air or nitrogen pressure and detection of fire, or a loss of air or nitrogen pressure and the activation of the alarm system by pulling the lever on a fire alarm pull station. Another advantage of the compressed air or nitrogen in a double-interlock preaction system is that it alerts the operator about branch piping or sprinkler head leaks through loss of pressure. In this case, the facilities maintenance technician (FMT) would know immediately if a construction worker damaged a head or piping without having to test the system.

6.7 SPECIALIZED FIRE SUPPRESSION SYSTEMS

Specialized fire suppression systems are designed to handle a specific hazard situation. Types of specialized fire suppression systems include:

- Early suppression fast response
- Water spray
- Foam/water sprinkler
- Fire-cycle

Early Suppression Fast Response Systems

Early suppression fast response (ESFR) systems were brought into use about 1990. According to Gross & Associates, these heads sense a fire in half the time that conventional heads take to sense the same event. ESFR heads put out about 100 gallons per minute (gpm), whereas conventional heads discharge 25 to 30 gpm. The water droplet size out of the sprinkler head is larger; thus, there is less evaporation and more of the water gets to the fire to extinguish it. Ceiling-mounted ESFR systems are generally used in place of in-rack sprinkler systems for warehouses that utilize shelving units for storage.

Water Spray Systems

Water spray systems are basically deluge systems that employ nozzles to attack the fire from different angles rather than just from the ceiling above. The nozzles are arranged around the machinery to be protected, and their spray patterns conform to the shape of the machine and the nature of the hazard. Water spray systems are also used to cool down a tank or machine to help prevent a fire from growing larger.

Foam/Water Sprinkler Systems

Foam/water sprinkler systems have a mixture of foam concentrate injected into the water supplied to the sprinkler head. Foam then sprays out of the heads and smothers the fire by filling the entire space. These systems are used when flammable liquids are present. Airports, for example, may use foam/water sprinkler systems in hangers because of the large amount of jet fuel present in jet aircraft.

Fire-Cycle Systems

Fire-cycle systems are placed in areas where cumulative water damage from the running sprinklers might be as bad as the fire damage. Fire-cycle systems secure the water flow to the sprinkler heads when the fire is extinguished through the use of a self-resetting, rate-of-rise heat detector. The heat detector senses a low space temperature, which indicates that a fire no longer exists. A timing circuit is activated and delays water shutoff to make sure that the fire is indeed extinguished. Should re-ignition of the fire occur, the heat detector senses the rise in temperature, reactivates the system, and water once again flows to extinguish the fire. The timing circuit is adjustable from 30 seconds to 15 minutes. According to Viking Corporation's website, when two heads are activated for 60 minutes in an ordinary hazard location during a fire, the amount of water used to extinguish the blaze is 3,120 gallons for a normal wet-pipe sprinkler system, whereas a fire-cycle system uses one-tenth the amount of water, or 312 gallons (3,120 gallons is about the same amount of water found in a 12-foot diameter, round swimming pool that is 4 feet deep). Fire-cycle systems are used in air traffic control facilities, museums, pharmaceutical plants, banks, historic sites, and so on.

6.8 WATERLESS FIRE SUPPRESSION SYSTEMS

Waterless fire suppression systems include halon, halon replacement gases, and CO. These systems use a chemical that either smothers the fire by excluding oxygen, or interferes with the radicals or chain reaction that allows a fire to continue and grow. Replacements for halon became necessary to meet the requirements of the Montreal Protocol and the U.S. Environmental Protection Agency's (EPA's) Clean Air Act.

Halon System

Halon (halogenated hydrocarbons for firefighting with numbers 1201, 1211, and 1301) were banned from production in 1994 in accordance with the EPA's Clean Air Act, but firefighting systems utilizing halon are still in existence today. It is still legal to purchase recycled halon to recharge an existing system.

Halon works by interfering with the fire's chain reaction, which is what keeps a fire going. Halon does not have to smother a fire in order to extinguish it. In fact, in engine-driven generator rooms, the engine(s) must be secured prior to discharging halon. Otherwise, the running engine(s) might pump the halon out through the engine's exhaust stack. Halon 1301 is effective at extinguishing a fire at a concentration of only 6 percent. At this concentration, a person in a compartment would still have plenty of air to breath. The proper procedure is to have a halon alarm sound prior to halon discharge so that anyone in the space has enough time to vacate. When halon contacts a fire, it produces two gases: hydrogen fluoride and hydrogen bromide, both of which are toxic. So it is imperative to leave immediately any space where halon has been discharged.

Carbon Dioxide Systems

Carbon dioxide (CO_2) smothers a fire by displacing oxygen found in the air. At a concentration of 28.5 percent, carbon dioxide is effective in extinguishing a fire. More CO_2 than halon (which requires only 6 percent concentration) is needed to fight the same size fire. The advantage of CO_2 is that there is no phase-out date, and it has worked well for many years. The disadvantage is that, at the concentration of 28.5 percent needed to fight a fire, the storage of CO_2 takes up considerably more space.

Montreal Protocol

After the discovery of the hole in the ozone layer in 1985, members of the United Nations met to regulate and phase out the substances that bear the major responsibility for this major environmental problem. In 1987, the Montreal Protocol on substances that deplete the ozone layer was written, and according to the United Nations Environmental Program (UNEP), Ozone Secretariat, as of August 25, 2010, 196 countries have ratified it. Some of the substances included in the agreement are chlorofluorocarbon (CFC) 11, 12, 113, 114, and 115, and halon 1211, 1301, and 2402. From both a diplomatic and an environmental standpoint, the Montreal Protocol has been a success. According to the U.S. National Oceanographic and Atmospheric Administration (NOAA), the CFC and ozone levels in the upper atmosphere are showing signs of leveling off, and some CFC levels have even decreased.

Halon Replacements

Several manufacturers produce halon replacement chemicals. All three chemicals— Novec™1230, DuPont™ FM-200®, and Halotron I (and II)—leave no residue and are suitable for applications in locations such as computer data processing and communication centers.

3M™ Novec™ 1230   According to the 3M Company, Novec™1230 is a fluoroketone that can replace halon as a fire-extinguishing agent. It is a sustainable replacement due to its zero ozone depletion and a global warming potential (GWP) of 1. A GWP of 1 means that, compared to a similar mass of carbon dioxide (carbon dioxide's GWP is standardized to equal 1), a gas will warm the earth the same amount. Novec™ 1230 has the lowest atmospheric lifetime for clean agent alternatives: 5 days (halon's is 29 years). It is a liquid at room temperature, but it works as a gas (boiling point of 120.6 degrees Fahrenheit). It is used in concentrations from 4 percent to 6 percent, and is nonconductive and noncorrosive.

DuPont™FM-200®   DuPont™ FM-200® is hydrofluorocarbon-277ea (HFC-227ea), a chemical that extinguishes a fire at concentrations of approximately 7 percent. HFC is a replacement refrigerant for CFC and hydrochlorofluorocarbon (HCFC). According to DuPont, it has zero ozone potential and has no phase-out date as mandated by the Montreal Protocol. FM-200® has a low toxicity. It is also used as a propellant in pharmaceutical inhalers.

American Pacific Corporation—AMPAC™ Halotron I® and Halotron II   Halotron I is a chemical based on HCFC-123. It is used in portable fire extinguishers and streaming applications. Halotron I is discharged as a quickly evaporating liquid with throw distances of from 6 to 45 feet. Halotron II is a chemical used for total space flooding. It is based on HFC-124a, HFC-125, and CO. It is used as a replacement for halon 1301 for total space flooding.

TABLE 6.1 Classes of fire and corresponding extinguishers

Class of Fire	Fire Components	Type of Extinguisher
A	Combustible solids: wood, paper, cardboard, organic cloths	Pressurized water or ABC dry chemical
B	Combustible liquids: petroleum and chemical liquids	Foam, ABC, or BC dry chemical
C	Electrical fire: energized electrical circuits	Carbon dioxide, BC or ABC dry chemical
D	Combustible metals: magnesium, titanium, zirconium, sodium, and potassium	Class D dry chemical (choose the type suitable for the metal; various types are available). The extinguisher's faceplate provides information on the effectiveness for a particular type of metal.
K	Kitchen grease fires, usually caused by deep-fat fryers	Wet chemical extinguisher that forms a soapy foam blanket over the grease (the soapy foam blanket both smothers and cools the grease below its ignition temperature).

6.9   FIRE EXTINGUISHERS

Several different types of fire extinguishers are in use today at properties and facilities throughout the United States. In office buildings, the most popular type of extinguisher is an ABC dry chemical extinguisher. The letters (A, B, C, D, and K) labeling the extinguisher come from the classes of fire as stipulated by NFPA. ABC extinguishers can handle most of the fire hazards encountered in this type of setting. Table 6.1 lists the type of fire extinguisher used according to the class of fire.

Fire extinguishers must be checked for proper charge levels monthly. This monthly check can be done by in-house maintenance staff and can be accomplished easily if the building is split up into four areas. One area can be completed every week along with the weekly alarm testing and operation of the emergency generator. Charge levels should be checked to make sure that the level of pressure charge has not been dropped by someone tampering with it or a faulty discharge valve. Some buildings monitor their fire extinguisher cabinet doors electronically and report to security when someone trips the door switch. Dry chemical and water fire extinguishers need to be serviced yearly by a trained technician. Extinguisher tanks also require periodic pressure testing with water; the interval of testing varies with the type of extinguishing agent.

6.10   CONTAINMENT OF FIRE AND SMOKE

Fire and smoke containment can be achieved through the use of heating, ventilation, and air-conditioning (HVAC) controls and equipment; fire dampers; smoke dampers; fire-rated walls; fire stop sealing materials; structural insulation; and fire doors. The FM should be thoroughly familiar with the operation, maintenance, and testing requirements of this equipment for effective emergency management. The FM also needs to understand how the HVAC systems respond to fire alarms and smoke detected in ductwork. Here are some questions to consider:

- How are the fire exit stairwells and elevator shafts pressurized in the event of fire?
- How does the HVAC/fire system secure air to the fire event area and pressurize surrounding spaces?
- How is over pressurization of ducts prevented when fire dampers are closed?

NFPA regulations require that fire and smoke dampers be inspected and tested one year after installation and then every four years. Hospitals are an exception; their fire and smoke dampers should be inspected and tested once every six years.

Heating, Ventilation, and Air-Conditioning Controls and Equipment

When fire is detected, the HVAC system, on a small scale of control, should be able to secure ventilation to prevent smoke from spreading throughout the building. On a larger scale of control, the HVAC systems should be able to manipulate the equipment to provide smoke control for the building. Both scenarios require interconnections between the fire alarm system and the building automation system (BAS).

Smoke control on a larger scale is provided in different ways, depending on the type of building and its HVAC equipment. Generally speaking, the system attempts to keep the floor where fire is present under a negative pressure condition while making sure adjacent floors are pressurized. Because gas flows from a place of high pressure to one with low pressure, placing the smoke-filled floor under a negative pressure keeps the smoke on that floor. Fire escape stairwells are kept under a positive pressure to ensure that smoke stays out of the stairwell when people open the door to the stairwell in order to evacuate a building. This means that exhaust and supply fans, and variable air volume (VAV) boxes need to be controlled in ways that meet the current fire situation. Some fire systems have their own smoke control panel so firefighters can start, stop, and override equipment to achieve effective smoke control.

Fire Dampers

Fire dampers prevent the transmission of fire from one space to another through ducting that penetrates fire-rated walls. A fire damper is a spring-loaded mechanism. It has a fusible link that melts when exposed to temperatures of about 165 degrees Fahrenheit or 72 degrees Celsius, thereby releasing the spring that shuts the damper. Where temperatures are normally elevated as is true in heated ducts, links that melt at a higher temperature (212 degrees Fahrenheit) may be permitted for use. The blades of the damper are generally in the plane of the firewall to keep the fire rating of the wall assembly.

Smoke Dampers

Smoke dampers are used to control the spread of smoke. They are operated by electric or pneumatic motorized dampers. Damper closing is triggered by smoke detectors located inside the ventilation duct.

Combination Fire and Smoke Dampers

Combination fire and smoke dampers have a fail-safe fusible link, for closing a damper in the event of a fire, and a damper motor for smoke management. The link closes the damper when activated by heat; it must also disconnect the damper motor that is used for smoke control. This disconnect feature ensures that the damper blades won't be restricted by the motor during a fire event. The link must be melted in order for smoke management to be initiated. Some dampers have indicator lights that indicate whether the damper is open or closed, a reset button, and a switch to open or close the damper locally.

Fire Doors

Fire doors can be composed of steel or solid-core wood. The doors may have glass sections in them; these glass sections must also be fire rated. Wired or ceramic glass is frequently used for this application. Fire doors must be either shut at all times, with a method of automatically closing them should they be opened, or they may be left open if a device is included that will cause them to shut during a fire event. A closing mechanism may be an electromagnet that holds the door open as long as power is supplied to it. The mechanism may also be a fusible link that melts and thus causes the door to close. Once closed, fire doors should latch shut via a listed door closer or spring mechanism.

Fire-Stop Sealants

When mechanical or electrical equipment must penetrate a fire-rated wall, the wall must be sealed with a fire-rated material. The general types of fire-stop materials are:

- Pillow
- Caulk
- Putty
- Mortar

The caulks and the putties are **intumescent materials** that expand when exposed to the heat of a fire (see Figure 6.5). Mortars are used around steel conduit and pipe. Mortars are not masonry mortars; they are formulated for fire sealing because they resist cracks and shrinkage when properly applied. Pillows are used when multiple conduits or cables that run through a fire rated wall are frequently disturbed through cable additions or eliminations.

Facilities personnel should be alert to and report any unprotected voids through fire-rated walls and floors.

FIGURE 6.5 Fire-stop sealant

Source: Photo courtesy of Oconomowoc Memorial Hospital.

Structural Fire Insulation

Building structural steel must be protected from the heat of a fire. The steel is sprayed with cement-type insulation. Insulation protects the steel for one hour, which is enough time for everyone to evacuate from the building. In the past, some of this insulation contained asbestos. Testing for the presence of asbestos in the insulation is necessary to provide the required disclosure to tenants and contractors who may disturb this material. For example, during construction, it may be necessary to remove some of this insulation in order to provide a good connection for wall support studs and hangers for suspended ceilings. Any insulation that is damaged or removed must be replaced.

6.11 SUSTAINABLE SECURITY MANAGEMENT

Sustainable security management focuses on four main areas: threat analysis, building design and infrastructure, personnel management, and continuous security review. Threat analysis is the same type of risk analysis one would employ for a disaster situation. Threat analysis looks at the building and its customers, tenants, and location to adjacent structures and the threats facing those buildings. Examples of questions to ask during an analysis include:

- How high a profile is the building, and could the building be used as a symbol if it were damaged or destroyed?
- Could tenants or governmental offices within the building be considered possible targets?
- What disruptions to other business or government operations could be achieved through a disruption in the building?

These same questions should be asked about surrounding buildings; even though the building the FM is managing might not be a potential target, the adjacent building(s) might be.

Threat analysis might reveal that the building is a low-priority target for terrorism, but it could be a high-priority target for vandalism. The next steps are to assess the likelihood of the threat and make the necessary modifications to address it. Modifications may include increased security presence at particular times, additional surveillance equipment, and decision-making equipment (such as locking doors when motion is detected or turning on cameras and lights).

Sustainable building security design is much easier now with three-dimensional building design software. The ability to perform a fly-through of a building model (as discussed in Chapter 5) can be applied to designing for security. Rather than trying to visualize a two-dimensional drawing, designers have a three-dimensional model that can provide a view similar to a person walking through the building while carrying a video camera. A fly-through also gives designers a good idea of how congested passageways may become during a disaster event, and may suggest alternative routes. The software files of these building models must be secured because of the enormous amount of data they would provide to someone seeking an unauthorized point of entry.

Building-security design can include the same elements of sustainable design, such as various land-scaping features: trees, planters, bike racks, and public and transport shelters, to name a few. However, the design of building security limits access to persons who have not been issued special access equipment or knowledge of how to get around these physical barriers. Exterior and interior lighting is a well-known way to increase building security. Utilization of intelligent lighting controls can enhance security while saving energy. Motion detectors can turn on or increase the amount of lighting. The unexpected illumination of an area in itself might signal that a potential threat has been detected.

Personnel management of the security force is a challenge. Security is an area that is frequently outsourced. Benefits to outsourcing are reduced costs, increased security personnel skill level, staffing flexibility, and reduced liability. Security personnel and security operational procedures should be reviewed regularly with the security supervisors assigned to the facility's account. To increase the sustainability level of the security operation, security personnel, including part-time temporary replacement, must receive training in security techniques and the equipment they will be using. Part-time temporary replacement night security personnel might not have adequate knowledge of the building. Having an appropriate security relief procedure that allows more knowledgeable personnel to transfer basic building knowledge to replacement personnel should be part of their standing orders. The FM needs to ensure that all security personnel are well trained and are knowledgeable in specific building equipment and procedures.

Traditionally, guards at office buildings who are present during the workday are full-time employees of the property or the security firm. Guards come to the building every day and become well known to tenants through greetings and minor assistance with tasks as the tenant enters the building and travels past the reception desk. This personal relationship has the added benefit of giving the guards a more direct knowledge of the people in the building and their habits. It makes it easier for a guard to detect when something or someone is out of place.

Continuous security review is a process whereby security procedures are examined and changes to increase security are made. The review looks at current procedures and examines whether the procedures are addressing the threat. The FM or the facility's head of security should also look at incident reports from other properties; access to these reports may be gained by participating in organized security groups that were formed to share information among property owners.

The performance of individual security personnel should also be reviewed to make sure they are meeting the requirements necessary to provide good security. For instance, the time it takes to complete the various rounds, the areas being patrolled, the equipment used, and the information they are writing in the security log are all items to be evaluated. Here are some items to consider when conducting performance reviews of security personnel:

- Security rounds should not be completed on a set schedule and with set routes because patterns will be detected by those wishing to breech security.
- If the guards walk the premises, perhaps implementation of equipment such as bicycles or motorized scooters might prove beneficial in terms of efficiency and ability to respond to an alarm.
- Guard logs should include suggested security improvements noted in a comment section. Valuable information can be obtained from security personnel on patrol.

Access Control

One of the oldest methods for increasing security is to lock the building and limit access using doors and locks. To limit access successfully requires a key system and key control. A key system is a system of master keys and submaster keys. For a large campus of buildings, a grand master key opens all the doors in all the buildings. **Building master keys** open all the doors in just one building. **Floor master keys** open doors on one particular floor. Tenant suites may have a suite master that opens all the doors in a particular suite. Suites may have doors that are **keyed alike**, meaning that their locks are pinned the same so that one key opens any of those doors. A **change key** is a term applied to a door that is on the master key system, but only one key beyond the master keys opens it.

Locksmiths use a keying chart, sometimes called a progression chart, to set up a key system. Progression charts allow the user to change locks without fear of cutting a key that will open another lock. As the number of keys in a keying systems increases and the possibility of cutting many keys also increases, the possible pinning combinations decrease. In this case, keys may be cut that vary only slightly from the keys used to unlock other doors. These keys are sometimes called *first cousin keys*. The fear here is that by manipulating a first cousin key in a different lock, it may actually open a door that that it wasn't intended to open. This is frequently discovered by accident when a key is used in the wrong door and yet opens the door. A way to prevent this is to add a different key blank to the key system that can facilitate more lock pinning combinations. In this case, the facility will then need two master keys.

A facility may also choose to increase security by having a dedicated key blank. For a fee, a lock company will issue to a property owner a key blank that differs in shape from all other blanks. The shape of the key prevents other keys from fitting in the lock's keyway. The owner is the only person who can purchase these key blanks, which prevents someone from duplicating a lost or stolen key.

Today, basic lock security is augmented with detection devices, security cameras, personnel verification devices, and alarms. Technology allows for access via a card with a magnetic strip that has information about the user. This magnetic strip provides management and security with knowledge about who is in the building and where they are in the building. This information is important, especially when managing the response to an after-hours fire or disaster situation. Elevators also require an access card that is either swiped across a reader or held up to a reader pad that allows the user to travel only to pre-designated floors within a building. Pressing the button for an unassigned floor might also trigger a security response.

Smart building technology can also take the information provided and turn on heating, lighting, and/or air conditioning for the space where the individual is assigned. Herein lies the opportunity to manage both facility security and the sustainable nature of the building and its systems.

Intrusion Alarm Systems

Intrusion alarm systems are designed to send an alarm to alert police and/or security personnel. When security personnel are present in the building, the alarm may be a silent notification to the security desk. Security personnel can then either train cameras on the area or immediately call for assistance. Most building security personnel are unarmed, so direct physical confrontation with intruders could put the security personnel in jeopardy. Audible alarms let personnel in the area know that a security breach has occurred. A loud alarm attempts to disorient the intruder so that the intruder ether leaves the premises immediately or doesn't stay as long as planned.

Alarms may be connected in a way that creates zones. Zoning alarms allow security personnel to disarm zones when after-hours activities occur in one part of a building. In this way, the rest of the building can maintain security while accommodating after-hours operations.

Many devices can be used to trigger an intrusion alarm, for instance:

- Door switches are the simplest type of alarm-triggering device. A door switch is a magnetic switch device with one half mounted on the door frame and the other half mounted on the door. Both halves are in close contact with one another. The two halves become separated when the door is opened, which triggers an alarm.
- Glass breakage detectors may rely on the sound that glass emits when it is broken.
- Motion detectors come in many varieties, for instance:
  - Passive infrared detectors can detect the rapid change in heat that occurs when someone enters a room. Passive infrared detectors detect the radiation given off by the skin temperature of 93 degrees Fahrenheit. These detectors look for a rapid change; otherwise, false alarms could occur as heated areas slowly cool off. Intelligent detectors automatically compensate for various parameters, so rodents and small pets, for example, won't set off the alarm. Glass is opaque to infrared light so an intruder moving on the opposite side of a glass wall won't be detected by a detector on the other side of the glass.
  - Microwave radar systems detect motion. The system sends out a pulse of energy and waits for the reflected energy to bounce back. An object entering the wave zone changes the amount of energy that is reflected back and sets off the alarm.
- Acoustical sensors are another type of sensor frequently used to detect breaking glass.

Closed-Circuit Television Surveillance

Closed-circuit television (CCTV) is a surveillance device that has been in use for over 40 years. Large facilities have numerous cameras mounted in areas where intrusions are likely to occur. Cameras are also installed to monitor parking areas and loading docks for the protection of visitors and service personnel coming to the facility. Additional CCTV cameras can give video information from different angles to help identify intruders or to see if they are trying to hide weapons or stolen items by keeping them against the opposite side of their body from the camera.

CCTV cameras can have built-in light-emitting diode (LED) lights to obtain detailed information at night, or they may be infrared cameras that can operate in total darkness. Replacing videotape, digital cameras and digital storage devices have greatly cut down on the labor and space needed to record and store video information. CCTVs may be activated by the same alarms used for intrusion detection. This has an added benefit of alerting security personnel to a problem that might otherwise go unnoticed.

Cyber Attack

Web-based computerization of building system control increases the possibility of cyberattack and the negative effects of such an attack. Building automation systems (BASs) have been controllable for many years from a laptop computer. Many building automation systems are protected by usernames, passwords, and firewalls. Should an unauthorized person gain access to the system's web address or its universal resource locator (URL), username, and password, that person can gain access at the level of the legitimate user. User levels can range from system administrator, a level that gives the person the ability to change everything including the control logic, to a student level, which gives the individual only the ability to view preselected display screens. An intruder armed with a moderate level of password at, for instance, the technician level gives the intruder the ability to make operational changes. With a laptop or other small computerized web-accessing device, the intruder can be inside the building making changes, such as turning off lights or setting off alarms in order to disrupt building security.

Some BAS installation/repair companies have a standard username and password that allows them access to the account at the system administrator level, which may be a holdover from the installation phase. Or perhaps the BAS provider offers the FM or other staff members a new password in the event of issues with system log-in. Although the provided password is a highly convenient solution, the existence of an access password outside facility personnel passwords compromises security. It would be better for managers themselves to have an alternate username and password that can be used should a password problem occur. Managers should give BAS companies access to the system only by assigning them a temporary username and password, and then deleting both once the work has been completed. The level of security of the BAS company should not become the level of security for the facility.

REVIEW QUESTIONS

1. Why are fire dampers important in a hospital building?
2. Describe a single-interlock preaction fire sprinkler system.
3. Describe the benefits of a fire-cycle sprinkler system.
4. What are the replacement extinguishing agents for halon?
5. What are ESFR systems?
6. What can an FM can do to ensure the quality of outsourced security personnel?
7. Where are fire-stop sealants used?
8. What is a deluge system?
9. What should an FM do if a jockey pump is starting and stopping frequently?
10. Explain the following terms: keyed alike, building master key, and floor master key.
11. The Montreal Protocol regulates _____.
12. A format and guidance that a manager can use when developing an emergency response plan can be obtained from _____.
13. What are some maintenance duties that security personnel may perform?

Essay

14. Discuss the procedures an FM should follow when a disaster event has occurred.
15. Discuss preparedness planning for disasters.

FACILITY AND GLOBAL ENVIRONMENTAL MANAGEMENT

INSIDE THIS CHAPTER

How Does Environmental Management Contribute to Sustainability?

The management of environmental issues is extremely important for any facility. Environmental management begins before a property is developed, extends through the life of the property, and continues through its recycling and redevelopment. The chapter discusses the different environmental concerns and the environmental personnel that can aid a facility with these concerns. The chapter defines the due diligence process for property acquisition and leasing, discusses the various environmental problems many facilities can face, and introduces the government regulations related to these hazards. The chapter ends with a discussion of two global environmental concerns—stratospheric ozone layer and the greenhouse effect—and the difference between carbon dioxide equivalent gases and biogenic carbon, and how they relate to a facility's carbon footprint.

7.1 ENVIRONMENTAL MANAGEMENT

A failure in environmental management can turn a valuable property asset into a major liability for many years. Correcting the problem can result in the loss of hundreds of labor hours of effort and huge amounts of capital. An environmental problem can happen through an unforeseen accident or disaster, or by a tenant's carless or illegal actions. As managers, facility managers (FMs) must perform the necessary investigative work and put the systems in place to reduce and hopefully eliminate all environmental risks. Sustainable facility management can be successful only if environmental management is practiced.

Environmental Consulting Personnel

Environmental management begins with a team. Two key members are an experienced environmental lawyer and an equally experienced environmental engineer. With the input of the FM, this team puts together an environmental management plan that meets all regulatory and strategic needs of the facility. Other professionals with even more specialized skills may be placed on the team as needed when specific environmental conditions are uncovered.

Environmental engineering consulting firms have several different job categories of professionals who can help the facility properly report and remediate environmental issues.

Some of these professionals are described in the following subsections.

Environmental Engineer or Environmental Scientist   Environmental engineers and environmental scientists utilize a variety of skills obtained from professional engineering and science disciplines such as chemistry, geology, hydrology, environmental science, civil engineering, hazardous waste management, wastewater control, storm water management, and air pollution control.

Hazardous Substance Sampling Technician   A hazardous substance sampling technician is skilled in taking, storing, and transporting samples that may contain hazardous materials. Persons need to be certified to sample specific hazards. For example, a lead-based paint sampling technician has skills specific to sampling lead paint and lead in drinking water. An asbestos sampling technician has skills for sampling asbestos.

Industrial Hygienist   An industrial hygienist is skilled in identifying hazards in the workplace and figures out how to minimize the hazards through various means. These hygienists utilize engineering controls to remove the worker from the hazard or to protect the worker. Hazards include hazardous chemicals, biological hazards, physical hazards, and environmental hazards.

Formulating an Environmental Management Plan

The *environmental management plan* is customized to meet the property's *current use* by tenants or in-house personnel if the property is owned and managed by the same entity. The plan must also reflect *past use* by previous owners and the *use of surrounding properties*, and it must be based on the property's current environmental condition. Discovering a property's current condition requires research and site investigative work by the environmental engineer. The engineer may review old aerial photographs and old Sandborn fire insurance maps of the property and surrounding businesses to discover previous use. (Sandborn maps were created by the Sandborn Company for the use of insurance agents in their assessment of a property's fire hazard.) Thousands of maps were created over a period spanning about 130 years, from 1867 to 1970, for the various urban centers across the country. These maps show the businesses in a particular area and the known hazards present (petroleum storage tanks, gasoline distribution centers, etc.), and provide a historic record of industrial development. Current owners of a property can then discover its past use by referring to these maps and the aerial photo taken in the past. This information is valuable when purchasing a property or trying to discover where contamination may have originated.

The environmental management plan also requires interviewing tenants and inspecting their spaces to discover and record environmental compliance issues and to review tenants' emergency action plans. Information on how to develop an emergency action plan in compliance with Occupational Safety and Health Administration (OSHA) standard 29 CFR 1910.38(a) can be found on OSHA's website. *Emergency action plans* define the steps to be taken during an emergency, such as a spill of a hazardous liquid. In the case of a hazardous material spill, the plan must define exit routes, emergency responders to contact (fire, police, hazardous material [hazmat]), the training requirements of in-house personnel responding to the spill, the personal protective equipment (PPE) required, and the regulatory agencies that must be alerted. In addition, a *hazardous waste plan* should define the required frequency of removal and treatment of tenant-generated hazardous waste to the appropriate hazardous waste landfill or recycling facility. Both plans must be available to ensure that the property remains in compliance with all applicable government regulations.

Environmental problems can occur on the land surrounding a building as well as inside the building itself. Some of these types of problems include:

- Soil contamination
- Water contamination
- Groundwater contamination
- Microwave and electromagnetic hazards

Some of the environmental issues that can occur inside a building include:

- Indoor air quality issues
- Hazards cause by building material(s)
- Hazards caused by waste generation
- Building habitation contamination
- Building use contamination

Air pollution is also an issue, and the FM must ensure that tenant operations are in compliance with all emission standards and regulations.

7.2 PROPERTY-RELATED ENVIRONMENTAL CONCERNS

The major areas of environmental concern are contamination of the property site and/or contamination of the building. A contaminated property site can have severe restrictions placed on it as to what type of building may be constructed on the property and for what purposes the building may be used. Such restrictions can affect the value of the property enormously. The authority having jurisdiction (AHJ) can also order the property owner to perform extensive costly remediation to prevent further damage to the environment and/or contamination of adjacent properties. Interior building environmental problems can decrease the value of the building and make it hard to lease. Both contamination of the property site and contamination of the building carry litigation risks, and professional environmental and legal representation should be sought by the property owner if one or both occur.

Site Contamination

When petroleum products, chemicals, and manufacturing wastes are dumped on the ground or buried in the ground, the result is soil and possible groundwater contamination. The contamination may remain on site in a localized area, or transport mechanisms may cause the contamination to move through the site and possibly even offsite. For example, if hazardous materials are dumped at a high point on the property, they may be naturally transported by gravity to a lower point down gradient. Material may be blown from one site to another by the wind, and surface runoff water may cause the materials to be washed to a new location.

Soil contamination may drift downward through the soil and reach groundwater, where it is then transported by the movement of groundwater. Groundwater can cause contamination to move in unexpected ways and change its direction altogether. Remediation of groundwater is often a much more difficult task than soil contamination. Generally, a number of wells need to be drilled on a site to determine the direction of groundwater flow because the topography of the surface may have little relation to the direction of the flow of water underground. Defining the flow direction of groundwater is important for defining where contamination is coming from or if contamination may be leaving the site and thereby contaminating adjacent properties. Allowing contamination to move offsite and onto an adjoining property can have serious legal ramifications for the responsible party.

Building Interior Contamination

Building materials and equipment can be sources of contamination. To discover these issues, the FM should engage the help of environmental professionals. Their review should be part of a yearly environmental audit, which in turn is part of the environmental management plan. The FM walks through the building spaces with the environmental professional(s) to uncover potential problems and identify solutions. Tenants are asked about their operations to ensure compliance with all government regulations. Some types of contamination that can occur inside a building include the following:

- Building construction material contamination (i.e., polychlorinated biphenyl [PCB], lead, asbestos, fiberglass insulation, volatile organic compounds [VOCs], formaldehyde, and mold)

- Mechanical equipment use contamination (fluorocarbons, waste oil, hydraulic oil, antifreeze, boiler chemicals, cooling tower chemicals, fuel wastes)

- Building habitation contamination (blood-borne pathogens, carbon dioxide [CO], odors, garbage, viral and bacterial contamination)

- Building use contamination (copy machine toners, printer ink cartridges, cleaning chemicals, chemical use in tenant equipment)

- Hazardous material storage for a tenant's business operations

7.3 ENVIRONMENTAL ISSUES PRIOR TO PROPERTY PURCHASE

Prior to purchasing a property, the prospective buyer puts together an acquisition team consisting of members of the organization who are skilled in the various aspects of property transactions. After performing market research for the property location, they then perform a financial evaluation of the potential purchase. If the financials look good, members of the team visit the property. If no significant issues that would prevent the property transaction from going forward arise at this stage, the team begins a series of costly investigations.

An environmental investigation is performed as part of an overall property assessment known as *due diligence*. Due diligence includes a visual assessment of a building's structural elements and building envelope. It evaluates items such as roofs and mechanical, electrical, and plumbing systems along with paved areas, drainage, and elevated parking garages. It looks for possible hazards such as suspected asbestos-containing materials inside the building, and oil-cooled transformers and light ballasts that may contain PCBs. The professional performing this part of the investigation also searches records for possible code violations and checks building permits. But this isn't enough. Due diligence includes an environmental site assessment (ESA) known as a Phase I. A Phase I is required by banks that lend mortgage money on a commercial property. A Phase I consists of visual site condition observations, and a search of current and historical records and maps of the site and surrounding areas. If suspicions arise after reading the Phase I, a buyer may request to perform Phase II work. Phase II work is the actual sampling of the soil, water, or materials that are suspect to determine if they are actually contaminated.

Phase I All Appropriate Inquiry

Phase I all appropriate inquiry (AAI) is performed by third-party environmental professionals hired by the property purchaser. Guidelines to be followed when performing a Phase I were first developed by the American Society of Testing Materials (ASTM) and were later revised to include the U.S. Environmental Protection Agency's (EPA's) new federal code requirement, 40 CFR Part 312—Standards and Practices for All Appropriate Inquiry. The result is ASTM's standard E-1527-05, which guides environmental professionals in meeting the requirements for a Phase I-AAI.

A Phase I-AAI goes a bit further than the old Phase I of the past by defining what constitutes an appropriate inquiry to identify the previous owners of the property and the current and historic uses of the property. This work includes asking questions of the owners, tenants, and operators—past and present. Certain industries are known generators of hazardous waste as a direct result of their factory operations; thus, it is important to identify the previous owners. Years ago, hazardous wastes were frequently dumped into pits on site and covered over with dirt. It was not uncommon for plant operators to dispose of waste oil and solvents by pouring them into a pit filled with wood and other combustibles, and setting everything in the pit on fire. This kind of activity is unethical by today's standards. It is a scary realization that at the peak of the industrial revolution in the United States, most industries were generally unaware that their actions could be responsible for the ultimate destruction of many of our natural resources, the elimination of wildlife habitats, and the degradation of human health.

The importance of a Phase I AAI doesn't stop at just learning if a property is likely to be contaminated or not. It provides the basis for a defense under the ***Comprehensive Environmental Response, Compensation, and Liability Act (CERCLA)*** (42 U.S.C. 9602-9675) as an ***innocent landowner*** of a property to limit the new owner's liability should contamination be found on the property now or in the future. CERCLA, a law enacted in 1980 and better known as the Superfund, was designed to correct the decades of unregulated and improper disposal of hazardous wastes. CERCLA has empowered the EPA to identify parties responsible for these threats to human health and force them to clean up the contamination. When the EPA cannot identify the party responsible, it uses government funds available to take on the cleanup job itself. If hazardous wastes are discovered on the property of an innocent landowner, the EPA won't make the property owner pay for cleaning up the wastes. Performing a Phase I AAI prior to purchase is an essential step for obtaining innocent landowner status. The data contained in a Phase I AAI also provides a basis for good environmental decision making with regard to purchasing the property, requesting Phase II testing, or abandoning the deal altogether.

Underground Storage Tanks

By performing a Phase I-AAI prior to property purchase, the presence of underground storage tanks (USTs) that may have been forgotten or abandoned can be identified. Leaky, abandoned underground storage tanks holding petroleum products can cause a tremendous amount of ground and water contamination. If USTs are found on the site, Phase II testing can be performed to determine if leakage has occurred, and if so, the extent of the contamination. If leakage is found, the location of the tank is now referred to as a LUST (leaking underground storage tank) site.

It is also important for FMs to know how close other property's LUST sites are to their property. Underground contamination can migrate from a LUST site and contaminate adjoining properties. Properties that are adjacent to and down gradient of a LUST site are of immediate concern. The Phase I- AAI can define these issues and provide the basis for action.

7.4   LEAD-BASED PAINT

The EPA's recent legislation, the Lead Based Paint Renovation, Repair and Painting Program Rule (RRP Rule) protects workers and occupants against lead-based paint hazards during renovation, repair, and painting. An EPA pamphlet (EPA-740-F-08-002) explains the rule and its requirements in detail. To summarize, the rule applies to operations that disturb 6 or more square feet of interior painted surface, or 20 or more square feet of exterior painted surface on schools, childcare facilities, and homes that were built before 1978. It requires that schools, property management, construction, repair, or any firms that do work that disturbs paint in pre-1978 buildings be certified by the EPA. A person representing the firm during the project work must be certified by taking an 8-hour course that covers lead-safe work practices, record keeping, testing, and the required information that must be provided to the occupants. The certified individual then trains the personnel performing the actual work.

An important exception is work done on windows: Whether their frames are painted, coated, or shellacked, the RRP rule is to be followed, whether or not the work constitutes disruption of 6 or more square feet or surface. Windows painted with lead-based paint are a major source of lead poisoning. As windows are opened and closed, the surfaces rub on one another producing, a fine lead-containing dust that is easily breathed into the lungs. From the lungs, it enters the bloodstream, poisoning the individual. Lead is most damaging to young children.

Several provisions and exceptions to this EPA rule are not stated in this chapter. It is strongly suggested that anyone interested or engaged in work that will disturb paint contact the EPA for complete and current information. Also, some states have added their own requirements to the EPA rule, so the relevant state's regulatory agency should be consulted as well.

Lead-Based Paint Testing

Chemical test kits are available that work by changing color when the material in the test reacts with lead in the paint. Any kit used must be approved by the EPA. The kits qualify the presence or absence of lead only on the surface of the material tested. Multiple layers of paint or other coatings can be problematic for these test kits. Chemical test kits sold in retail stores do not tell the user how much lead is present.

The two quantitative methods for testing for lead-based paint are sampling with laboratory testing and X-ray fluorescence (XRF) (see Figure 7.1). Paint sample testing involves having a licensed sampling technician remove a sample of paint and sending it to an approved lab. The lab pulverizes the sample, puts it in acid, and performs tests to define the amount of lead present.

An XRF tester is a handheld device that is smaller than a shoebox. The XRF machine is placed directly on the painted surface to be tested. A shutter door opens in the device, and high-energy gamma and X-rays are emitted and penetrate the painted surface. The energy causes electrons to be ejected from their atoms. The amount of electron energy can be read by the XRF tester and converted to the amount of lead in the paint. XRF machines are used only by trained professionals. They provide results in about a minute, and their use does not damage the painted surface.

FIGURE 7.1 Testing for lead-based paint with a Niton XRF tester. Testers such as this one give immediate results without disturbing the finished surface.

Source: Photo courtesy of Thermo Fisher Scientific.

XRF machines are relatively expensive, at around $15,000 to $30,000, but there are options for renting. XRF machines can tell the investigator the location of the lead and the lead concentration in parts per million (ppm). XRF machine users must receive training in device operation, and follow state and local licensing and registration rules. The machines must be calibrated periodically. The speed, accuracy, and total lack of any damage to the surface being tested make XRF machines an attractive testing option, especially when multiple tests need to be performed quickly and discreetly.

7.5 ASBESTOS[*]

Asbestos is a high-tensile strength mineral fiber that resists heat and corrosive chemicals. Asbestos has been used in floor tiles, acoustical ceiling tiles, gypsum board, gypsum joint compound, roofing felts, roofing materials, pipe insulation, spray-on fireproofing, electrical equipment, glues and mastics, gaskets, and many other materials. Asbestos is most dangerous when it is friable and, when disturbed, becomes airborne. It can remain suspended in the air for a long time and can easily be breathed into the lungs.

Asbestos-Related Diseases

Asbestos fibers lodge themselves in lung tissue, which can result in any one or more of the following:

- Asbestosis, a progressive, long-term disease of the lungs that causes scarring of lung tissue, making it difficult to breathe. There is no cure.

- Lung cancer causes the largest amount of deaths associated with asbestos exposure. People involved with mining of asbestos and the manufacture and use of asbestos products have a greater risk of getting lung cancer than the general population. Other factors, such as smoking, may increase the chance of getting lung cancer.

- Mesothelioma is a rare form of lung cancer that may appear years after exposure to asbestos. It is found in the thin lining (membrane) of the heart, lungs, chest, and abdomen, and is almost always associated with asbestos exposure.

Setting up an Asbestos Operations and Maintenance Program

As part of the U.S. EPA's Operations and Maintenance (O&M) Program, the first step in managing asbestos is to appoint an asbestos program manager (APM). The APM then has the building inspected by a person licensed to inspect for asbestos to determine if asbestos-containing materials are present. Samples are taken of all suspected materials, and the condition of the materials and their potential for disturbance are noted. An O&M plan is then developed based on the inspection and assessment data. The condition of the asbestos-containing materials (ACMs) may necessitate that certain abatement actions be conducted by qualified personnel. A special requirement for schools is that all regulated ACMs that are identified must be inspected by the APM every 6 months to verify their condition.

Notification and Preventing Contact with Asbestos

Tenants of a building who may come in contact with regulated ACM must be notified of its presence to prevent the disturbance of regulated ACM. Facility workers and housekeeping staff must also be notified, as required by OSHA. Building owner should inform occupants about the presence of ACM by distributing written notices, posting signs or labels in a central location where affected occupants can see them, and holding awareness or information sessions. Some states and localities have right-to-know laws that may require all occupants, workers, or visitors to buildings with ACM to be informed about the presence of asbestos. OSHA also requires warning signs to be displayed at each regulated area so that an employee may read the signs and take necessary protective steps before entering the area.

In 1994, OSHA issued its final rule with regard to the safety of workers exposed to ACM. Suspected materials that were installed prior to 1980 must be tested. Drywall and drywall compound and tape must be tested prior to demolition activities. The rule also provides guidelines for housekeepers who clean and maintain asbestos-containing vinyl floor tile. Among other requirements, the rule requires that floors be stripped using only wet methods, and only floor burnishing pads on floor buffers at speeds of 300 rpm or less are to be used. The floor cannot be sanded to even out surface imperfections or for any other rea-

---

[*]The information in this section on asbestos has been provided courtesy of the U.S. Environmental Protection Agency and the Illinois Environmental Protection Agency.

son. The floor finish can be burnished at any speed if there is a sufficient number of floor finish coatings (at least three or more) to prevent floor contact with the burnishing pad.

Training Requirements

The training requirements discussed in this section are intended for workers who work in an environment where regulated asbestos-containing materials are present and there is a possibility of disturbing these materials through their work activities.

Occupational Safety and Health Administration Requirements   OSHA requires employers to institute and ensure participation in a worker training program for employees exposed to fiber levels (either measured or anticipated) at or above the permissible exposure limit of 0.1 fibers per cubic centimeter [f/cc] as an 8-hour, time-weighted average (TWA) and/or the excursion limit of 1.0f/cc as a 30-minute TWA. This training program consists of the initial training period, the duration of which is determined by the type of work the employee performs, and annual refresher training. For additional information about these training requirements, see the OSHA regulations in 29 CFR & 1910.100 (j)(7).

EPA's Asbestos Hazard Emergency Response Act Requirements   School maintenance and custodial staff who perform work that may disturb asbestos-containing building materials (ACBMs) must be trained according to the provisions in the Asbestos Hazard Emergency Response Act (AHERA). According to EPA regulations, these workers must take 16 hours of O&M training consisting of 2 hours of asbestos awareness training and 14 hours of special operations and maintenance training.

State and Local Requirements   Some states and municipalities may also have specific training requirements for workers who could be exposed to asbestos or who work in a building with ACM present. Follow the requirements of the jurisdiction having authority (JHA) and the federal requirements.

National Emissions Standard for Hazardous Air Pollutants Regulations   National Emissions Standard for Hazardous Air Pollutants (NESHAP) rules are federal regulations under the Clean Air Act (CAA) that apply to the facility owners and contractors who perform work in public and commercial buildings. Asbestos NESHAP regulations address common small-business activities such as milling, manufacturing and fabricating, demolition and renovation, , active and inactive waste disposal, and asbestos conversion. A privately owned home or apartment with four or less units is exempt from the regulations unless the building either was used or is planned for use as a commercial or public facility.

Compliance with NESHAP   The FM must ensure that an inspection of the entire facility or, at a minimum, the affected portion of the facility for the presence of regulated ACM takes place prior to any demolition or renovation activities. Friable, regulated ACM is present in many public and commercial buildings built before the mid-1970s and in some buildings built after the mid-1970s. If he or she is uncertain about the presence of regulated ACM at the facility, the FM should hire a licensed asbestos inspector who can obtain the necessary samples for confirmation via laboratory analysis.

Although the FM should hire a licensed asbestos contractor to conduct regulated ACM removal work, the FM should be aware of the following requirements for contractors during removal activities so that the contractor fulfills the required duties:

- A NESHAP-trained person must be present.
- The area from which the asbestos will be removed should be sealed off, and any forced-air heating systems should be shut off.
- Workers should wear a respirator fitted with cartridges that filter out asbestos fibers.
- The asbestos should not be broken down into small pieces because this can increase the amount of airborne asbestos fibers.
- The asbestos should be kept wet during the entire removal process.
- The asbestos waste should be containerized and labeled for disposal at an approved landfill.
- The area from which the asbestos was removed should be cleaned thoroughly with a wet mop, rags, or sponges.
- Asbestos removal workers should decontaminate themselves.

7.6 ADDITIONAL ENVIRONMENTAL CONCERNS

The following subsections list three additional areas of environmental concern. Facilities may have their own additional areas of concern that may require the services of an environmental engineer for their discovery. Biological concerns from Legionnaires' disease, radiofrequency (RF) radiation from cellular telephone equipment, and electromagnetic from high voltage wiring are discussed. Although cellular telephone equipment provides some challenges, it also provides a possible opportunity to gain revenue from rooftop areas that traditionally generate only a maintenance expense.

Legionnaires' Disease and Its Prevention

In 1976, the first identified outbreak of Legionnaires' disease occurred in Philadelphia, Pennsylvania, striking conventioneers who attended an American Legion convention. Legionnaires' disease is frequently caused by aerosolized warm water that contains legionella bacteria (*L.pneumophile* or *L.micdadei* bacteria are most common). The disease can also be caused by drinking water contaminated with the bacteria. Not all people who are exposed to the bacteria become sick; in fact, only five out of 100 people who are exposed develop the disease, and 10,000 to 40,000 in the United States are stricken with Legionnaires' disease annually. The disease-causing bacteria are not transferred by person-to-person contact.

Hospital plant engineers increase the temperature of the heated water inside water heaters in order to kill the bacteria. Water that leaves the heater must be cooled by mixing it with city water before it is sent throughout the hospital to prevent scalding. Bacteria may also be killed by the C-band of ultraviolet light UVC light treatment, filtration, electric pulse treatment, and chemical treatment. It is important to contact professionals to address this problem by developing a plan of action tailored to the specific facility.

Cellular Telephone Sites

The proliferation of cellular telephones has greatly increased the number of cellular telephone antenna sites across the country. Cellular antennas require an elevated location that is free from interference from surrounding structures. This provides a leasing opportunity along with some additional challenges for the FM. The opportunity lies in obtaining rental income from cellular telephone equipment installation in locations inside and outside the building that are frequently unused space. Locations include rooftop areas for antenna installation and building interior areas for the necessary electronic equipment (if the roof has sufficient support, the electronic equipment may be mounted on a platform situated on the roof deck). The challenge lies in mounting the equipment so that it doesn't spread radiation and so it does not preclude other potential sources of income (such as by mounting antennas in a way that prevents additional cellular carriers from mounting their equipment). The three areas where cellular telephone antennas and equipment are installed include the following:

- Raw land sites
- Co-location
- Existing building

Raw land sites provide a small fenced-in space where a cellular tower can be erected and the telecommunication switching equipment can be housed. These sites must be accessible by road and have access to electric power and telephone lines. Three types of towers can be built on raw land: guy wire towers, lattice towers, and monopole towers. Guy wire towers can extend several hundred feet up. Guy wire towers are metal lattice structures that receive their strength from guy wires attached to the upper portion of the tower and angled down and away from the tower. These wires are attached to a concrete foundation embedded in the earth. Lattice towers are like the Eiffel Tower in one respect: They are constructed of a lattice of metal angles and other shapes. Monopoles are like big steel flanged pipes that are reduced in diameter in sections as they rise above the earth. Monopoles are somewhat cleaner looking because they hide the antenna wires inside the monopole. The owner of the property receives rental income for the land used in conjunction with the tower.

Co-location is when other carriers choose to locate their antennas and equipment on an existing tower. If the lease is structured to protect the owner's interest, these new cell phone carriers provide additional rental income to the property owner.

An existing building can provide an opportunity for cellular telephone companies if the height and location of the building is suitable. But *existing building* is a relative term because it does not adequately describe the other structures used to mount cellular antennas, such as farm silos, water towers, and church steeples. Existing buildings sometime have antennas mounted on the outside wall of penthouse mechanical rooms, or on padded skids located directly on the roof deck close to the edge of the roof. Electric lines and phone lines are run to the base equipment either through fireproof chases that exist inside the building or by mounting conduit to an exterior wall of the building. The base equipment might weigh 5,000 lbs, so building structure must be evaluated.

Cellular antenna equipment, being so high up, poses no known health risk to people on the ground. Cellular antennas broadcast parallel to the earth with some downward scatter, so most of the energy is directed away from people on the ground. People on the ground receive levels thousands of times less than the Federal Communication Commission's (FCC's) safe exposure level. But cellular antennas on a roof might expose a person standing directly in front of the antenna to levels that may approach or exceed the FCC's safe exposure level. This should be a consideration for persons performing maintenance on the roof or if a building gives access to the roof for tenants.

Electromagnetic Fields

As electric current moves through a wire (as is the case with power lines carrying electricity at high voltages), an electromagnetic field is created around the wire. As voltage is increased, the strength of the field is also increased. Voltages of 100,000 to 300,000 volts can be found traveling through overhead transmission lines. According to the National Institute of Environmental Health Sciences (NIEHS) website, the few studies on adult exposures showed no link between electromagnetic field (EMF) exposure and adult cancer. Research studies performed during the 1980s showed a link between EMF exposure and childhood leukemia. But after reviewing decades of research, the NIEHS has concluded that only a weak association exists between exposure and childhood leukemia. Facilities wishing to locate near high-voltage transmission lines should consider the possible effects of electromagnetic fields.

Radon

Radon is a colorless, odorless, radioactive gas that is the result of decaying radioactive radium found naturally in the soil. It can also be found in drinking water when well water is used. Radon gas exposure has been linked to lung cancer. Radon mitigation for radon contamination in soil prevents the radon from getting into the building. Techniques for mitigation include caulking any cracks in the foundation, and collecting and venting off any gas that accumulates in sump pump crocks. Radon can be removed from drinking water with an air stripper system or activated carbon filtration.

The U.S. EPA produces maps, available on their website, that show the likely concentration of radon present in soils for every state. Actual radon levels can be determined only through testing.

7.7 GLOBAL ENVIRONMENTAL CONCERNS

A multitude of issues are of global environmental concern: air and water pollution, loss of natural habitat, nuclear waste, freshwater availability crisis, food production and food health crisis, ozone layer depletion, and the greenhouse effect and global warming (global weather instability). As caretakers of the built environment, facility managers must understand the issues at hand and respond to them in the most sustainable ways.

Facility managers should also understand the interrelated aspects of their decisions. For example, purchasing power from a nuclear plant may reduce a facility's carbon footprint, but it simultaneously adds to the crisis of nuclear waste disposal and increases the possibility of nuclear catastrophe.

Stratospheric Ozone Layer

The ozone (O_3) layer protects the inhabitants of the earth by shielding them from harmful ultraviolet radiation emitted by the sun. Ultraviolet radiation not only makes the planet hotter, its harmful rays are a leading cause of skin cancer. The Montreal Protocol, designed to protect the ozone layer, is a diplomatic triumph, with 170 nations signing the treaty legislation. The goal of the legislation is to reduce ozone depletion in the upper atmosphere by phasing out the manufacture of halogenated hydrocarbons used as refrigerants, aerosol propellants, and fire suppression agents. According to the U.S. EPA, efforts have been successful, and it estimates that the ozone layer has not grown thinner over most of the earth since 1998. The arctic ozone layer is projected to return to its pre-1980 levels somewhere between 2060 and 2075.

Greenhouse Effect

The *greenhouse effect*, also known as *global warming*, is the trapping of heat in the earth's atmosphere by carbon dioxide gas, a variety of other gases, and water vapor. A gas molecule must have at least three atoms to be an infrared heat absorber. Nitrogen and oxygen molecules, which make up the bulk of our atmosphere, have only two atoms per molecule. The amount of water vapor (H_2O) in the atmosphere is a function of atmospheric temperature of the earth and is not normally thought of as a greenhouse gas. Rather, increases in the amount of atmospheric water vapor are a result of global warming.

Carbon dioxide enters the atmosphere when materials that contain carbon are burned. The atomic weight of carbon is 12 and that of oxygen is 16. Therefore, CO_2 has a molecular weight of 44 (12 + 16 + 16). The ratio of carbon dioxide to carbon in CO_2 is 44/12 = 3.667. One pound of carbon therefore requires 2.667 pounds of oxygen to burn, resulting in the combination 1 + 2.667 weighing 3.667 pounds. So the combustion of a fuel that contains one pound of carbon results in 3.667 pounds of CO_2 being emitted.

Combustion of fuel also adds other gases to the atmosphere that in turn add to pollution, but they are not considered as being greenhouse gases (GHGs). Nitrogen oxides and sulfur dioxide (SO_2) are also emitted during combustion. Nitrogen dioxide is a reddish brown gas commonly known as smog, and sulfur dioxide combines with water vapor to form acid rain. Carbon monoxide is another pollutant emitted. Carbon monoxide (CO) results from incomplete combustion. NO_x, CO, SO_2, and nonmethane volatile organic compounds (NMVOCs) are considered to contribute indirectly to the greenhouse effect.

Prior to the industrial revolution, carbon dioxide concentrations in the atmosphere were about 280 ppm. Today carbon dioxide concentrations in the outdoor air are approximately 385 ppm. The result has been that the average temperatures across the globe have risen 0.74 degrees Celsius ± 0.18 degrees Celsius since the late nineteenth century, according to National Oceanographic and Atmospheric Administration (NOAA).

This rise in temperature around the world is known as global warming. Although global temperatures are rising, the phenomenon should more accurately be called *global weather instability* because this is its most damaging effect, and it is a direct result of air pollution.

Carbon Dioxide Equivalent Gases

The Kyoto Protocol lists six greenhouse gases that are targeted for reduction through international agreement. (The United States has not signed the Kyoto Protocol.) The gases have a carbon dioxide equivalent (CO_2E) rating. The ratings reflect the gases' potential for atmospheric damage over a 100-year period. According to Table 7.1, 1 pound of methane has the same damaging effect that 23 pounds of carbon dioxide has on the environment. Conversely, using methane landfill gas to power a co-generation plant at a facility to make heat and electricity produces CO_2, which when compared to methane is much less damaging to the environment and utilizes a source of energy that otherwise would be wasted.

The U.S. EPA has formulated regulations that, *if enacted*, would regulate commercial operations that produce 25,000 metric tons or more of CO_2 per year. At this writing, some facilities are required to report their CO_2 gas emissions to the U.S. EPA, which is scheduled to provide a final rule sometime in 2013 after public comments. More information can be found on the EPA's website.

Example: Calculating CO_2 Emissions

According to the U.S. EPA's Greenhouse Gas Equivalencies Calculator, if a facility's boilers consume 5,000,000 therms of natural gas in one year, the resulting CO_2 gas equals 25,000 metric tons. One therm of gas is the amount of gas that contains 100,000 Btu of heat energy. Gas varies in heat content, so it is better to be billed based on therms than on a volumetric measurement such as cubic feet per minute. The threshold for reporting CO_2 emissions from stationary combustion equipment is 30 mm Btu/hour, or 30 million Btu/hour. Below this level, no reporting is necessary.

One way to report emissions is to calculate the amount of CO_2 emissions based on the amount and type of fuel consumed. The following example illustrates such a calculation.

The BTU input can sometimes be found on the boiler's nameplate. If it isn't, the boiler horsepower rating and efficiency can be used to calculate Btu/hr of heat input using the following formula (BHP stands for "boiler horsepower"):

Heat input in Btu/hour = boiler capacity in BHP × 33,446 Btu/hr/BHP × 1/(efficiency as a decimal)

TABLE 7.1 The six greenhouse gases regulated under the Kyoto Protocol

Greenhouse Gas	Chemical Designation	Carbon Dioxide Equivalent	Source
Carbon dioxide	CO_2	1	Burning fuel containing carbon
Methane	CH_4	23	Landfill gas, livestock digestion, wetlands
Nitrous oxide	N_2O	296	Naturally, through the breakdown of nitrogen in soil and in the ocean; fertilizers
Hydrofluorocarbons	HFCs	12–12,000	Air-conditioning and refrigerant equipment (unless destructive chlorofluorocarbons [CFCs] are replaced)
Perfluorocarbons	PFCs	5,700–11,900	Aluminum production, semiconductor industry
Sulfur hexafluoride	SF_6	22,200	Electrical transmission equipment

For example, what is the horsepower rating of a boiler that is 80 percent efficient that has a heat input of 30,000,000 Btu/hr?

$$30,000,000 \text{ Btu/hr} = \text{boiler capacity in BHP} \times 33,446 \text{ Btu/hr/BHP} \times 1/0.80$$

The result is that a boiler with a BHP rating of 717 BHP or below and an efficiency of 80 percent or more would not require reporting. This example assumes that the boiler will be operated at maximum capacity 365 days per year. In reality, this is never the case because boilers are designed to operate to match the heating load. The formula above is simply another way to eliminate reporting data for equipment whose heat input falls way below the U.S. EPA's requirements. Above the 30 mmBtu/hour level, the actual amount of therms of natural gas consumed over a one-year period by the boiler is input into the EPA's Greenhouse Gas Equivalencies Calculator and the actual amount of CO_2 emitted is calculated for reporting purposes.

Biogenic Carbon

Biogenic carbon, is contained in biomass which consists of plants, plant materials and wood. The combustion of biogenic carbon does not count toward greenhouse gas emissions because it is assumed that, during the life of the plant, the plant absorbs as much carbon dioxide from the atmosphere as it gives off when it combusts. This assumption explains why several power companies are investing in wood waste–electric power conversion plants. Some of these plants will actually gasify the wood at high temperatures, which produces a synthetic natural gas used to run diesel or gas turbine generators. Other plants will use the wood directly as a fuel source for a boiler. It is hoped that someday this diesel engine technology will be replaced by fuel cell technology that will convert the hydrogen gas directly into electricity without the mechanical engine losses of today.

Diesels can also operate on biodiesel made from waste cooking oil and soybeans. Flex fuel vehicles can run on E85 fuel, which is 85 percent ethanol (C_2H_6O) and 15 percent gasoline. This combination, when burned, is considered to be carbon neutral, meaning that it does not add to greenhouse gas emissions from burning carbon. Using flex fuel vehicles is one method of reducing a facility's carbon footprint.

Carbon Footprint

Carbon footprint is a summation of all a facility's activities that cause CO_2 and CO_2E to enter the atmosphere. The two most significant contributors to GHG are the energy a building consumes and the source of that energy; both affect the amount of carbon entering the atmosphere. When comparing two identical buildings that consume the same amount of electric power, but one building uses power provided by a coal-fired plant and the other uses power from a nuclear plant, the resulting carbon footprint is vastly different. The building that has its power provided from coal has a much larger footprint. Consideration as to how power is produced is as important as reducing consumption.

The production of electricity from nuclear power presents enormous challenges. Electricity from nuclear power, once thought to be too cheap to meter, suffers from the enormous cost to build nuclear plants and the huge cost of nuclear waste disposal. The safety of nuclear plants, especially in seismically active areas, is an area of such concern that both Japan and Germany are moving away from this power source and shifting instead to renewable sources of energy.

Of additional importance for facilities is the carbon generated by occupants of the building when traveling to and from work. To reduce the overall carbon footprint, facilities should consider public transportation options and the addition of charging stations for electric vehicles, especially when that charging power is from a renewable source. Other contributors to a carbon footprint are a facility's emissions of HFC and CFC refrigerants, off-gassing of building furnishings and finishes, and the solid wastes generated by the building's occupants. Organic wastes eventually decompose and emit methane, a greenhouse gas that is twenty times more damaging than CO_2 over a 100-year period.

The U.S. EPA provides online resources in the form of a *carbon calculator* to aid businesses in discovering their carbon footprints (see the U.S. EPA website). Power usage and power-generating plant type are input into the calculator to achieve an amount of carbon that is emitted into the air yearly. Having this actual carbon emission baseline data is important in determining the level of success of any carbon reduction plan. For additional information, the International Facility Managers Association (IFMA) Foundation has produced a series of free sustainability white papers, which are downloadable from its website. One of these white papers is the "Sustainability Guide—Carbon Footprint."

Carbon Footprint Reduction

A multistep approach to reducing a facility's carbon footprint is an effective way to achieve the greatest success. The approach begins by performing an evaluation of energy consumption, energy production, building operation, recycling, trash, and green chemicals. Then a plan is developed to achieve energy reductions. The manager must also look outside the facility (such as transportation and heat island effect) in order to factor in all the components that make up the facility's carbon footprint.

Energy Consumption   Power plants produce a tremendous amount of CO_2 in the combustion of fossil fuels, so the obvious first step is to reduce energy usage. One of the easiest and most effective methods is through the use of efficient lighting technology from energy-efficient lamps and intelligent lighting controls. The use of day lighting provides energy-free lighting for building inhabitants.

Electric motors are another big consumer of building power. To reduce this load, energy-efficient motors can be installed. If the load on the motor varies, having a variable frequency drive that controls motor starting and running can provide significant savings.

Energy is not always just electric power. Many buildings have boilers for heating, and these buildings often use natural gas or number two diesel as their fuel source. Boilers can be retrofitted with energy-saving devices, or new high-efficiency condensing types of boilers can be installed.

Elevators are another significant consumer of energy, but several strategies are available to reduce consumption, including more intelligent controls and regenerative elevators. These approaches can return power to the grid because the elevator turns the elevator winch motor into a generator as gravity moves the elevator downward.

A source of energy loss is lack of insulation. The building itself might lose heat due to poor insulation or transfer heat through leaky windows with a low insulation (R) value. (The R-value is the resistance to the flow of heat.)

Energy Production   Renewable energy sources such as photovoltaic, wind, solar, hydroelectric, and geothermal (see Chapter 12) can be added to a facility's utility infrastructure to reduce the amount of kilowatts from nonrenewable sources or Btus from carbon fuels. No renewable strategies reduce the carbon footprint as well. Fuel cell technology can be part of a building's microgrid (see Chapter 12) to convert natural gas into electricity. Microturbines can be fueled by landfill gas, thereby preventing methane from being released into the atmosphere.

Innovative Operational Strategies   An area of great concern to green building organizations is the failure of some buildings to perform as designed. Chapter 8 offers several operational strategies for HVAC systems that reduce a facility's carbon footprint. Facility managers can ensure that operational strategies are working through retrocommissioning, energy audits, and use of the U.S. EPA's Portfolio Manager software program.

Water   Using water consumes a precious resource. Water is provided by huge electric-driven pumps at a water utility. Wastewater must be pumped and treated. Rainwater storage provides a way to reduce the power consumed and eliminates waste treatment costs entirely. Sewage can also become a source of water if it is extracted from sewer mains or captured from the building's drains and treated (see the section called Sustainability by Design in Chapter 13, which showcases the Council House 2 project in Melbourne, Australia).

Green roofs (see Chapter 12) reduce the amount of water entering the storm sewer, the heat island effect, and the amount of air conditioning required. The plants themselves remove CO from the atmosphere. Water can also be kept out of storm sewers by using permeable paving materials that allow for the recharging of subsoil with water.

Another method of water conservation is to install low-flow restroom fixtures or no-flow waterless urinals in restrooms.

Recycling and Trash   Recycling keeps materials out of the landfill. Using recycled materials to create new materials can reduce the amount of energy required for their manufacture. Construction projects should strive to use recycled building material from renewable sources. Trash can be managed in innovative ways. Facilities spread out over many acres may consider the installation of smart trash compacting receptacles that are solar-powered and can transmit data to ensure efficient routing of trucks for pickup.

Green Refrigerants and Chemicals   Newer refrigerants are more friendly to the environment when leaks occur. Refrigerants such as HFC-134a, HFC-407c, and HFC-410a were considered to be green refrigerants. These refrigerants have a lower potential to harm the ozone than older refrigerants do, but unfortunately they are potent greenhouse gases. An alternative has been developed for mobile use (trucks and autos): HFO 1234yf has a global warming potential (GWP) that is 335 times less than HFC-134a.

Some types of chemical cleaning solvents contribute to GHG emissions. Green housekeeping and maintenance chemicals should be used to prevent gas emissions through evaporation.

Outside Factors   Consideration must be given to the commuting habit of building occupants. To attain the smallest carbon footprint, public transport should be encouraged by installing convenient and comfortable shelters and other amenities. This approach may actually tie in nicely with heat island reduction through landscaping and the installation of more solar reflective pavements. Purchasing locally produced material reduces a carbon footprint because it reduces the GHG that results from truck, train, ship, or aircraft emissions, which would be produced if materials had to be transported.

REVIEW QUESTIONS

1. Why is it important to complete a Phase I environmental assessment prior to a property transaction?
2. Describe the due diligence process in property transactions.
3. What is NESHAP?
4. What is a major source of lead poisoning in an old apartment building?
5. What is a LUST site?
6. What is biogenic carbon?
7. What is a carbon dioxide equivalent?
8. Name factors to consider when a manager is attempting to define a facility's carbon footprint.

Essay Questions

9. Write a paper about the various methods of environmental remediation that can be performed to clean up groundwater or soil contamination.
10. Research and describe the carbon regulations enforced in Australia. What has been the effect of these regulations on the people and industry?
11. Write an essay describing how calculating a facility's carbon emissions can lead to improved operational efficiencies.

BUILDING SYSTEMS AND CONTROLS

INSIDE THIS CHAPTER

Is Heating, Ventilation, and Air-Conditioning (HVAC) Control and System Knowledge Necessary to Be an Effective Facility Manager (FM)?

The main thrust of this chapter is HVAC ventilation systems and control. The chapter begins with building control systems, both the older pneumatic system and the newer, direct digital control system. It talks about the language of this system that allows different equipment to "speak" to each other and work together. The chapter then talks about the components of a healthy, comfortable indoor environment and the various types of air supply systems and equipment used to produce it. Emphasis is placed on reducing energy consumption through systems and control.

8.1 MANAGING BUILDING MECHANICAL SYSTEMS

Mechanical systems deliver heating, ventilation, or cooling media to the spaces within the building. These systems can be based on water, air, steam, electricity, or a combination of these heat transfer media. A building's mechanical systems are its heart, and a building control system is its mind. HVAC management provides comfort for building occupants by providing excellent interior conditions with regard to thermal comfort and healthy air. The HVAC manager oversees this by managing the indoor air quality (IAQ) in the built environment. IAQ is a combination of air temperature, airflow, air movement, outside air ventilation, and purity. IAQ is important for the health and productivity of building occupants. Additional attention must be paid to the building's energy consumption and carbon footprint while striving to achieve a quality indoor environment for building occupants.

A facility manager needs a working knowledge of building systems, equipment, and operation. Strong management and financial skills are valuable attributes for a facility manager. What makes facility managers essential to the organization is the combination of managerial and engineering skills, the FM's ability to function in the real world of getting a job done with all the operational, nuts and bolts knowledge that goes with it, and the alternate reality of the corporate financial world. This dual functionality is what makes FMs so valuable, and why mechanical and operational knowledge is essential. Operational knowledge provides the manager the ability to give direction and guidance to the technically skilled facility maintenance workers under his or her command. A manager with this ability has confidence and can respond quickly and effectively to emergency situations. The manager knows his or her building and knows how to get the job done.

8.2 BUILDING CONTROL SYSTEMS

Building control systems are devices that control HVAC equipment and lighting. The newer direct digital control (DDC) computerized systems have the ability to control a variety of ancillary equipment, such as sun shading devices; photovoltaic, solar heating equipment; wind turbines; and water-harvesting equipment. They can tie into other systems to view status information from elevator controls, and fire and security alarm systems. DDC systems can also make decisions on how equipment should be operated and can closely monitor energy use. Control systems fall under three categories:

- Pneumatic systems
- Electro/pneumatic computerized systems
- Direct digital control (DDC) systems

FIGURE 8.1 Molecular sieve desiccant dryer used to remove moisture from the compressed air in pneumatic systems, thus ensuring clean dry air to HVAC pneumatic controls and actuators

Source: Photo courtesy of Oconomowoc Memorial Hospital.

Pneumatic Systems

Early building control systems were pneumatic systems that utilized air pressure produced by a compressor and stored in tanks. This main air pressure needed to be filtered and the entrained moisture removed. Moisture removal was accomplished by refrigerating the air exiting the storage tanks. Today, a more energy-efficient method of removing moisture can be achieved with a molecular sieve desiccant dryer. A molecular sieve (see Figure 8.1) actually strains the moisture from the air by providing passage ways through the desiccant that are too small for water molecules to pass through but are large enough for air to pass through easily.

Air pressure from these early systems was sent to a variety of pneumatic controls. The controls provided air signals to operate equipment such as heating valves, cooling valves, pneumatic damper motors, and air bladders found inside early diffusers. These early systems weren't very flexible in their operation, and changes were sometimes made through the use of manual bypass valves. Although these systems were automatic, they often required mechanical adjustments from facility maintenance personnel in order to operate efficiently.

Electro/Pneumatic Early Computerized Systems

Control systems in the early 1970s got a boost in their functionality from the introduction of electronic controls. Through the use of devices such as transducers that could convert a pneumatic signal to an electric signal (see Figure 8.2), operators could now command simple operations from a computer terminal. Electric sensors such as a resistance temperature detector (RTD) could now send signals to a controller that could make corrective decisions, and the data could be viewed by a computer. Electronic

FIGURE 8.2 Pressure transducer

Source: Photo courtesy of Oconomowoc Memorial Hospital.

control also allowed equipment to be started and stopped from the computer terminal. If control points varied from the set-point, an alarm was generated. These alarms were recorded on a dot matrix printer and stored in the computer's memory. The system also had a time clock function that allowed for scheduling the on/off operation of equipment.

Direct Digital Control (DDC)

DDC systems consist of a computer terminal, a router, a server and software, a controller, and sensors. DDC systems are in use today, and they are a major improvement over prior pneumatic systems. DDC systems can process information and make decisions within parameters set by the programmer or maintenance technician. A major improvement was having a graphic interface that depicts building systems. Rather than navigating through lines of text to find the device that needed to be adjusted, the technician views a computer rendering of the system (see Figure 8.3). The computer rendering shows all the devices that can be controlled, and the technician simply clicks on the picture of the device. Another new benefit that these systems offered maintenance technicians was to provide actual position information of the controlled device. In the past, all that the technician knew was that something was open or closed, or on or off. Now, the maintenance technician can view a device that is opened or closed stated as a percentage.

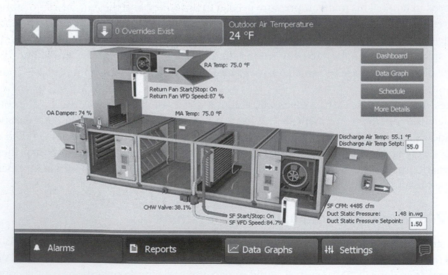

FIGURE 8.3 Trane Tracer DDC air handler control screen

Source: Materials courtesy of Trane.

This information provided an important troubleshooting tool. For example, if the supply air temperature was much too warm, the maintenance technician could see what percentage (from 0 to 100) the cooling valve was open. If the valve were open only a small percent, that information would alert the maintenance technician that the valve might be stuck.

The advent of smaller, more powerful computer chips allowed DDC systems to make decisions and take corrective action automatically. It caused a shift in equipment logic from "It's runtime, so let's turn it on" to "It's time to run, but is the equipment needed yet?" If indoor conditions did not warrant the equipment being turned on, the equipment remains off. In fact, if HVAC equipment is needed, the equipment provides only enough heating or cooling to meet the current need.

8.3 MAJOR COMPONENTS OF A DDC SYSTEM

The major DDC system components are as follows:

- Computer terminal
- Controllers
- Software, server, and router
- Points
- Sensors

See Figure 8.4 for an illustration of a DDC system and its components.

Computer Terminal

A building manager interacts with the DDC system via a desktop or laptop computer terminal's keyboard and mouse. The computer's display screen instantly displays the information necessary to operate

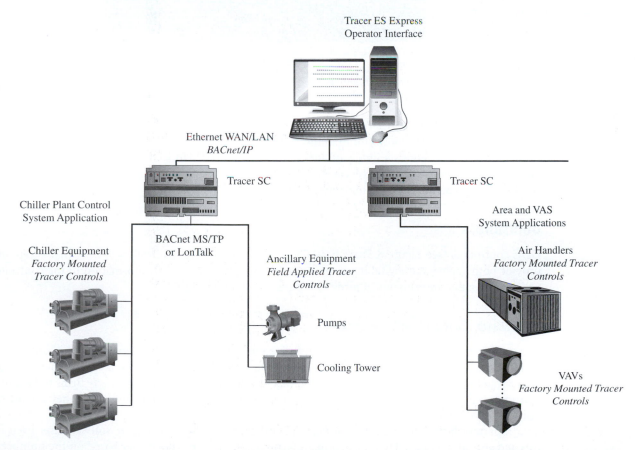

FIGURE 8.4 DDC system

Source: Materials courtesy of Trane.

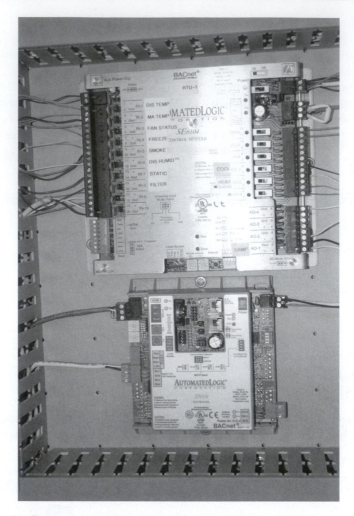

FIGURE 8.5 DDC controller

Source: Photo courtesy of Waukesha County Technical College.

and control the building's mechanical equipment using two common types of display: graphic and line of text. A graphic display shows, for example, a picture of building floor areas or mechanical equipment. A line of text display consists simply of lines of text, as in "AHU No.1 OCCUPIED ON." Both of these displays report the equipment operating data, internal building conditions, and even external weather conditions that can affect the building occupants' comfort.

Controllers

DDC controllers (see Figure 8.5) are basically programmable logic controllers (PLCs) that accept and provide both analog and digital inputs and outputs. Sensors provide the information to the controller, which acts on that information according to the software programming.

Software, Server, and Router

Control logic resides in the software stored in a computer server. Controllers utilize this software by being connected to the server through a router. If the system utilizes a web server, then access to the system can be obtained anywhere in the world that has Internet access. Most property firms today prefer that their building engineers have the ability to access the system at home and when they are off-site visiting other properties.

Important factors for FMs to consider when purchasing a DDC system are software user licensing fees and cost of upgrades. DDC system manufacturers often charge the owner based on the number of persons and/or computers accessing the software, which can be a financially burdensome surprise after spending a six-figure amount on a new system and then having to pay for additional licenses so all the facility personnel can use it. The other unfortunate surprise is that the new system may be due for a software upgrade before the warranty period has expired. The FM needs to include provisions in the bid specifications that cover the number of persons and computers needing access and that require the inclusion of any system software upgrades occurring throughout the warranty period. It is also a good idea to have, at a minimum, a two-year warranty on DDC systems. In areas of the country that experience four seasons, the system should go through each season to verify that the system operates properly when switching from one season to the next. It is also recommended that the warranty start when a third-party consultant has verified that the system is fully operational; otherwise, the warranty period will start on a system with faulty programming and/or faulty parts.

Points

Point is a term control technicians use to describe a single item of information in a control system used in a data storage location. Points are either input or output, and they can provide digital or analog signals to the controller. An address is used to identify points in a digital control system. The address is a unique numeric or alphanumeric identifier. Points are known as internal if the data is created by control software, external if the data is generated by an external device. Global points provide data that is transmitted to the network so that other controllers can use it.

Sensors

Sensors are instruments that are in contact with the item or medium to be measured. Sensors come in many different types, including position sensors, motion or occupancy sensors, temperature sensors, light sensors, pressure sensors, pressure differential sensors, electric load and voltage sensors, and equipment condition sensors.

Quality sensors measure data such as:

- Space temperature
- Humidity
- Carbon dioxide
- Filter status
- Ambient light

Safety sensors measure data such as:

- Presence of smoke
- Rise in temperature
- Airflow
- Presence of water
- Water flow
- Level (of a fluid, for example)
- Low voltage
- Flame failure
- Freezing conditions

Operational sensors measure items for system control. The following are examples of operational sensors:

- Pressure
- Equipment status
- Static air pressure
- Airflow
- Air temperature

Some examples of sensors will be described in more detail in the following subsections.

Pressure Sensors   Pressure sensors display pressure in pounds per square inch or inches of water. An example of a pressure sensor is a fuel oil pressure sensor for a boiler. It might also be a differential pressure sensor across an air filter. Filter sensors can tell the operator that air filters must be changed or that they are getting dirty but are still serviceable. Air filter sensors must be very accurate because the pressure differential is generally in tenths of an inch of water. The *pressure differential* is the difference between the pressure of the air entering and leaving the filter.

Static Air Pressure Sensor   *Static air pressure* is the weight of air in a duct measured in inches of water. By generally accepted convention, it is measured either at the fan outlet or at roughly three-quarters the way down a main duct run. For example, on a twenty-story building that had a sixth-floor mechanical room, the static pressure sensor would be found on the sixteenth floor.

The static sensor is used to provide a 4 to 20 milliamp (ma) signal to the frequency drive controller. The signal causes the supply air fan to speed up or slow down. Static air pressure is the weight of air in a duct and provides a more useful signal than interior space air temperature for controlling the speed of the supply air fan. As a building interior heats up, the building's variable air volume boxes dampers open wide. The supply air pressure in the duct has less resistance on it, and the static pressure in the duct drops. The fan needs to speed up to supply more air into the duct to increase the static pressure. Static pressure therefore is the sum of everything that influences building comfort.

The building engineer should be aware of the static pressure reading that provides a comfortable interior environment for the building occupants. Static readings differ somewhat because different buildings have different ventilating equipment, have different heat loads generated by tenants and equipment, and are constructed from different materials that absorb and reject heat differently. One building might be comfortable at ¾ inch of water column (w.c.) static pressure at the static sensor location. Another building needs 1½ inches w.c. of static pressure, while another might require an entirely different amount. The static air pressure signal is converted to a 4 to 20 ma electric signal to control fan speed via a frequency drive. On older systems, the static pressure signal was a pneumatic signal that controlled the opening of vortex vane dampers on the supply fan suction, which controlled the CFM (cubic feet per minute) output of the fan.

Airflow Sensors   There are two basic types of variable air volume (VAV) boxes: pressure dependent and pressure independent. Both control the amount of air flowing into a space for heating, cooling, and ventilation. Today almost all VAV boxes are pressure independent. Pressure-dependent VAV boxes vary damper position based on space temperature alone. Fluctuating duct pressures cause different amounts of air to flow into the space for the same space temperature reading. The greater the air pressure in the duct, the more air will flow into the space. For this reason, changing duct pressures result in poor space temperature control. Pressure-independent VAV boxes compensate for varying duct pressures by using a combination of duct pressure and space temperature to control damper position. Duct pressures vary as outside environmental conditions or internal building heat loads change during the day.

For example, solar energy heats the building's perimeter spaces exposed to the sun. VAV boxes that serve these building spaces compensate for the increased load by opening more, in other words, by changing the angle of the damper inside the box to provide the required flow of cooling air. But if the building's demand for cooling air is great, duct pressure drops, which results in less airflow into the space. Airflow sensors mounted on the inlet to pressure-independent VAV boxes compensate for changing duct pressures by increasing or decreasing the VAV's damper position beyond what would normally be called for based simply on space temperature.

To measure airflow, airflow sensors need to obtain a velocity pressure reading in order to register airflow velocity. Velocity pressure is the force of moving air. Static pressure is the weight of air. The combination of velocity and static pressure is called total pressure. It is easy to measure total pressure directly, but it is difficult to measure velocity pressure directly.

The sensor consists of multiport devices that are generally configured like a ring or a cross. This configuration averages the varying air pressures created as air flows through the duct. The sensor is located in the airflow stream leading to the air damper in the VAV box. The sensor has ports that face into the airflow and ports that face downstream. The sensor ports that face into the airstream are called total pressure pickups because these sensors register total pressure. The sensors that face in a downstream direction compared to the airflow are static pressure sensors; these sensors are not affected by velocity pressure. The difference between the total pressure and the static pressure gives a direct reading of velocity pressure, which, in the case of a pressure-independent VAV box, adjusts the damper position so that the required airflow is maintained no matter how high or low the static pressure is in the duct.

Airflow sensors can also be calibrated in feet per minute. If you know the cross-sectional area of the duct in square feet, airflow volume can be calculated using the following formula:

$$V = 4004.5 \times \sqrt{hv}$$

$$Q = A \times V$$

hv = velocity pressure in inches of water for standard air

Q = flow in cubic feet per minute(CFM)

A = cross-sectional area of the duct in square feet

V = velocity of the air in feet per minute

Air Temperature Sensors   There are several types of air temperature sensors. Old pneumatic systems relied on bimetallic elements: two strips of metal that expanded at different rates. When connected to a linkage, these elements could move a needle that would give a temperature reading.

Today, most air temperature sensors are either thermistors or resistance temperature detectors (RTDs). Thermistors and RTDs are composed of materials whose resistance changes with a change in temperature. Thermistors are composed of a semiconductor material; RTDs are composed of platinum. A voltage is connected to the RTD, and the change in resistance causes a voltage drop that can be measured on a voltmeter calibrated to read temperature.

Thermocouples are another type of temperature sensor and are often used to read higher temperatures such as emergency generator exhaust temperatures. These sensors are comprised of two dissimilar metals that are in contact with each other. Physics tells us that a small voltage will be generated and current must flow when two dissimilar metals are in contact with each other. Temperature variations cause a change in the voltage produced. A voltmeter that is sensitive to very low voltages is calibrated to read the temperature.

8.4   CONTROL PROTOCOL: THE LANGUAGE OF DDC

The two structural types of protocols or language structure for programming and networking controls are:

- Proprietary protocol
- Open protocol

Protocols enable equipment to communicate with controllers and facilitate the networking of numerous pieces of HVAC equipment together.

In the past, HVAC equipment manufacturers used a proprietary protocol for the HVAC equipment and the control devices on their network (the protocol wasn't available to other manufacturers). A protocol is a type of computer language that allows components on the network to communicate with one another. Deals had to be struck between manufacturers to gain the communication information needed, which greatly limited a customer's ability to install equipment manufactured by different vendors.

Open Protocol

An open protocol is a communication standard that is published so that all manufacturers can use it to establish communication between devices on the network. The two major open protocols are BACnet™ and LonWorks.™ BACnet™ (Building Automation and Controls Network) is an American National Standards Institute/American Society of Heating Refrigeration and Air Conditioning Engineers (ANSI/ASHRAE) nonproprietary, open protocol communication standard that doesn't require additional hardware. LonWorks™ uses a protocol called LonTalk. Most LonWorks systems utilize a proprietary computer chip known as a Neuron® chip mounted to the control circuit board of each piece of equipment on the network that is to be monitored and controlled. The Neuron chip was designed by the Echelon Corporation.

BACnet™ and LonWorks™ devices cannot communicate directly with one another. To establish communication between them, it is necessary to install a device called a *gateway* so that data from one device can be translated into a readable form. Issues and costs related to protocols need to be investigated thoroughly by the FM prior to making equipment purchase decisions.

8.5   COMFORTABLE AND HEALTHY BUILDING ENVIRONMENT

IAQ and space quality are the two main components of a healthy and comfortable building environment. IAQ components include:

- Air temperature
- Relative humidity
- Air movement and purity

Space quality components include:

- Noise and vibration
- Lighting

Air Temperature

Air temperature is the most important component of IAQ. The majority of buildings attempt to achieve indoor temperatures of 72 degrees Fahrenheit, plus or minus 2 degrees. A space temperature of 70 to 74 degrees Fahrenheit has been generally accepted as a pleasant comfort range for most people. To save energy on air conditioning, buildings, especially government buildings, may set their thermostats at 76 degrees and allow summertime interior temperatures to rise to 78 degrees. In many commercial buildings, a 78-degree interior space temperature would be considered excessively hot, resulting in complaint calls and a possible violation of the requirements of tenant leases that specify a 72 ± 2 degree space temperature.

Relative Humidity

Humidity is very important because it relates to the way humans regulate body temperature. Humans sweat to reduce their skin temperature. When sweat evaporates on a person's skin, it creates the effect of evaporative cooling. Humidity that is too high restricts the evaporation of sweat and inhibits this biological cooling, causing building occupants to feel hot, moist, and sticky. Relative humidity (RH) is the percentage of the total moisture that air can hold at a specific temperature. Forty percent relative humidity means that the air can hold 60 percent more moisture until it hits 100 percent and becomes saturated. The comfort range for humidity inside a building is 30 to 60 percent relative humidity; about 45 percent RH is optimum.

Cooling coils in an air-conditioning system that operate at a temperature below the dew point of the incoming air cause condensation of moisture on the coils. (The dew point is the temperature at which moisture will condense out of the air, forming droplets of water on the coils.) This lowers the relative humidity and results in a more comfortable environment. During the cooler months, as outdoor temperatures drop, so does the air's ability to hold moisture. In the wintertime, the drier air is then expanded when heated, further reducing the amount of moisture per cubic foot of air delivered.

Low RH accelerates the evaporation of moisture on a person's skin, making him or her feel cold. Too low an RH allows static charges to build up, causing shocks to occupants and possible damage to computer devices if touched. Supplemental humidity must be added for these reasons. A humidifier injects steam into the supply airstream.

An efficient humidity control system balances the need to remove moisture in the cooling season with the energy needed to reduce the chilled water to a temperature below the dew point. The **dew point** is the temperature at which moisture condenses out of the air and onto a surface that is at or below the dew point temperature. The chilled water temperature supplying cooling to the air-conditioning coil might actually be less than what is required to produce a comfortable space air temperature. This lower temperature ensures proper dehumidification by having the metallic surface of the cooling coils below the dew point. If the chilled water temperature set-point is raised to save energy, then the RH will go up. Too much moisture remains in the airstream, which can lead to uncomfortable conditions and unhealthy mold growth. During the heating season, increasing RH above 60 percent by steam injection wastes energy. So the FM needs a system that can balance comfort and energy efficiency.

Measuring RH   Relative humidity can be measured using a sling psychrometer and the psychrometric chart, or handheld direct-reading digital meter. A sling psychrometer gives the user the wet bulb and the dry bulb temperatures. When these two temperatures are plotted on the psychrometric chart, the intersection point rests on top of a curved line representing the relative humidity as a percentage. Relative humidity data can also be provided through humidity sensors connected to the building management computer system. An FM should periodically review RH levels to ensure a comfortable, healthy, and energy-efficient building environment.

Air Movement and Purity

The movement of air is important for the thermal comfort of occupants. Movement accelerates the cooling effect on a person's skin and eliminates unpleasant air stagnation. Air movement prevents gaseous contaminants from becoming concentrated in an area, and it removes these airborne contaminants out of the space.

Pure air is a combination of gases that consist mostly of nitrogen and oxygen. Contaminants that effect air purity are:

- Manmade gases; examples include volatile organic compounds (VOCs), oxides of nitrogen, carbon dioxide, sulfur dioxide, petroleum-based and chemical gases such as chlorofluorocarbons (CFCs).
- Natural gases; examples include methane and ozone.

Other contaminants of air include:

- Dirt, dust, dander, smoke, soot, and ash
- Pollen, spores, mold, viruses, bacteria
- Odors
- Radioactive particles

Filtration   Removing particulates from the airstream is one way to improve air purity. Dirt, dust, pollen, insects, and airborne plant material are all removed through filtration. Several types and designs of filters are available on the market.

ASHRAE has standardized filter ratings through its Minimum Efficiency Reporting Value (MERV) rating system. MERV ratings from 1 to 16 tell the user how small a particle the filter can stop. A rating of 16 stops the smallest particle measured in microns. A filter with a MERV rating of 5 to 8 protects areas in commercial applications where the filter needs to remove particle sizes of 3 microns and larger. MERV-rated filters of 9 to 13 remove fine dust particles of 1 to 3 microns, and MERV-rated filters above 13 are used in hospitals to trap bacteriological contamination because they can trap particles that are 0.3 micron in size. They are typically installed downstream of lower-rated MERV filters that act as prefilters; otherwise, the more expensive, higher-rated MERV filter becomes clogged quickly.

High Efficiency Particulate Air (HEPA) is another filter rating; it was established by the U.S. Department of Energy. Simply stated, it describes a filter that can remove 99.97 percent of airborne contaminants that are 0.3 micron in size. The 0.3 particulate size was determined as the most difficult to trap. In an airstream containing 10,000 particles of 0.3 microns in size, only 3 particles will get through. To get an idea of the size of a micron, 25,400 microns equals 1 inch, and about 600 microns can fit in an area the size of a period (.). The human eye can see particles as small as 10 microns, and the size of pollen grains range from 5 to 100 microns. Filters rated 16 to 20 MERV are considered to be HEPA filters, with higher number corresponding to greater capture percentages.

Ultraviolet Light   The C-band of ultraviolet light (UVC) has been proven effective in killing airborne and surface viruses and bacteria. Ultraviolet light is generated by high-intensity lamps located inside the air-handling unit. Human exposure to UVC light can quickly cause eye and skin damage. Electrical lockouts, which secure power to the lamps when access covers are removed, must be installed. Ultraviolet lights can be installed downstream of cooling coils and above cooling coils and drain pans to control microbial growth. They can also kill bacteria trapped on the surfaces of filter elements. UVC light units are frequently installed in hospital air-handler units to help prevent the spread of viruses and bacteria through the air supply system.

Noise and Vibration

Noise and vibration directly affect the indoor environment. There are many strategies for reducing the amount of noise within occupied spaces. Fiberglass bat insulation may be placed inside walls and laid on top of the tiles of an acoustical ceiling. Demising walls can be added where only partition walls are required to prevent sound from moving through the plenum space above the ceiling. Open floor plans where cubicles are used is a more challenging situation because of the ease at which sound can travel throughout the space.

Sound-masking equipment (which is discussed in Chapter 13) can be installed to help reduce the sound of people speaking. Noisy mechanical equipment can sometimes be encased in sound-absorbing enclosures.

Vibrations can be generated by rotating machinery and transmitted for long distances through the building structure. Sometimes a glass of water sitting on a desk will start to vibrate with round visible waves evident on the surface of the water. Vibrations of this nature can be annoying to someone subjected to them for long periods of time. Vibrations can also generate noise when objects vibrate against one another. The solution is to isolate the vibration at its source. Rubber isolators and springs can be installed at points where rotating equipment is attached to its mounting to help prevent vibrations from being transferred to the building's structure. Metal ventilation ducts that can carry vibrations are isolated by installing short canvass duct sections to prevent the transfer of vibrations further down the duct.

Lighting

Pleasing glare-free lighting sufficient for the task is a necessary component of a quality indoor work environment. The use of day lighting and automatic shading can provide the proper mix of interior and exterior lighting without glare.

Electrical code requirements in many states mandate that lighting in offices be controlled by occupancy sensors that turn off the lighting when the space is unoccupied. Lighting might also be turned off or reduced when there is sufficient ambient lighting entering the space through windows. A more detailed discussion of lighting is found in Chapter 10.

8.6 AIR SYSTEMS FOR AIR SUPPLY AND MOVEMENT

The two general classifications of air supply systems for delivering air to spaces inside a building are:

- Forced mechanical ventilation (overhead and underfloor)
- Natural ventilation

Forced Ventilation: Overhead

Standard ventilation systems supply air through ducted ceiling diffusers and exhaust air through unducted grills in the ceiling. The open space above the ceiling is called a **plenum return**. Ceiling plenum returns are created by having a central return grill on each floor of a building that pulls return air through the space created between the dropped acoustical ceiling and the floor above (see Figure 8.6). This structure eliminates having unnecessary return air ducting.

Locating both air supply and air return grills in the ceiling can create problems. In small office spaces, this arrangement inevitably results in some freshly conditioned air being sucked back into the return system before it can benefit the occupants. Also, should the occupant be sitting too close to the supply air diffuser, cool air hitting the forehead can be uncomfortable. This cold, forehead-chilling breeze may cause the occupant to install a homemade cardboard air deflectors taped to the diffuser to block air from flow in their direction. The turbulence created interferes with the coanda effect created by the diffuser. The **coanda effect** happens when flowing air or liquid adheres to a smooth, curved surface. Ceiling diffusers are constructed of a series of curved plates so that instead of the air coming straight out of the diffuser, it adheres to the curved surface and travels horizontally. By doing so the air has a better chance of mixing with the existing air in the space. This mixing prevents cold air from hitting the occupants of the room. The coanda effect can be witnessed when a liquid is poured out of a glass that is tipped slightly. Instead of the water flowing directly downward out of the glass, it flows down the side of the glass and sometimes down the arm of the pourer. The solution to the above problem is to move the diffuser a sufficient distance from the occupant to allow proper mixing of ventilation air and space air to take place before it comes in contact with the occupant.

Forced Ventilation: Underfloor

Underfloor ventilation is a newer concept that is more logical from a thermal air movement standpoint. Underfloor systems frequently provide greater comfort than standard ceiling-mounted ventilation. Underfloor ventilation consists of a raised floor where the supply ducts are run. The floor consists of $2' \times 2'$ panels that fit in a channel. With the installation of carpet tiles on top, below-floor access can be gained from almost anywhere. Just like raised floors used in computer data centers, data and power cabling can also be run through an underfloor system.

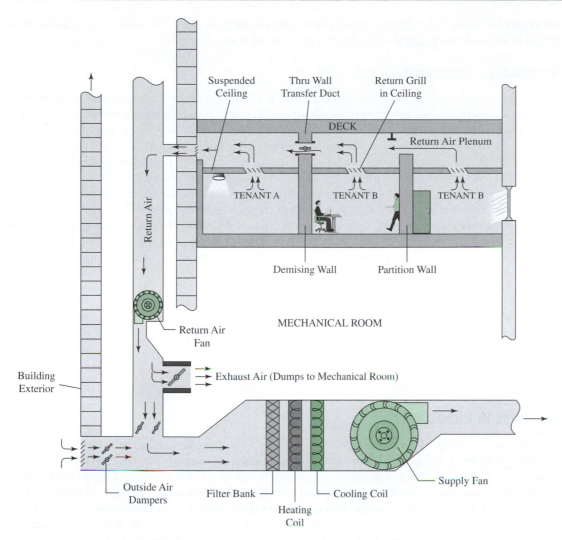

FIGURE 8.6 Return air plenum system, showing return air traveling between the suspended ceiling and the deck of the floor above it. No return ducting on the floors is required. Demising walls between tenant spaces require a transfer duct.

During the cooling season, cool air enters at floor level, where it warms and rises upward to the ceiling return, creating a natural thermal circulation. During the heating season, circulation is still upward. Warm air mixes with the cool air at floor level and rises upward to the diffuser. This natural movement transports contaminants upward to the return grill.

Natural Ventilation

Since the 1970s, the majority of comfort control systems for buildings relied on a totally sealed building envelope. The only source of outdoor ventilation was via infiltration through the exterior walls, through opening and closing doors, and via mechanical ventilation provided by an air-handling unit.

Natural ventilation is sometimes provided through operable windows relying on wind pressure and cross ventilation through a space to provide a healthy and comfortable environment. For natural ventilation to be successful, an open floor plan is necessary to allow air to move freely across the floor of the building. Natural ventilation can also be created by using the stack effect. The physics involved in the stack effect are the same as those used to obtain draft in early boilers. Prior to mechanical ventilation for boilers, boilers were fed with air for combustion by the *stack effect* created by a tall masonry chimney. In buildings, warm air rises through a shaft or atrium area in the building, causing a suction draft that pulls fresh air in through vents at the bottom level of the building.

Natural ventilation has many limitations, but the energy-saving benefits can be significant. Natural ventilation systems cause a greater variance of space temperature throughout the day. These temperature swings can be uncomfortable, and during hot humid months it can be intolerable. Natural ventilation eliminates fan noise, but it frequently substitutes street noise in its place. Fire code issues relating to the spread of fire and smoke can also be a major challenge for these systems

Ventilation Standards

Air is used to heat, cool, or both heat and cool, and at all times it keeps the spaces within the building properly ventilated. The air system must supply air and, in most cases, recirculate a portion of the air sent to the spaces to be reused again. This recirculated air is known as return air. The current ASHRAE standard for calculating outside air requirements for proper ventilation is ASHRAE Standard 62.1-2007. This standard applies to most buildings except single-family residential and multifamily residential up to three stories (these residential buildings are covered by ASHRAE Standard 62.2-2007).

It is important to understand that, at any particular time, the fresh air ventilation requirements are fixed because they are largely based on the estimated number of people who will occupy the space, but supply air requirements constantly vary with the thermal conditions of the space because air is required to heat or cool it. In large existing buildings, two sets of dampers are provided for the intake of outside air. Minimum air dampers open to provide the air required for ventilation, and main air dampers open to provide the air needed for cooling the building.

8.7 VENTILATION SYSTEMS

Mechanical ventilation systems can be divided into two categories:

- Constant volume systems
- Variable air volume systems

Constant Volume Systems

Constant volume (CV) systems feed air into the space at a constant rate. To meet the heating or cooling demands of the space, the air first flows over the fins of heating and cooling coils. The amount of heating or cooling media must be varied in temperature or amount in order for constant volume systems to meet space temperature requirements. So the air temperature supplied to the space varies according to the space load. Two methods for accomplishing this are *direct expansion (DX) systems* and *face and bypass dampers*.

For DX air conditioning in CV systems, the refrigerant liquid is sent directly to a refrigerant expansion valve located at the entrance to the evaporator coil. Cooling capacity needed to meet space demand is controlled by the expansion valve varying the amount of refrigerant allowed to enter the evaporator with airflow remaining constant.

Face and bypass dampers are used in constant volume chilled water systems to control the amount the air is cooled. A set of dampers is placed in front of the cooling coil, and another set of dampers is placed on top of the cooling coil. Opening and closing these dampers regulates the airflow through and around the cooling coil, which in turn regulates cooling capacity without having to change the cooling media's temperature or volume.

Space heating, when using constant volume systems may be achieved, by using a gas-fired heat exchanger, electric resistance heat, steam or water coils or renewable methods such a geothermal or solar heat.

The disadvantage of CV systems is that they waste fan energy by always having the fan operate at the same airflow output. When less ventilation is necessary to meet IAQ standards and space temperature requirements, energy is wasted by supplying too much air. Any air supplied must be filtered and either heated or cooled. This unnecessary heating and cooling wastes additional money and energy.

CV Rooftop Units   Rooftop units (RTUs) are packaged heating/air-conditioning units that sit on top of a raised curb constructed of $2' \times 8'$ pressure-treated lumber or a prefabricated metal roof curb from the unit's manufacturer. Both types of curbs are set on the roof around the opening in the roof deck and flashed over with roofing material to make them watertight. RTUs have one supply fan that provides conditioned air to the space. A second fan is strictly for venting out combustion gases used in the heat exchanger when heat is required. The supply fan also recirculates air from the space and draws in fresh outside air through economizer dampers. The economizer damper provides ventilation air and free cooling for the space. In the past, most RTUs provided the supply air at a constant volume to the space.

Some RTUs today utilize variable frequency drives to control fan speed and therefore also control the amount of air supplied to the space. Higher-tonnage RTUs frequently have two air refrigeration compressors that are staged in their operation to satisfy the space load.

Rooftop units are especially useful in multitenant buildings such as strip shopping malls. Mall tenants generally have the responsibility to heat and cool their rented space, and frequently they must maintain the equipment as well. RTUs are easy to operate and require little maintenance. Tenants can operate the unit from a thermostat and can program the unit's operation from there as well. Preventive maintenance most often consists of a spring and fall servicing, with quarterly filter changes by an HVAC contractor. An important part of the service is to check the unit electrically, check refrigerant levels, and spray-clean the air-cooled condenser with coil cleaner.

Should unit replacement be necessary; the first steps are usually just a matter of disconnecting the ductwork and the electrical, thermostat, and the gas line. The unit can be lifted off the roof in one piece via a crane and a new unit set in its place and reconnected. The disadvantage of rooftop units is lower energy efficiency, which is sometimes half or less than the efficiency achieved by water-cooled chiller units.

Variable Air Volume Systems

The two components of variable air volume (VAV) systems are supply system control and terminal unit control. The terminal unit control varies the amount of air to the zone in order to control zone temperature. The supply system control varies fan output (see Figure 8.7).

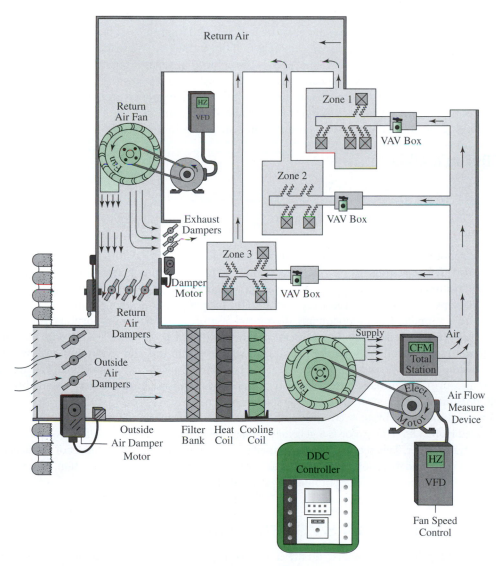

FIGURE 8.7 Variable air volume system, showing how the system receives inputs and controls airflows for an energy-efficient and comfortable building environment. The controller in the diagram is also known as a field controller in the system. *Hz* stands for "Hertz," which is cycles per second, or frequency. *VFD* stands for "variable frequency drive."

Supply System Control

Supply system control is achieved by controlling supply fan output. Early designs placed a duct static pressure sensor two-thirds the way down the supply duct. As VAV boxes opened in response to increased cooling demand, the air pressure sensed by the static sensor also dropped. This drop sent a signal to the main supply fan, which resulted in the speeding up of the fan through the use of a variable frequency drive (on pneumatic systems, vortex vane dampers in fan's suction were opened). Increasing fan speed or vortex vane opening provided greater air pressure supplied to all zones of the building.

Vortex vane dampers are wedge-shaped dampers that basically restrict the amount of air coming into the circular opening of the supply fan. Positioning these wedge-shaped dampers to lie flat starves the fan of air completely and can effectively drop the load on the fan to provide for smooth, low-current-draw startups.

Terminal Unit Control

Terminal units, as the name implies, are the last piece of ventilation equipment prior to the air entering the space(s) in the zone. The volume of air (calculated in CFM) admitted into a zone is controlled in order to achieve the desired space temperature. All VAV boxes are devices that have two basic parts: a rotating blade damper and a damper motor for airflow regulation. A thermostat sends a signal to the damper motor to open or close the damper to meet the temperature requirements of the space. In this configuration, the VAV box is known as pressure-dependent. The amount of air pressure in the duct has a direct effect on the amount of air coming out of the diffuser because duct pressure is not considered to be in damper positioning; only space temperature is considered in *pressure-dependent* VAV systems. For example, assume that tenants start leaving an office building at the end of a workday. VAV boxes close in the vacated spaces, which causes duct static pressure to go up. The tenants remaining in their offices will receive more air than they need. To compensate for changes in duct pressure, a *pressure-independent* VAV box is used. An airflow sensor consisting of a multiported ring or cross arrangement is installed just upstream of the damper. Damper position is then regulated by a combination of actual airflow and space temperature signals so that the correct amount of air is delivered to the space.

VAV Box Actuation

There are two types of actuators for VAV boxes:

- Pneumatic
- Electronic, or DDC

The damper motor in a **pneumatic VAV system** is a cylindrical can that has an actuating shaft protruding from one end. The actuating shaft is connected to the rod of the damper blade. A spring-loaded rubber air bladder, piston, or diaphragm is used to move the actuating shaft. The spring provides a force against the bladder or piston. The spring returns the shaft to the proper position when air pressure has changed. Air pressure is supplied from a pneumatic thermostat at a pressure that is proportional to space temperature. For pressure-independent systems, the airflow sensor and a volume reset controller are installed. The volume reset controller allows the operator to set the minimum and maximum airflow settings. Because pneumatic VAV boxes are controlled at the boxes themselves via thermostatic pressure signals and flow measuring, pneumatic boxes cannot relay information for use by the control system directly. As a result, pneumatic VAV terminal units are employed only in early systems.

In systems controlled by an electronic thermostat or by DDC system control, an electric damper motor is used (see Figure 8.8). The motor is a generally a small, synchronous, shaded-pole motor powered by a step-down transformer. Two sets of contacts are used to control motor direction. One set of contacts rotates the motor in the closed direction, and the other set rotates the motor in the open direction. The amount of movement depends on a low-voltage or milliamp signal that is proportional to the signal(s) received.

8.8 VENTILATION SYSTEM CONTROL STRATEGIES

It is important for the FM to understand the control strategy employed by the building. The FM should recognize the limitations of the control strategy resulting from the type and amount of control equipment. Developing this knowledge of building systems and controls is exceedingly important as organizations strive to achieve energy efficiency or work toward gaining Leadership in Energy and Environmental Design (LEED) certification points.

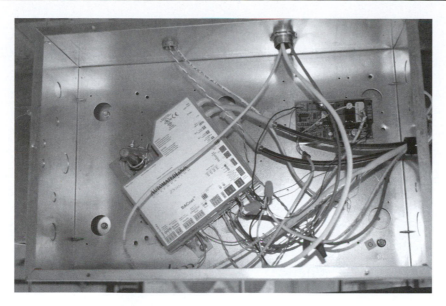

FIGURE 8.8 DDC-controlled VAV box

Source: Photo courtesy of Waukesha County Technical College.

Ventilation Reset of Outside Air

In this control scheme, information on the amount of airflow delivered to every zone is sent to the building automation system (BAS) controller. The BAS has the design information about the specified airflow requirement for each zone stored in its memory. The BAS can then calculate the amount of ventilation air required. The BAS also monitors the actual amount of outside ventilation air coming into the building. With these two pieces of information (outside air and zone ventilation requirements), the BAS can calculate the required airflow and then reset the outside air damper to ensure that all spaces are properly ventilated.

Supply Pressure Fan Optimization

The first step in supply pressure optimization is installing DDC-controlled VAV terminal units that can provide damper position information to the BAS. The BAS system continuously searches all the VAV boxes in order to discover which VAV box is open the widest. Duct static pressure is reset until the box requiring the highest inlet pressure is almost completely wide open. The speed of the main supply fan is adjusted by its variable frequency drive to vary the output CFM of air to meet the demand. This control method can save a significant amount of fan energy.

Demand Controlled Ventilation

The amount of fresh air supplied to a building is measured in cubic feet per minute (CFM) or liters per second (L/s). The figure is based on space use, amount of occupancy, and square footage formula developed by ASHRAE. Proper ventilation levels can be achieved through the use of demand-controlled ventilation (DCV) using carbon dioxide (CO_2) levels. The maximum amount of CO_2 in commercial buildings, as required by ANSI/ASHRAE (62-1989), is 1,000 parts per million (ppm). However, CO_2 readings taken outside a building may range from 400 to 450 ppm, and in congested urban areas, readings they may be higher by an additional 50 to 100 ppm.

The source of most of the elevated CO concentrations in the air of commercial buildings is from people exhaling. So readings higher than the outdoor level that occur inside a building have a direct relationship to the number of people in the building. DCV systems also need to know the outside CO_2 reading so they can calculate the amount of outside air necessary to reduce the level of CO_2 in the inside air. These readings are done by either monitoring carbon dioxide level with a handheld meter for several days and using an average level for the calculation, or installing an outdoor CO_2 sensor (see Figure 8.9) for a direct reading. The individual amount of CO_2 that a person generates is largely a function of his or her activity level and physical health, so an average amount is used. CO_2 sensors can tell the building automation system to increase or decrease ventilation to meet the 1,000 ppm standard.

FIGURE 8.9 CO_2 sensor

Source: Photo courtesy of Waukesha County Technical College.

DCV's potential to save energy is greatest when occupancy levels vary greatly from the design level. For example, when an event is over and everyone leaves the auditorium, the DCV's sensor senses the drop in CO_2 levels and automatically adjusts the amount of outside air coming into the building to a lower level by closing the outside air-intake air dampers. Lowering the amount of cold outside air to be heated or hot outside air to be cooled saves energy. DCV is most effective for places where occupancy changes greatly from one time period to the next, for example in auditoriums, conference rooms, cafeterias, theaters, and gymnasiums. See Figure 8.10.

DCV doesn't provide the same value in areas where the number of persons and occupancy times are constant and well defined. The energy-saving potential is lower in these areas because of the cost to purchase, install, maintain, clean, and periodically calibrate the CO_2 sensor. In these locations, a simple occupancy sensor may be sufficient to detect occupancy. Another solution is simply to adjust the time schedule on the DDC computer to put the space into occupied or unoccupied mode and thus match the actual occupancy condition. This solution may have the greatest cost benefit overall for spaces with reliable, fairly constant occupancy loads and times.

Combined CO_2-Based DCV and Ventilation Reset

Combined CO_2-based DCV and ventilation reset seeks to limit the high costs of CO_2 sensors in every zone. Cost reduction is done by not including those zones that are lightly populated and whose population does not change very much over the course of the day. Even if a CO_2 sensor is present in the zone, it would likely tell the system to provide the designed ventilation rate for most of the day. CO_2 sensors are then placed only in areas where occupancy is high and could change rapidly. The capability to provide ventilation reset is present in all zones of the building.

Free Cooling

When the outside air is cool and dry, building mechanical cooling systems can be shut down or operated at a low level. Comfort for the interior space is provided by opening the outside air dampers and drawing in fresh air, a process referred to as *free cooling*. A factor limiting the use of free cooling is the moisture content of the incoming air. There is a lot of heat energy in moist air, which is known as *enthalpy*. This energy must be removed by the cooling coil in order to drop the temperature of the air. Even without mechanical cooling, air that is too moist can promote mold growth and is uncomfortable for building occupants. An enthalpy controller is added to the system that closes the main dampers when the air is too moist. Enthalpy control is frequently found on air economizer dampers for rooftop air-conditioning units. See Figure 8.11.

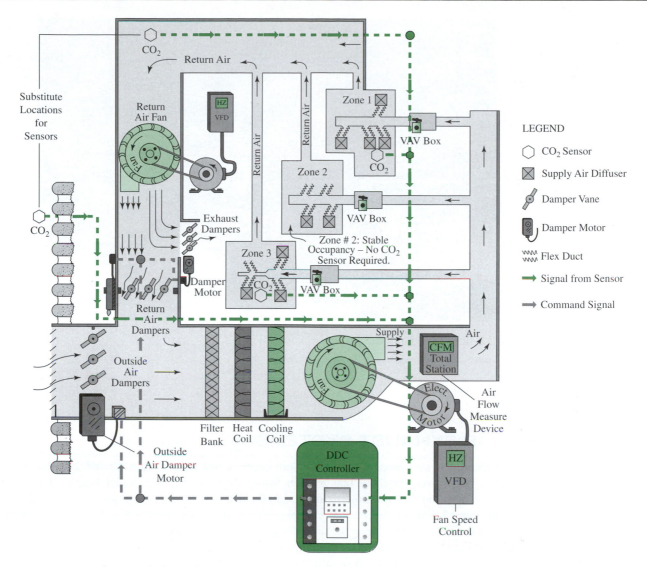

FIGURE 8.10 Demand-controlled ventilation based on CO_2

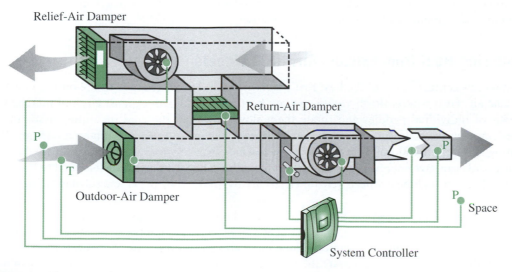

FIGURE 8.11 Air economizer system
Source: Material courtesy of Trane.

FIGURE 8.12 Rooftop exhaust fans

Source: Photo courtesy of Oconomowoc Memorial Hospital.

8.9   AIR EXHAUST SYSTEMS

Just as fresh outside air is needed to dilute indoor contaminants, there is also a need to eliminate contaminants at the source. Processes that generate fumes, dust, smoke, or excessive heat are targeted for removal by exhausting the fumes and/or air around them before they can contaminate the building ventilation air. Fume hoods are devices that sit above the source of contamination and pull a negative pressure around the source, discharging the contaminated air to the outside.

When exhaust equipment (for an example, see Figure 8.12) pulls out more air than can be replaced by base building systems, a makeup air unit (MAU) is installed. The MAU supplies air and it has the ability to heat the air prior to delivering it to the space. This heating prevents people from being blasted by cold replacement air when the outside air temperature is low. Conversely, some MAUs can cool the air if it is too hot with a vapor compression refrigeration system, or they can boost system efficiency by using a geothermal heat pump with a water-based condenser.

Capturing Heat from Exhaust Air: Heat Wheels

Heat wheels capture some of the heat that otherwise would be discharged to the outdoors with the exhaust air. Heat wheels slowly rotate, pick up heat from the warm exhaust air, and thereby warm dozens of individual paddles and rotate them slowly through the cold incoming airstream. The wheels are placed halfway in the exhaust airstream and halfway in the intake airstream. When heating is required and building exhaust air is warmer than incoming outside air, heat wheels (see Figure 8.13) can save energy by absorbing heat from the exhaust air and transferring it to the incoming air. Even though heat wheels rely on electricity to rotate, they provide a substantial net gain in energy.

A makeup air unit (MAU) with a heat wheel, also known as an energy recovery ventilator (see Figure 8.14), is not only used to bring in fresh outside air when large volumes of air are exhausted from a space to remove excess heat or chemical fumes, but it also recaptures some of the heat contained in the exhaust air. The heat wheel preheats the incoming air, reducing the amount of supplemental heat required to bring the air to an acceptable supply temperature.

FIGURE 8.13 Heat wheel
Source: Courtesy of Greenheck Fan.

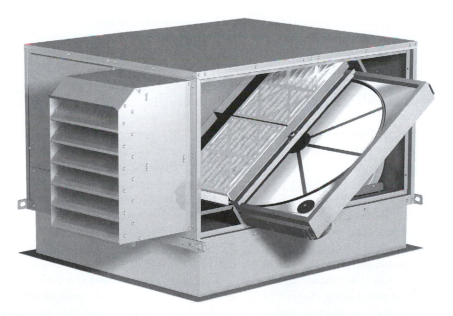

FIGURE 8.14 Energy recovery ventilator with heat wheel and integral water-based heat pump
Source: Courtesy of Greenheck Fan.

8.10 WATER SYSTEMS

Water has the ability to move energy into or out of a building space (heat transfer). Water systems convey either hot or cold water to the heat transfer coils inside terminal units located in or near the space to be conditioned. The water terminal unit may be a fan coil unit located in a distribution air duct, a baseboard heating convector, or a unit ventilator. A major advantage of this system is that relatively small water pipes can replace large metal ventilation ducts and still achieve the same amount of heating or cooling. A ⅜-inch-diameter tube supplying cooling water has the same capacity to cool as does an 8-inch-diameter duct carrying air.

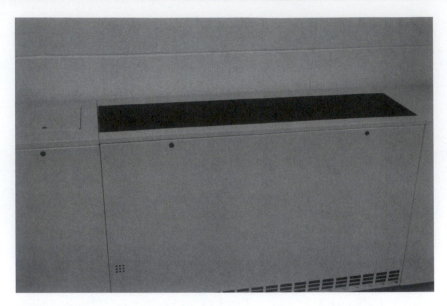

FIGURE 8.15 Unit ventilator

Source: Photo courtesy of Waukesha County Technical College.

Unit Ventilators

Unit ventilators (see Figure 8.15) are terminal units mounted on the exterior wall of a building. The vast majority of these units use water as the transfer medium inside the heating and/or cooling coils of the unit. Air flows over the coils typically by an internal supply fan run at low, medium, or high speed. Unit ventilators may have one set of coils for heating and another for cooling. Automatic water-flow-regulating valves control the amount of heating or cooling to be provided based on space temperature. Attached to the unit ventilator is a metal sleeve that goes through the wall to the outside. This opening allows the unit ventilator's fan to drawn in fresh outside air and to exhaust a proportional amount of room air.

Most new units are DDC and require little monitoring once they are programmed. These units have electronic damper motors that open dampers to provide the proper amount of outside air. Dampers also close when the space is unoccupied, which prevents freezing of the water coils when outside air temperatures fall below freezing. If space cooling is provided, drain piping must also be included to remove water that accumulates in the condensate pan. Some units employ a heat wheel for heat recovery from exhaust air. This equipment is frequently used in K–12 educational facilities because of its ability to provide fresh outside air without the need for opening windows.

To improve their electric and operational efficiency, unit ventilators may be equipped with electronically commutated motors (ECMs). ECMs can be used to drive the fan (see Figure 8.16). The ECM is a direct current (DC) motor with a permanent magnet rotor that has a built-in inverter (see Figure 8.17) so that alternating current (AC) can be supplied to it. ECM motors are significantly more efficient than the typical capacitor start fractional horsepower motors that have been installed in the past. Unit ventilators may also have a desiccant heat wheel in the airstream that removes heat and moisture from the incoming outside airstream and exhausts it to the outside in the summer. In the winter, the heat wheel uses waste heat from the exhaust air to help heat the incoming outside air in the winter.

Chilled Beam Systems

To bring the cooling equipment into the individual spaces inside the building, additional ventilation air, which normally would be required to condition the space temperature, can be eliminated. The direct result is that ventilation air duct sizes can be reduced as much as 50 percent. A 1-inch water pipe can provide the same cooling or heating as can be provided by an 18-inch-square air duct. This reduction in air duct size can result in lower ceiling heights, thereby allowing more floors in a shorter building, which in turn allows an increase in available space. There are two types of chilled beam systems: active and passive.

FIGURE 8.16 Unit ventilator with an electronically commutated motor (ECM)

Source: Material courtesy of Trane.

FIGURE 8.17 An engine control board for an electronically commutated motor (ECM) provides solid-state commutation for motor control.

Source: Material courtesy of Trane.

Active Chilled Beam Systems   Active chilled beam systems (see Figure 8.18) consist of copper tubes with attached aluminum fins mounted in a rectangular box, which in turn is mounted in the ceiling. When used for air conditioning chilled water flows through the copper fin tubes from the building's chiller plant. Ventilation air is fed to the box by a duct from the building's primary air system. The air is directed to a series of nozzles. As the air streams out of the nozzles, warm air is pulled into the coils,

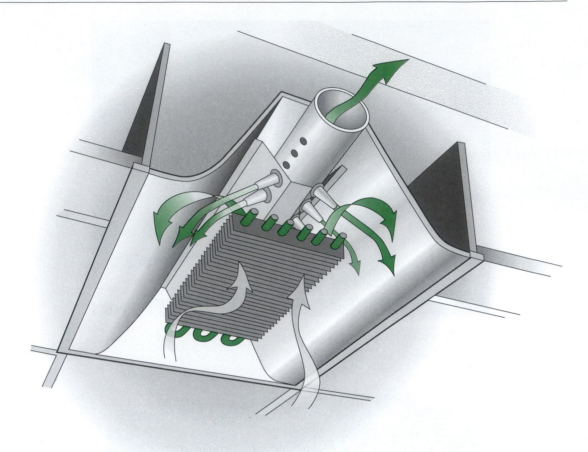

FIGURE 8.18 Active chilled beam

Source: Courtesy of USEPA—Labs for the 21st Century.

where it is cooled. The space air, which has been cooled, mixes with the ventilation air and is blown into the space, thereby providing a thorough mixing of supply and return air. Active chilled beams may also employ a second set of coils for heating the space.

The cooling capacity of chilled beams is limited by condensation that forms on the fin tube surfaces and drips into the space. Therefore, the temperature of the chilled water flowing through the tubes in the chilled beam must be above the dew point. Water temperature is generally somewhere between 55 degrees and 63 degrees Fahrenheit. The air supplied to the chilled beam must also be above the dew point temperature. As a result, chilled beams remove only sensible heat from the space. *Sensible heat* is heat that raises the temperature of the air and can be measured with a thermometer. This provides occupants with a comfortable space temperature. But more must be done to control humidity for overall comfort. Ventilation air is supplied through a dedicated outdoor air system (DOAS). Because the DOAS is providing only ventilation air, it can be much smaller than a system that provides air for a typical VAV system that must provide both outside ventilation air and return air.

Dehumidification in chilled beam systems can be handled in several ways. Some systems employ desiccant wheels in the primary air system to absorb the moisture in the air. Other systems remove latent heat from the moisture in the air. *Latent heat* is the heat required to change the state of a substance (from a solid to a liquid, or a liquid to gas) but does not change its temperature. Removing latent heat does the reverse and allows moisture to condense out of the air thereby dropping its relative humidity. These systems employ a technique referred to as a *runaround coil* (see Figure 8.19). The DOAS air handler has two coils: a runaround coil and a dehumidification coil. The dehumidification coil may be a DX coil connected to a refrigerant compressor. The DX coil is completely sandwiched between the two halves of the chilled water coil. The warm outside air blows on half of the chilled water coil first, and as the water flows through it, the water inside the coil picks up the heat. The air then flows into the DX coil where it is subcooled. Subcooling occurs when the air is cooled a few degrees lower that what is required to remove moisture and latent heat. This ensures effective moisture removal. But this cold air could cause moisture to condense on the chilled beam and drip into the space. To correct this, the warm water exiting the first coil section *runs around* the DX coil and into the second half of the coil,

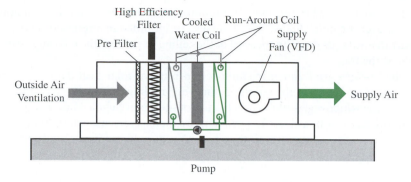

FIGURE 8.19 Runaround coil

Source: Courtesy of USEPA—Labs for the 21st Century.

thereby reheating the subcooled air and providing the required chilled water temperature for the chilled beam to cool the space and prevent water condensation on the surface of the chilled beam.

Active chilled beams may also be used in underfloor ventilation systems, known as ***underfloor air distribution (UFAD)***, which utilize the underfloor space as an air supply plenum. The air supply sends ventilation air through the active chilled beam, cools or heats the air, and directs it out through a rectangular grill in the raised floor. When located along the building perimeter, under the windows in the floor, the chilled beam can take in cold air falling from the exterior glass, heat it, and redistribute it into the space. This prevents cold air falling off the windows from migrating along the floor and chilling the feet of the occupants.

Passive Chilled Beam Systems   Passive chilled beam systems (see Figure 8.20) are not connected to the building's ventilation system. They are constructed basically the same as active chilled beams but they lack heating coils or air nozzles. They rely on the natural circulation of convection currents of air. As cool air flows out through the center of the beam, warm air flows up the sides of the beam and in

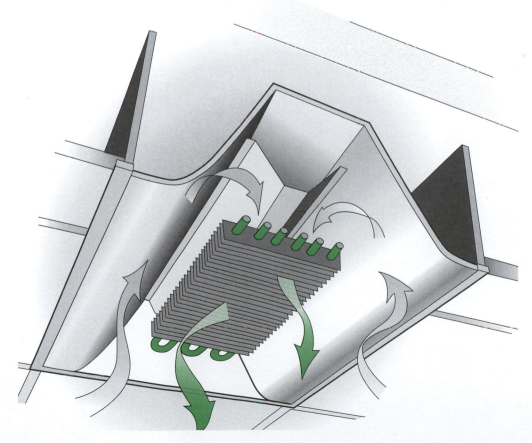

FIGURE 8.20 Passive chilled beam

Source: Courtesy of USEPA—Labs for the 21st Century.

through the top. Passive chilled beams must be mounted several inches from the ceiling to allow warm air to flow up and over the unit. Chilled beams can also be mounted in a recessed ceiling enclosure with a separation skirt that extends down to the level of the ceiling to prevent the cold airstream from returning to the intake at the top.

Passive chilled beams are frequently employed to remove solar heat gain in exterior zones. By mounting passive chilled beams in the ceiling around the building's perimeter and above the windows, warm air rising off the window glass flows into the top of the passive chilled beam and travels as cool air out the bottom. Although costing about 15 percent more than a conventional system, chilled beams can provide a payback in energy savings over time and are a unique alternative to VAV systems.

REVIEW QUESTIONS

1. What is an open protocol?

2. Why are unit ventilators sometimes used in classrooms?

3. Demand control ventilation is based on being able to sense the concentration of _____ in an interior space.

4. Three types of temperature sensors are _____, _____, and _____.

5. Relative humidity can be measured using _____.

6. Free cooling is limited by the _____ of the incoming air.

7. What are the main components of a DDC system?

8. The main components of a comfortable and healthy building environment are?

Essay Question

9. Name and describe three different ways to improve the quality of air supplied to the building interior.

10. Discuss the possible advantages of chilled beam systems over conventional main building cooling systems that use cooling coils in a main air handler.

MAJOR BUILDING EQUIPMENT SYSTEMS AND SUBSYSTEMS

INSIDE THIS CHAPTER

In the Face of Changing Technology, Can Physical Plants Ever Be Considered Sustainable?

This chapter reviews basic vapor compression and the absorptive refrigeration process for air conditioning. The chapter discusses commercial air-conditioning equipment consisting of electric centrifugal chillers, steam absorption units, and heat recovery chillers. It then goes into chiller operational strategies, ice storage systems, and cooling towers, including their operation, maintenance, and water treatment. The chapter moves into the topic of building heating, with a discussion of boilers, combustion, and efficiency basics. Heating plant operation and licensing is the next topic, followed by boiler water treatment and testing. Condensing boilers (whose increased efficiency has made them extremely popular) and the need for proper design to maximize their efficiency are discussed. Plant sustainability is an extremely important topic for achieving an overall sustainable facility, and this topic ends the chapter.

9.1 REFRIGERANT COOLING BASICS

The basic vapor compression refrigeration cycle uses a refrigerant gas as the cooling medium. The cooling gas works directly, as is the case in a direct expansion system, or indirectly, by chilling water in the evaporator section of a chiller. In the vapor compression cycle, a compressor pumps the refrigerant through the system. A condenser utilizing either air or water for refrigerant cooling changes the refrigerant from a hot gas to a liquid. Then the liquid refrigerant enters an expansion device that causes a drop in temperature of the refrigerant; in the process, it converts some of the refrigerant back to a gas. The mix of liquid and gas enters the evaporator section, which removes the heat from the building air by circulating water through a cooling coil in the air handler, and the air handler supplies the building with ventilation air. As the air passes through the hundreds of fins on the coil, the chilled water flowing through the inside of the coil becomes warmer, and the building ventilation air becomes cooler. This now warmer "chilled water" goes back to the evaporator, where it heats the refrigerant, turning it to a gas; in the process, the cold refrigerant once again chills the water. The refrigerant gas that collects in the evaporator goes into the suction of the compressor, and the refrigeration process repeats.

Compressors

Several types of compressors are in use today: reciprocating, screw, scroll, and centrifugal. Compressors raise the temperature and pressure of the refrigerant gas. Reciprocating, screw, and scroll compressors directly push the molecules of refrigerant gas together. The friction of molecules rubbing together generates heat as the pressure is increased. Inside centrifugal compressors, the refrigerant gas is spun by a centrifugal pump that throws the gas outward at a high velocity, giving the gas a high kinetic energy. The kinetic energy is converted to pressure as the gas travels through diffuser passages that increase in diameter. This slows the gas down and increases its pressure. The gas, now increasing in area (which further increases its pressure), then enters a volute chamber. The increasing pressure and slowing of the gas converts the kinetic energy to potential energy. As pressure increases, so does molecular friction between gas molecules, which increases gas temperature.

Condensers

The heat of compression must be rejected so the refrigerant gas will be able to absorb heat from the building. Heat is rejected in a condenser. The condenser rejects enough heat to convert the hot high temperature refrigerant gas into a subcooled liquid. *Subcooled* means that the refrigerant liquid is a few degrees lower than the condensing temperature.

Air-Cooled Condensers
The simplest type of condenser is the air-cooled type. In this type, cooling coils have outside air blown over them. The air flows over the fins of the cooling coil, taking away the heat, cooling the gas, and changing it to a liquid. Typical home central air conditioners have a cylindrical condenser mounted outside.

Water-Cooled Condensers
Water-cooled condensers require a source of cool water to drop the temperature of the refrigerant gas in the condenser. The most common way to obtain cool water is through the use of a cooling tower. Water-cooled condensers greatly improve the efficiency of alternating current (AC) systems, with efficiencies approaching 0.55 kilowatts per ton (kW/ton).

There are two types of water-cooled condensers. One type relies on city water as the source of cold water to cool the condenser. City water arrives at a temperature of 55 to 60 degrees Fahrenheit. These water-cooled condensers are generally small tonnage systems and are used mostly when specific equipment in a space requires cooling around the clock. Such a requirement arises, for example, when a computer data center is added to a building after it is built. The central plant for the building shuts down at 6:00 p.m. and, without cooling, the computer equipment overheats. If the data center is located in the middle of the building, cooling equipment generally cannot be mounted on the roof and with piping run down through numerous tenant spaces to reach the data center. Attempting to do so may cause disruptions to other tenants and be prohibitively expensive. One choice for cooling a space that doesn't have access to the building exterior is to use municipal water to cool the condenser. But such systems are costly to operate because they incur charges for both water and sewerage. Instead, most buildings utilize cooling towers to provide cool condenser water for the main building air-conditioning equipment.

Evaporator and Expansion Device

Expansion devices may be orifice plates, nozzles, or thermal expansion valves. Expansion devices convert 25 percent of the liquid refrigerant to a gas, thus dropping its temperature as it enters the evaporator. The evaporator is a heat exchanger that cools the building supply air as it flows over its coils. As the building air is cooled, it gives up its heat to the refrigerant, changing the refrigerant back to a gas. Refrigerant must enter the compressor as a gas or serious damage could occur.

9.2 WATER-COOLED ELECTRIC CENTRIFUGAL CHILLER SYSTEMS

Most large buildings employ electric centrifugal chiller plants to provide air conditioning (see Figure 9.1). Centrifugal chillers use the vapor compression refrigeration cycle and take over where air-cooled equipment leaves off. Units can range into the thousands of tons of cooling capacity. Package units are available up to about 3,000 tons, and field-erected units can be three times that size. Other types of chiller compressors are used to meet the demands of smaller loads. These compressor configurations are reciprocating, helical-rotary (also known as a screw compressor), and scroll.

Centrifugal chillers don't have to use water to cool the hot refrigerant gas in the condenser. They can use fan-cooled coils located outside, just like residential air-conditioning units. But water cooling is much more efficient than air cooling. It achieves an efficiency increase through lower condenser temperatures

FIGURE 9.1 Electric centrifugal chiller

Source: Photo courtesy of Oconomowoc Memorial Hospital.

and pressures. By using cooling towers to supply water that is below ambient air temperature, efficiencies as low as 0.55 kilowatts per ton have been achieved.

Electric centrifugal compressor chillers today employ a combination of vortex vane dampers (which have been used for decades on centrifugal chillers) and variable frequency drives to control the machine's capacity. Vortex vane dampers control refrigerant flow and preswirl the refrigerant as it enters the suction eye of the impeller. The compressor motor can be slowed down or speeded up by using a variable frequency drive. The compressor should have to work only as hard as necessary to meet the demand for cooling. Building engineers today can easily view all pertinent data (pressure, temperature, hours of operation, alarm conditions, and run status) from a computer screen mounted on the chiller (see Figure 9.2).

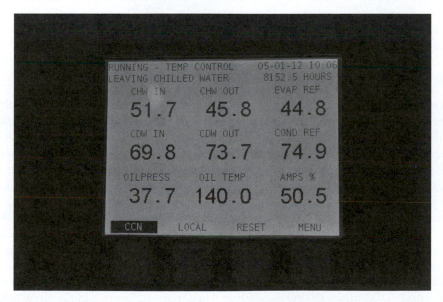

FIGURE 9.2 Centrifugal chiller control screen

Source: Photo courtesy of Oconomowoc Memorial Hospital.

9.3 CHILLER ENERGY STRATEGIES

Several well-known chiller strategies are available for efficient chiller operation. The older strategies are using a swing chiller, operating with a modified strainer cycle, changing the load on a chiller through chilled water reset, and allowing a natural circulation process rather than a mechanical compressor to move refrigerant through the system via refrigerant migration. The newer strategies are adding ice storage systems to make and store cooling, and using a heat recovery chiller that recovers heat that would otherwise be wasted and simply ejected into the surrounding atmosphere.

Swing Chiller

The name *swing chiller* comes from the practice of using new employees to work the *swing shift* at a factory. For chiller plants, when a large chiller is employed to meet the building's air-conditioning load, an additional smaller-size chiller is used to meet partial load conditions. Centrifugal chillers generally operate most efficiently when they are producing chilled water at 80 to 100 percent of their capacity.

Example: A 100-Ton Swing Chiller

A 100-ton chiller is used as a swing chiller (see Figure 9.3) at a plant where the main chiller is 320 tons. If the swing chiller is operating at 80 percent capacity, it is operating at its highest efficiency. If the 320-ton chiller operates at 80 tons, it is operating at only 25 percent of capacity and its efficiency is considerably lower. If the swing chiller at the correct size for handling the lower load conditions, for example, in the spring and fall, it can make a big difference in the amount of energy consumed.

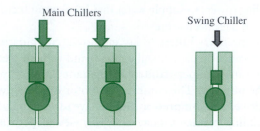

FIGURE 9.3 Swing chiller arrangement

Swing chillers operate on a daily basis as follows. The swing chillers start up first and run up to full capacity. Then the main chiller starts and begins to take the load. The swing chiller quickly backs off, giving the main chiller as much load as possible. The swing chiller then shuts down. Should the load actually exceed the main chiller's capacity, the swing chiller will start up again to take the additional load.

Modified Strainer Cycle or Waterside Economizer

A strainer cycle is used when the cooling tower water temperatures are below the required chilled water temperature necessary to cool the building. Tower water is used to cool the building air directly. A disadvantage of this approach is that the tower water is dirty; even though it goes through a strainer (hence the term *strainer cycle*) it is still not as clean as the water in a closed loop. In a modified strainer cycle (see Figure 9.4), the tower water is used to cool the chilled water through a plate and frame heat exchanger. The centrifugal chiller is completely shut off. The plate and frame heat exchanger separates the relatively dirty tower water from the clean, treated chilled water in the cooling coils.

One detail to consider before using a strainer cycle is the ambient air temperature. If the ambient temperature is low enough to cool the tower water, it is often cool enough to be used directly, and the building is cooled by simply taking in more outside air. In fact, if ambient air is sufficiently cool, the building control system may opt to shut down the chillers completely and use cool outside air directly to cool the building without operator input. This process is called ***free cooling***. A restriction on free cooling is when the outside air contains too much moisture. High moisture content in the air, which can cause mold growth in the ducts, also means that the air has a high enthalpy (energy content) due to the latent heat contained in the water vapor in the air. Under these conditions, using the strainer cycle for cooling the building has a definite advantage by providing only enough moist outdoor air for ventilation, and not the massive amount of air required for building interior cooling.

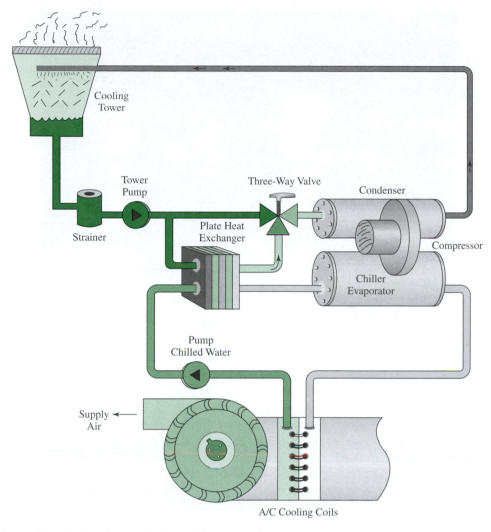

FIGURE 9.4 Modified strainer cycle/waterside economizer

Chilled Water Reset

In constant flow chiller evaporator systems, the chilled water temperature set-point is raised during periods of high load. This reduces the load on the chiller, thus saving energy, and might be necessary in systems without sufficient capacity to keep the chiller from overloading. There are limits on how high the set-point can be raised: Raising the set-point reduces the amount of moisture that can be removed from the airstream by the fan's cooling coil.

In variable flow systems, chilled water reset might not achieve the desired result of saving energy because the chiller's coefficient of performance (COP) is about ten times greater than the COP of the chilled water pump. COP is a dimensionless efficiency rating that is the ratio of cooling effect expressed in British thermal units (BTU) per hour (BTU/hr) divided by the energy input in BTUs. For example, a chiller that consumes 0.55 kilowatt per ton has a COP of 6.4. At higher temperatures, the chilled water pump has to pump more, and this increase in pumping power may cancel out any gain achieved by resetting the chilled water temperature.

Refrigerant Migration

The chiller system, valves, and piping must be designed specifically for refrigerant migration. Refrigerant migration is achieved when tower water is colder than the desired chilled water temperature. The compressor is turned off, and refrigerant vapor naturally migrates to the condenser from the evaporator as building water warms the liquid refrigerant, turning it to a gas. The gas then enters the condenser, where it condenses to a subcooled liquid. It is claimed that 40 percent of chiller capacity can be achieved using refrigerant migration and without turning on the compressor.

FIGURE 9.5 Calmac ice storage tanks

Source: Material courtesy of Trane.

Ice Storage Systems

Ice storage systems do not save energy directly, but they improve the efficiency of the power utility's delivery of power. The systems work by making ice for storage (see Figure 9.5) during off-peak demand power periods when electricity is cheaper. The ice is then used the following day to provide cooling for the building. Depending on the amount of ice stored, the system (see Figure 9.6) might need to be supplemented with a chiller. In one system design, the chiller tonnage can be reduced by

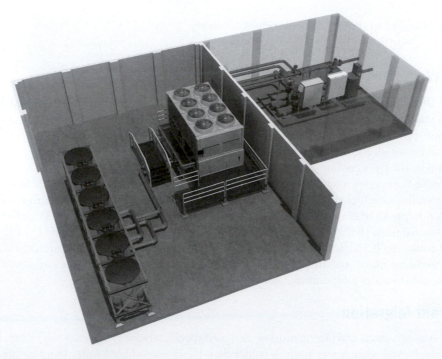

FIGURE 9.6 Ice storage system

Source: Material courtesy of Trane.

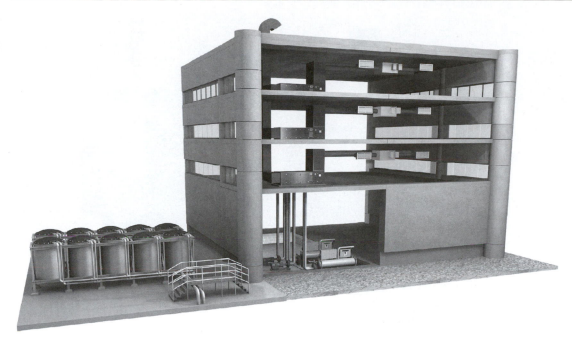

FIGURE 9.7 Ice storage system elevation

Source: Material courtesy of Trane.

40 to 50 percent. But this reduction means that, during the evening hours, the chiller must produce enough ice to cover 50 percent of the daytime load. When making ice, the chiller must operate at a chilled glycol/water temperature below the freezing point of water, or about 25 degrees Fahrenheit. During ice making, chiller efficiency suffers because the chiller has to operate at such low temperatures. But chiller efficiency recovers when the chiller is operated during the discharge cycle the next day. During the day, the chiller is required to make chilled water at only 52 degrees Fahrenheit because the temperature is further reduced to 44 degrees Fahrenheit by the ice. The advantages are that a smaller chiller can be used, and electric utility costs are lowered by shifting load to off-peak periods. Figure 9.7 shows how an entire office building's heating, ventilation, and air-conditioning (HVAC) system can incorporate ice storage with the flexibility, efficiency, and reliability of a plant with two centrifugal chillers.

Heat Recovery Chiller

Chillers require a place to reject the heat of refrigerant compression. Chillers either reject the heat through a cooling tower, transferring the heat to a ground source, or use a river or other natural water source for cooling.

Heat recovery chillers (see Figure 9.8) can be used capture this heat and use it to preheat domestic hot water or to supply heat to air-reheating coils in the air ducts. Heat can be recovered by installing a second condenser on the chiller. There are other methods of recovering condenser heat, but this method actually improves chiller efficiency by having a greater condenser surface area. Newer building codes, which normally restrict ventilation reheat, may allow this recovered heat to be used to supply hot water for reheat coils in ventilation supply ducts.

A heat recovery chiller may also be a small additional chiller that runs all the time, providing cooling for the building and hot water for a variety of uses. The system works well when the building has a constant demand for both hot water and air conditioning, which is frequently the case for a hospital. These units achieve a comparatively low kW/ton of cooling level (approximately 0.38 kW/ton integrated part load value (IPLV), because much less energy is lost to the atmosphere. Some units employ magnetic bearings in their compressors, which eliminates the need for compressor oil because they operate without bearing friction.

FIGURE 9.8 Heat recovery chiller

Source: Material courtesy of Trane.

9.4   COOLING TOWER BASICS

Towers are rated in tons, similar to ratings for air-conditioning units. The tonnage of the cooling tower must be greater than the tonnage of the chiller because a cooling tower removes both internal building heat and the heat of refrigerant compression. The tower is sized for a tonnage that is 25 to 50 percent greater than the chiller it serves. Cooling towers use the principle of evaporative cooling. When water evaporates, it creates a cooling effect that reduces the temperature of the remaining water. One pound of water evaporated releases 1,000 BTUs.

Evaporation is most effective when the surface area is increased; that is, the water is spread out into a thin film. Cooling towers (see Figure 9.9) have several nozzles that spray water on a material called tower fill. The ***tower fill*** spreads out the water and increases its surface area. A fan(s) is employed to blow air or pull air across the wet fill material. Some of the water evaporates, but the remaining cool water falls to the bottom of the tower. This process reduces the temperature of the water coming into the tower about 10 to 15 degrees Fahrenheit. The area in the bottom of the tower that collects the water is known as the tower sump. Drift eliminators cause the vapor leaving the tower to change direction, which causes some water droplets to fall back into the tower. This saves some of the water, which otherwise would be lost (about 3 percent is lost due to drift). Makeup water must be added to the tower sump to compensate for the water lost through evaporation. A tower evaporates about 12 gallons of water per minute for every 500 tons of cooling tower capacity. An electronic or simple ball float valve is used to keep the sump at operating level by adding makeup water (city water).

Fan operation is very important to tower capacity and the efficiency of the equipment the tower serves. Towers employ a variety of capacity control methods. Two or three speed fan motors, controllable pitch fan blades, or variable frequency drives are used to control fan output. Towers may also use variable frequency drives on the tower pumps to simultaneously reduce the pumping energy when lighter loads dictate that less tower capacity is needed.

A much overlooked area is improving tower efficiency to boost chiller plant efficiency. Old towers may have poorly designed spray nozzles and antiquated fill media. Managers of older chiller plants might consider tower rebuilding as a possible method to increase cooling capacity of the chiller plant.

Cooling Tower Water Treatment

Incidents of improper cooling tower maintenance and treatment can result in serious illness and even death. Different types of towers, the water/chemical contact time, changing water quality, or the prevailing conditions in various geographic areas are factors that influence which method works best. Making appropriate equipment and chemical choices should be a detailed and informed decision involving

FIGURE 9.9 Cooling tower on the roof of a physical plant building

Source: Photo courtesy of Oconomowoc Memorial Hospital.

consultation from industry experts, expert consultants, and appropriate government agencies, and through the benchmarking of best practices of successful cooling tower operators.

Chemicals are added to the tower to control microbial growth, convert scale to a nonadhesive sludge, and prevent corrosion. Chemicals that attack microbial growth fall into two categories: oxidizing chemicals (chlorine, chlorine dioxide, and bromine compounds) and biocides (chemicals that are toxic to bacteria). The goal of bacterial water treatment is to reduce the amount of colony-forming units per milliliter (CFU/ml) below industry standards.

Chlorine and Bromine   Chlorine and bromine have been used for many years. Both can be delivered in solid or liquid form. Chlorine reacts with water and forms hypochlorous acid and hydrochloric acid. Bromine reacts with water to form hypobromous acid. Chlorine reacts with ammonia compounds and amines to form chloramines. Chloramine is the chemical in swimming pool water that burns swimmers' eyes and has a potent chlorine smell. Bromines also react with ammonia compounds to form strong biocides called bromamines. Chlorine also reacts with some scale-reducing chemicals and corrosion inhibitors. Organic material can react with chlorine to form trihalomethanes, which at high levels are harmful to humans. The efficiency of chlorine drops as pH* increases above the neutral reading of 7.0. A pH of approximately 8.0 is necessary for tower corrosion control. Bromine works better at a higher pH. Both chlorine and bromine work well but produce harmful byproducts.

Chlorine Dioxide   Chlorine dioxide is actually a weaker oxidant than chlorine or hypochlorus acid on the oxidation-reduction potential (ORP) scale,[†] but it can transfer five electrons compared with a maximum of two for chlorine or hypochlorus acid. This electron transfer ability gives chlorine dioxide a greater oxidation/disinfection capacity. Being a weaker oxidant, chlorine dioxide reacts more selectively with other chemical compounds. In comparison to chlorine, chlorine dioxide does not chlorinate organics, react with ammonia, or oxidize bromide to bromate. Chlorine dioxide exists as a dissolved gas in water, and it is stable enough to build a measurable residual disinfectant level that remains for several hours. Chlorine dioxide (ClO_2) is largely unaffected by the pH of the water and works well with systems that require elevated pH levels, such as cooling towers.

Chlorine dioxide (ClO_2) doesn't react with most other water treatment chemicals and is more effective at controlling biofilm buildup and algae growth. Also chlorine dioxide won't degrade chemicals such as phosphonates used for scale control and triazole corrosion inhibitors. Chlorine dioxide has been

---

*pH is an acronym for "potential hydrogen." The pH scale ranges from 0 to 14, and a pH of 7 is neutral. A neutral solution has the same concentration of hydrogen ions (H^-) and hydroxyl ions (OH^-). An alkaline solution contains more hydroxyl ions; an acid solution contains more hydrogen ions.

[†]Oxidation occurs when a molecule loses an electron. Reduction occurs when a molecule gains an electron (becomes more negatively charged). A solution's ability to give up or accept electrons is its potential. The ORP scale ranges from −1,000 mV to +1,000 millivolts (mV). More information can be found at the Georg Fischer Signet LLC website.

found particularly effective against legionella bacteria. With the proper equipment, chlorine dioxide can be made in the mechanical spaces of many facilities.

Microbe Resistance   Microbes have the ability to develop a resistance to some methods of chemical attack. To counter this ability, cooling tower water is sampled and placed in a closed container with a food material for molds. If mold starts to grow immediately with vigor, it is time to switch the chemical used to combat microbe growth. Once the microbes are removed from the cooling tower, the operator can return to using the original chemical.

Scale Treatment   Cooling towers must also be treated for scale. When water evaporates, it leaves chemicals and soluble minerals behind. The scale-producing minerals are mostly calcium and magnesium carbonate. Evaporation builds up the concentration of these contaminants in the tower water, which is generally referred to as the hardness of the water. When the concentration of calcium and magnesium carbonate exceeds the saturation point, they precipitate out of the water solution and onto the cooling tower's internal surfaces, covering them with scale. Makeup water added to the system to replace the evaporated water also contains scale-causing minerals, and the concentration is raised even higher.

Hardness of Water   Towers generally operate at or below 350 parts per million (ppm) to 500 ppm of scale (hardness). The incoming water may contain about 100 ppm of hardness. Every cycle leaves an additional 100 ppm, so the tower is limited to 3.5 to 5 cycles of concentration before it must be blown down. Phosphonates, organic chemicals, and polymers are chemicals used to control scale. Acid may also be injected to reduce the hardness of the water, but acid is highly corrosive to tower metal. It also reduces the pH of the water, which can increase the chance of corrosion. A pH of 8.5 or above helps to reduce the corrosion of copper and galvanized steel.

Nonchemical Treatment

Nonchemical biocides and scale prevention methods are gaining in popularity because they are environmentally benign. Chemical treatment products and their byproducts are a source of pollution when they are introduced to the environment. Several nonchemical methods that have been used with varying outcomes and degrees of success include the following treatments:

- Water softening
- Ozone
- Ultraviolet (UV) light
- Electric pulse and electrostatic
- Hydrodynamic cavitation
- Ultrasonic
- Magnetic

Water Softening   Before the water enters the cooling tower, it flows through a zeolite bed inside a softener. The zeolite gets coated with minerals from the water, which greatly reduces the hardness of the water and the need for additional scale chemical treatment. Once the zeolite has been coated, it is recharged by flushing a salt solution through the bed. One disadvantage of this water-softening method is that, although the treatment of the water stream is nonchemical, regeneration with a salt solution introduces salts into the wastewater stream.

Water softeners reduce the amount of scale-reducing chemicals needed. If the makeup water contains 100 ppm of hardness, cutting it in half to 50 ppm through softening yields ten cycles of concentration ($50 \times 10 = 500$ ppm), which keeps the hardness of the tower water to the upper 500 ppm limit. This cuts down the frequency of tower sump blowdown (the removal of a portion of the water from the bottom of the tower sump to reduce the concentration of dissolved minerals and sludge) in half, which saves water and chemicals.

Ozone   Ozone treatment utilizes ozone gas, which is harmful to humans if breathed in and is a powerful biocide. Ozone is frequently used to treat drinking water at municipal plants. Ozone (O_3) is an unstable gas with a half-life of just minutes. With such a short life, ozone must be produced on site, and it dissipates in water so quickly that it cannot be found in blowdown water. Using ozone thus avoids any restrictions on discharging water containing chemicals. Ozone introduces only a fraction of the dissolved solids that chemical treatment causes; therefore, towers using ozone can be operated at higher cycles of concentration. At higher cycles of concentration, the pH of tower water is raised, which helps to prevent corrosion. Ozone is made near the cooling tower via an ozone generator. According to the United States Environmental Protection Agency (U.S. EPA), ozone works best at water temperatures below 104 degrees Fahrenheit; above 110 degrees Fahrenheit, the solubility of ozone is effectively zero. This characteristic of ozone precludes its use in absorption refrigeration plants, which heat water to higher temperatures.

Ultraviolet Light   Ultraviolet light reduces the number of microbes in water by exposing them to electromagnetic radiation, which penetrates the cell membrane, damages their DNA, and makes them unable to reproduce. To be effective, UV light must be applied continuously to the water, and the water must be filtered. The water must also have a disinfectant added to it because UV has no residual effect and thus cannot reduce the level of bacteria colonizing on surfaces of piping and other materials not exposed to UV light. However, UV light decreases the amount of chemical disinfectants required, thereby increasing the sustainability of the treatment system overall.

Electric Pulse and Electrostatic   The electric pulse and electrostatic process operates with high-frequency electromagnetic energy that is pulsed through the water entering the cooling tower or the water entering the refrigeration condenser. The high-frequency pulse creates seed crystals for minerals such as scale-forming calcium carbonate to attach themselves. This prevents scale from plating out on cooling tower surfaces. The seed crystals then drop out of the solution and fall to the bottom of the cooling tower sump or on tower fill material as a powder-like material. The mineral power also encapsulates the bacteria floating in the water, preventing it from reproducing. The pulse system is similar to one of the cold pasteurization processes (electroporation) used to destroy bacteria in dairy products. The electrical pulse is strong enough to breech the cell membrane of bacteria. The short-lived bacteria (with a lifespan of only one or two days) must then spend time and energy to repair themselves rather than reproducing, which causes bacteria counts to fall.

Electric pulse systems also claim to operate at higher cycles of concentration, which means that water can remain in the system longer before the concentration of scale-producing minerals must be reduced by blowdown and thus decreases the amount of water used. Because no chemicals are added to systems using electric pulse, blowdown water may be utilized for a nonpotable use. The higher pH of the water in the cooling tower that is a result of fewer blowdowns aids in preventing corrosion.

Electrostatic treatment units are similar to electric pulse systems except that they use a static electric charge, and the charge is applied on a continuous basis rather than pulsed. An electrostatic field charges the minerals and keeps them in suspension so that more collisions between molecules occur. The increase in the number of collisions forms precipitates that can be removed easily through blowdown.

Hydrodynamic Cavitation   During the rapid high-pressure changes that occur as a result of hydrodynamic cavitation, micrometer vapor bubbles form and collapse rapidly, causing locally high temperatures and pressures. This process may cause an inactivation of surrounding organisms near the bubbles due to shock waves and high temperatures. High temperatures cause inversely heat soluble calcium carbonate ($CaCO_3$) to precipitate out of solution as a nonadhering colloidal crystal.

Ultrasonic   Ultrasonic energy results in cavitation, which in turn produces resonance and pressures. This ultrasonic energy can damage various parts of algae and bacteria, particularly cell walls and cell membranes. The ultrasonic waves can also prevent algae from settling on surfaces and creating a biofilm. Ultrasonic water treatment devices can reduce the amount of chemicals used.

Magnetic   In a magnetic treatment process, water is sent through a magnetic field. Magnetism alters the form of calcium carbonate ($CaCO_3$) and causes it to precipitate as suspended particles in the water. Precipitation prevents the calcium carbonate from adhering to heat transfer surfaces and forming scale. The precipitated particles are then removed by filtration. The effectiveness of magnetic systems relies on the velocity of the flow through the magnetic field and the strength of the magnetic field.

Cooling Tower Maintenance

Cooling tower maintenance is seasonal in cooler climates. The unit is opened during the heating season, spray-washed, cleaned, and inspected. In warmer climates, the cooling tower is shut down for

FIGURE 9.10 Sand filter, which removes suspended solids from a cooling tower
Source: Photo courtesy of Oconomowoc Memorial Hospital.

maintenance once or twice a year. Scale, rust, and other debris collecting in the sump should be removed. Strainers and filtering devices (see Figure 9.10) must also be cleaned. Floats and water-purging valves need to be inspected and cleaned. Fan belts must be checked and bearings greased. The cooling tower's steel housing should be inspected for corrosion; if the corrosion is excessive, the tower must be sandblasted and recoated.

The overall effectiveness of the various nonchemical treatments for cooling tower water has been the subject of a great deal of debate. Questions arise about whether treatment equipment should be combined with chemical treatment and/or other nonchemical methods to ensure effectiveness. A detailed research report on the topic from the American Society of Heating, Refrigerating, and Air Conditioning Engineers, ASHRAE Project No. 1361-RP, was submitted by the University of Pittsburgh, Department of Civil and Environmental Engineering (principal author Radisav D. Vidic, contributing authors Scott M. Duda and Janet E. Stout, "Biological Control in Cooling Water Systems Using Non-Chemical Treatment Devices," April 2010). The report can be obtained at the Special Pathogens Lab website.

9.5 ABSORPTION AIR-CONDITIONING UNITS

Absorption chillers do not have a machine-driven compressor like vapor compression air-conditioning refrigeration systems do. Absorption units (see Figure 9.11) utilize the absorption refrigeration cycle. Absorption units, like vapor compression units, use refrigerant and have a condenser, expansion device, and evaporator. Absorption units differ by replacing the compressor with a device called an absorber that holds the absorbent liquid. They also use a circulation pump and a device called a generator that acts like a nonmechanical compressor.

Lithium bromide is a nontoxic salt used as an absorbent. Lithium bromide has a high affinity for water, and water is the refrigerant in a lithium bromide absorption unit. From an environmental standpoint, lithium bromide absorption units are the most earth friendly. But water at normal atmospheric pressure is a terrible choice as a refrigerant. Water turns to a vapor at 212 degrees Fahrenheit, much too hot for air conditioning. Water in a vacuum is a different story. In an absorption unit, water vaporizes in a vacuum of 0.15 pounds per square inch absolute (psia) at a temperature of 45 degrees Fahrenheit. This temperature is cool enough for most chilled water needs. In the condenser section, the water is at a higher pressure of 1.5 psia and condenses at 115°F. In these conditions, the refrigerant vapor created by the generator can be condensed with a typical cooling tower.

The generator section acts like a compressor, raising the pressure and temperature of the refrigerant. But it does this in a way that is similar to how a boiler raises temperature and pressure: Heat is added via steam, hot water, or a direct-fired natural gas burner, raising the temperature and pressure of the refrigerant.

Absorption units are not nearly as efficient as air- or water-cooled vapor compression air-conditioning systems. Some units operate at around 3 kW/ton of refrigeration, whereas many water-cooled chillers

FIGURE 9.11 Steam absorption chiller

Source: Material courtesy of Trane.

are designed to operate at 0.5 kW/ton. Firing an absorption unit with a costly, nonrenewable energy source makes the unit unsustainable. Instead, absorption chillers can utilize a source of waste heat from a factory operation, heat that would otherwise have to be released to the atmosphere or a water source (river, lake, the ocean). Capturing this heat for air conditioning makes these chillers sustainable; using water instead of toxic chemicals as a refrigerant makes them earth friendly as well.

9.6 HEATING AND STEAM POWER SYSTEMS

Depending on the type of plant and state and/or local regulations, a person operating a plant may have to be licensed. The license is provided either directly from the state, city, other local jurisdiction, or through a recognized third-party licensing organization that provides testing and licensure. The FM wishing to obtain a boiler license must find out which license is acceptable for the jurisdiction. Two third-party licensing organizations are the National Institute for the Uniform Licensing of Power Engineers (NIULPE) and the American Society of Power Engineers (ASOPE). Information on these third-party licensing organizations can be found at their websites.

For someone who wants to become a director at a hospital or other healthcare facility, an upper-level boiler license at the first-class or chief-engineer level for high-pressure boilers may be a requirement for employment. Requirements for boiler licensure vary, but generally it is necessary to have a combination of experience and education prior to taking the examination. Typical boiler licensing levels are as follows:

- Chief engineer
- First-class engineer
- Second-class engineer
- Third-class engineer

Various licensing authorities often have different licensing levels based on plant size and type. Also the requirements placed on the plant and the license holder can vary for operations, maintenance, and inspection. Frequently, the lowest class of license (third class) is suitable for the operation of only low-pressure boiler heating plants producing steam at 15 pounds per square inch (psi) or less.

A facility manager may not be directly responsible for the day-to-day operation of a boiler plant. For example, a director of facilities may manage a large university campus that has a physical plant building and a plant manager. The ultimate responsibility for the safe and efficient operation of the entire facility lies with the director, and he or she needs to be able to oversee the important aspects of boiler plant operations, even if he or she has no day-to-day responsibility. Important boiler operational issues for boilers include the following:

- Air emission control regulations
- Boiler inspections
- Fuel efficiency
- Boiler deterioration and maintenance issues
- Boiler chemical treatment and water discharge permits
- Monitoring and staffing requirements
- Safe operation

Boiler Basics

The two major types of boilers are fire tube boilers (see Figure 9.12) and water tube boilers. The basic systems of any boiler are the draft system, steam or hot water system, feedwater system, combustion/fuel system, and boiler control system. The draft system provides the air required for combustion and moves the combustion gases through the boiler to achieve the necessary heat exchange between the flue gases and the boiler water. The steam or hot water system is all the piping, heat transfer equipment, valves, and safety devices necessary to provide heat where it is needed. The feedwater system provides water for the boiler. Water must be treated and, as is the case with high-pressure boilers, heated so that it does not damage the boiler. The boiler control system ties all the systems together and adjusts them for efficient operation.

The fuel system provides the method of delivering the fuel to the boiler in a condition whereby it can be burned efficiently. Liquid fuels such as fuel oil need to be pressurized, residual fuel oil number six needs to be heated to ensure proper viscosity, and both need to be atomized for efficient burning. Atomization is the spraying of compressed air or steam through small holes in a boiler nozzle to break the fuel up into thousands of tiny droplets to gain increased fuel surface area for more efficient combustion. Gas fuels such as natural or LP gas must be provided at the correct pressure, and solid fuels such as coal are frequently pulverized to a talcum powder–like consistency and sprayed into the boiler.

The combustion control system controls the efficiency of combustion. When fossil fuel burns completely, combustion products consisting of gases, water vapor, and ash are produced. On cool days, white clouds can be seen exiting a boiler stack. The white part of the cloud is water vapor. Unfortunately, combustion also emits other gases, which vary depending on the fuel used, fuel quality, and the combustion efficiency. A light, yellowish brown haze may also be seen being emitted into the atmosphere. The majority of this haze is nitrogen dioxide gas, which is mixed with other emission products and reacts with UV light to produce the yellowish brown color. The more common name for this haze is smog. Boilers with

FIGURE 9.12 Fire tube boiler in a hospital's heating plant

Source: Photo courtesy of Oconomowoc Memorial Hospital.

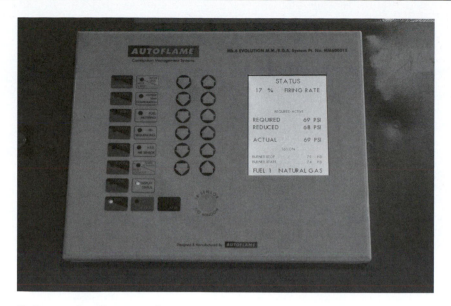

FIGURE 9.13 Boiler combustion control

Source: Photo courtesy of Oconomowoc Memorial Hospital.

burners and combustion control systems (see Figure 9.13) that produce low levels of nitrogen oxides (NO_X) and can greatly limit these emissions while improving boiler efficiency.

When fuel burns incompletely, black smoke and soot are produced, a more troubling situation. Soot coats boiler heating surfaces and effectively insulates them from the boiler water. Because heat is restricted by the soot from heating the water in the boiler, the heat leaves the boiler by traveling out the stack. This situation means that more fuel must be burned to get the same results as would be achieved with clean boiler heating surfaces. The boiler's combustion control system provides the correct amount of fuel and air (draft) to provide clean, efficient operation and avoid incomplete combustion.

The U.S. EPA is required by the Clean Air Act Amendment of 1990 to regulate air emissions of hazardous air pollutants (HAPs) using a technology base standard. The National Emissions Standards for Hazardous Pollutants (NESHAP) and the Maximum Achievable Control Technology (MACT) are the two standards used. NESHAP deals with hazards from existing and new sources; MACT requires facilities to meet emissions levels based on the best-performing facility of a similar type. The U.S. EPA is also required by the Clean Air Act to provide the National Ambient Air Quality Standard (NAAQS) for six criteria pollutants: carbon monoxide (CO), lead (Pb), nitrogen dioxide (NO_2), ozone (O_3), particulate matter (PM), and sulfur dioxide (SO_2). These pollutants are of particular importance because all six are produced by boiler operation. Plant managers need to be aware of current air and water pollution regulations.

Boiler Efficiency

Annual fuel utilization efficiency (AFUE) is a measurement of the amount of fuel energy that is converted to heat. When using natural gas as a fuel, therms are use to describe the heat content of the gas. One therm is equivalent to 100,000 BTU of energy. If one therm of energy is consumed in a natural gas boiler that has an AFUE of 80 percent, 80,000 BTU are converted to useable space heat, and 20,000 BTU of fuel energy is lost in the combustion and heat transfer processes. AFUE ratings do not include heat loss due to inefficient distribution through ducting and piping systems.

9.7   MAKING STEAM

Raising the temperature of water to the boiling temperature of 212 degrees Fahrenheit at atmospheric pressure requires 1 BTU per 1 pound of water. Taking 1 pound of 60-degree water to the boiling point requires 152 BTU. But the water still isn't boiling. Only sensible heat has been added to the water. *Sensible heat* is heat that can be measured with a thermometer. Adding sensible heat doesn't physically change the material that is being heated; it just makes it hotter.

To change the water into our new substance—steam—we have to add ***latent heat***, which is the energy needed to change the physical state of a substance without a decrease or increase in temperature.

FIGURE 9.14 Steam pressure gage

Source: Photo courtesy of Oconomowoc Memorial Hospital.

If enough latent heat is added, water can be changed from a liquid to a gas, which is steam. Nine hundred seventy BTU per pound of water must be added to change water into steam at 212 degrees Fahrenheit. With a single source of heat, it takes longer to boil water than to simply raise its temperature. When the water boils at 212 degrees Fahrenheit and at atmospheric pressure, the steam is known as saturated steam. If additional heat is added to saturated steam, it becomes superheated steam; the degrees of temperature added above saturation are known as degrees of superheat. The concept of latent heat is important for understanding efficiency gained by using a condensing boiler. The basic concept is that heat can be extracted when steam is condensed because latent heat was added to water to create steam.

There are two major types of boiler plants: steam plant and hot water plant. A commercial boiler that produces only hot water is commonly referred to as a hot water *boiler*—even though nothing boils! When a home is heated by hydronic heat, the heating unit is called a hydronic boiler. When water is heated to make steam, the units are referred to as steam boilers (an accurate term). As pressure in a steam boiler goes up, the temperature at which water will boil also goes up. Steam at 15 pounds per square inch gauge (psig) is at a temperature of 250 degrees Fahrenheit and is saturated steam. Every steam temperature has a corresponding steam pressure for saturated steam. If heat is continually added to saturated steam, the steam becomes superheated. Superheating is a way to increase the energy content of the steam.

When boilers develop 15 psig or less steam pressure, they are considered to be a low-pressure boiler; above this pressure level, boilers are classified as high-pressure (see Figure 9.14). Hot water boilers that operate at or below 250 degrees Fahrenheit and 160 psig water pressure are also considered low-pressure, according to the American Society of Mechanical Engineers (ASME) boiler-pressure vessel code. The pressure in a hot water boiler is allowed to be more than ten times that of a steam boiler because all the pressure is generated by the boiler pump.

As stated earlier, the two major types of boiler designs for fossil fuel boilers are fire tube boilers and water tube boilers. Each will be discussed next in the following subsections.

Fire Tube Boilers

To envision the generating section of a fire tube boiler, picture a large, cylindrical steel tank that is enclosed at both ends and looks something like a big soup can. The top and bottom of the "can" are called tube sheets. Matching holes are punched in either end of the tube sheet, and generating tubes are run through the holes, fitted tightly, and expanded into the metal. One very large hole is located below the smaller tubes for a large tube several inches in diameter; this large tube is the furnace flue. The tank is filled with water, the outside of the tubes get wet, but the insides remain dry.

On the front end of the tank is a cap that houses the burner and combustion air supply equipment and the front smoke box chamber. On the back end of the tank is another cap that makes a flue gas turnaround chamber (also known as a smoke box) to turn the flue gas and force it through the generating tubes. If the chamber contains refractory brick or other types of insulation to insulate against the high heat of the flue gas at this point, the boiler is referred to as a *dry back boiler*. If the chamber also has a water jacket to absorb heat, it is called a *wet back boiler*.

For the first step in heating the water, the boiler's burner makes hot combustion gas that travels through the furnace flue. The gas goes out the flue at the back of the boiler and turns to go through the tubes. Hot combustion flue gases pass through these tubes. Heat is transferred to the tube metal, which becomes hot and heats the water in the tank (the temperature of the water goes up). Hot gases traveling the full length of the boiler tube is known as a pass. Fire tube boilers have at least two passes. The first pass is when the hot gas travels the length of the furnace. The second pass is when it travels from the back of the boiler and through the tubes to the front. Baffle plates (if installed) add additional passes. Baffle plates limit the number of tubes into which the gas can travel. Hot flue gas can then be directed to enter another set of tubes and thus travels in the opposite direction.

One of the ways to guess at the number of passes a fire tube boiler has is to look at the stack location. If the stack is in the back of the boiler, it is a three-pass boiler. If the stack is in the front, it is either a two-pass or a four-pass boiler.

Because the steam-generating tank is enclosed, the pressure of the steam rises when steam is made. The steam is trapped between the water level and the shell. More and more pounds of steam are pushed into this fixed volume and the pressure goes up, causing an increase in steam pressure.

Fire Tube Hot Water Boilers

Fire tube boilers can generally tolerate poorer water quality. A fire tube boiler's straight tubes make them easy to retube, clean, and repair. Fire tube boilers hold large quantities of water, so small leaks at the tube shell aren't as large a problem as the same leak in a water tube boiler. If a fire tube boiler is used to make hot water, the boiler and associated piping are completely filled with water. Expansion tanks are used to provide a place for water to move into because, when water is heated, it expands. Steam boilers are filled below their normal operating water level (NOWL) before starting to compensate for water expansion.

In commercial installations and most residential installations, electric-driven centrifugal pumps are used to move the heated water throughout the heating system. Pumps may also be used to help return water to the boiler. Small hot water supply pumps are located at individual fan coil units to make sure that hot water doesn't bypass the heating coil. The pump forces the heated water through the heating coil, thus overcoming the additional frictional resistance created by the coil. An electronic or pneumatic valve regulates the amount of water allowed into the coil based on space temperature.

All of the pressure in a fire tube boiler that is generating steam must be contained by the large tank surrounding the fire tubes and furnace flue. In this situation, a lot of surface area is directly exposed to pressure. Force = pressure × area, so the shell thickness must be increased as the pressure goes up. This limits fire tube boilers to a steam pressure of 350 psi. Fire tube boilers also take longer to start up than most other types of boilers because of the large volume of water that must be heated.

After operating for a heating season, fire tube boilers are drained and opened up if the period of nonuse will last ninety days or more. The outside of the tubes are washed down as soon as the boiler has cooled to prevent scale from sticking to the tubes. The inside of the tubes are brushed to remove soot. The boiler inspector is contacted and, if required, arrangements are made for an internal inspection. A water-absorbing material such as quick lime or silica gel is placed on a tray inside the boiler to prevent corrosion due to moisture. Always refer to the manufacturer's manual for specific maintenance requirements.

Water Tube Boilers

Water tube boilers have water in the tubes surrounded by the hot gases of combustion. Most commercial water tube boilers used for heating buildings keep the water as water; they do not apply heat to change the water to steam. These boilers use pumps to circulate water through the building.

Hospitals and power utilities need the additional heat and energy that can be gained from generating steam. Steam has the added advantage of being able to flow through pipes, without the need for pumps, strictly because of the pressure difference. It also provides a reliable source of moisture for humidification. Many types of water tube boilers are further defined by their tube shape. Early water tube boilers were straight tube boilers with water headers on either end of the tubes. This design was simple, which meant that tubes could be replaced easily, but it was inefficient from the standpoint of energy.

A D type water tube boiler is a great improvement. It has a steam drum at the top and a water or mud drum at the bottom, with the curved water tubes running in between.

Water tube boilers of all tube configurations expose steam pressure to a smaller area than do fire tube boilers. These boilers can be made with pressures that range from 15 psig to 1,500 psig or more. In fact, water tube boilers in some new power-generating plants use an idea that was engineered and patented by Mark Benson in the early 1920s. Benson boilers can be made in pressures exceeding 3,208 psig, which is the supercritical pressure of steam. The supercritical pressure of steam is when the pressure is so high that the density of steam and water is the equal.

Ultra-supercritical steam generators produce a pressure of 310 bar at 650 degrees Celsius. A bar is equal to approximately 14.5 psi, which means that 310 bar is equal to 4,496 psig, and the temperature is equal to 1,202 degrees Fahrenheit. At the supercritical pressure of 218 bar, the pressure is so great to that no boiling occurs.

The reason power companies use supercritical boilers is that pressure and heat equals energy. The more energy contained in the steam, the more electric power it can produce by spinning a turbine generator. It also nets a fuel savings of from 3 to 4 percent over conventional boilers. Most ultra-supercritical boiler pressure ratings today are in the range of 3,400 to 4,200 psig, with steam temperatures of approximately 1,000 degrees Fahrenheit.

9.8 STEAM HEAT PLANTS

In the 1950s, most heating plants produced steam for heating. At that time, the large population increase of the baby boom generation required many schools to be built, and many included steam heat boilers.

Steam has several advantages over hot water and some notable disadvantages. For large multibuilding facilities, steam is generated in a central boiler plant building (also known as the physical plant building). The steam travels in pipes underground in tunnels large enough that a person can walk through. These tunnels also serve as a maintenance technician's personal highway for moving quickly from one building to another and avoiding the weather outside. As long as the pressure of the steam is high enough to overcome frictional losses in the piping, the steam travels solely based on the pressure difference. The steam moves from a high-pressure area to a low-pressure area. The steam enters a building and then travels through branch lines into heat convectors, radiators, and fan coil heating units. At the outlet of each type of heater is a steam trap. The job of the **steam trap** is to hold back the steam until it condenses inside the heating unit to form water. This water is now called condensate. The condensate is collected in a condensate drain tank and then fed back to the boiler via a pump. At this point, the condensate water is now called boiler feedwater.

Steam traps are a big maintenance issue for steam heat systems. All traps operate automatically, but almost all are mechanical in their operation. Steam traps are affected by corrosion, dirt, and wear. Leaky traps can be a major source of lost plant efficiency because they allow steam to leak through them and out the vent in the condensate receiver tank. Steam traps are frequently checked by using a noncontact infrared thermometer. An infrared thermometer's laser dot is pointed at the inlet and outlet of the steam trap. The temperature difference before and after the trap for a heating system should be 10 to 20 degrees Fahrenheit because condensed water is cooler than incoming steam. If the temperature is about the same as the normal steam temperature entering the heater, the trap is leaking and must be serviced. If the temperature difference is too great, then the trap might have failed in the closed position. The failure of a heating coil to provide heat might also be the result of water flooding the heater after a trap has become plugged. A recommended interval for testing low-pressure steam traps is twice a year. At facilities with hundreds of traps, this maintenance task can be quite an endeavor.

9.9 PRETREATMENT OF BOILER FEEDWATER

Pretreatment of boiler feedwater is necessary to limit the amount of chemicals added to the boiler water to prevent scale buildup on and corrosion of the heating surfaces inside the boiler. These chemicals eventually find their way into the environment when boilers are drained or blown down.

Water Softeners

As mentioned previously with regard to cooling towers, water softeners can remove dissolved minerals by softening the water. Water softeners require salt for regeneration of the zeolite bed that removes the calcium and magnesium from the water. Some facilities use bags of salt; others purchase salt in bulk and

FIGURE 9.15 Reverse osmosis (RO) machine in a boiler plant

Source: Photo courtesy of Oconomowoc Memorial Hospital.

make a brine solution bath for regeneration. Softening removes some, but not all, of the minerals. Removing dissolved minerals before they enter the boiler reduces the amount of phosphate chemical scale treatment needed.

Reverse Osmosis

To remove more minerals from softened water, the softened water is sent to a reverse osmosis (RO) machine (see Figure 9.15). At this point the mineral content is very low, but chemicals are still needed to reduce scale. The water is forced under pressure through a membrane where most of the molecules of the minerals cannot pass through the membrane. The highly clean and softened water is forced out the other side of the membrane. The clean water goes to the boiler, and the mineral-containing water is drained to the sewer.

De-aeration by Feedwater Heating

A direct contact de-aerating feedwater heater (see Figure 9.16) is a device that removes air and some of the boiler metal-damaging oxygen without the addition of chemicals. The heater is essentially a vented tank located above the feedwater pump. Both feedwater and steam are sprayed so that they contact one another inside the tank, which immediately heats the feedwater. Heating the water drives off oxygen by reducing its solubility. The heater is vented to the atmosphere, which allows venting of the oxygen. The de-aerated feedwater flows into the pump's suction, and the pump sends the feedwater into the boiler. De-aeration helps prevent oxygen corrosion inside the boiler. Corrosion in a boiler is a very serious condition that can lead to tube failure.

Boiler Water Chemical Treatment

All commercial boilers require their water to be treated with chemicals to prevent the formation of scale and to reduce the damaging effects of the remaining oxygen after de-aeration. Chemicals are fed into a boiler system through chemical feed pumps whose output volume is readily adjustable. Boiler water contains scale-forming minerals: calcium, magnesium, and silica. Boilers also need chemicals to remove oxygen. Removing oxygen from boiler water prevents rusting and pitting of boiler metal. Boilers also need chemicals added to the steam lines to prevent carbonic acid formation in condensate.

Some typical boiler chemicals for low-pressure boilers and boilers of a few hundred pounds pressure are:

- Phosphates turn dissolved minerals into a nonadhering sludge.
- Sodium sulfites scavenge oxygen from the boiler by turning sulfites into sulfates. A residual level of 30 ppm of sodium sulfite ensures that no oxygen is present in the boiler water.

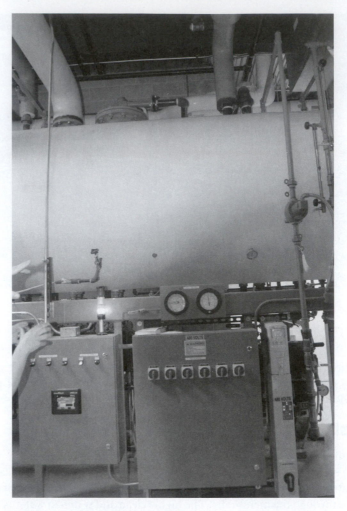

FIGURE 9.16 Direct contact de-aerating feedwater heater
Source: Photo courtesy of Oconomowoc Memorial Hospital.

- Hydrazine, which is used only in high-pressure utility boilers, removes oxygen.
- Amines, also called steam line treatment, prevent carbonic acid from forming in condensate.
- Alkalinity builders keep boiler water at a pH of 8.3 or more, as required by the boiler manufacturer.

Boiler Water Testing

In high-pressure steam heating plants, boiler water should be tested daily. A variety of instruments and chemical test kits are available for this purpose. Although many strides have been made toward using chemicals of low toxicity, the chemicals used to treat boiler water are an issue for sustainability because they eventually end up in drinking water supplies. It is essential that the smallest amount of chemicals possible is used for boiler water treatment. To achieve this, proper boiler water testing should indicate the precise amount of boiler treatment chemicals needed.

9.10 MANAGING BOILER AND COOLING TOWER CHEMICAL TREATMENT

Many facilities utilize the services of a water treatment chemical contractor. The contractor's services usually include monthly (or some other agreed-on period) testing of the water in the boiler and cooling tower. The contractor fills out a treatment record sheet and gives a copy to the boiler operator. The contractor then makes recommendations about dosage rates or chemical use. The contractor supplies the chemicals on a regular basis. In some cases, the facility personnel do not handle the chemicals or the pumps that supply chemicals. This is generally done for the safety of facility personnel.

Facility managers sometimes develop a mistrust of water treatment contractors because they seem to be constantly increasing dosages that use up a lot of expensive chemicals, or they recommend more costly chemicals. In this situation, knowledge of chemicals and their use can be very valuable. Simply call the water treatment contractor and ask for an explanation about their recommendations. Also ask how they arrived at their dosage levels. If their answer is unsatisfactory, seek advice from a third party. But beware: The new water treatment contractor may paint a much darker picture of the actual situation in order to get new business. To combat this situation, benchmark the amount and type of chemicals used by other organizations with similar equipment that is of similar age.

Example: Water Treatment Service

A business has used the services of one particular water treatment contractor for many years. For the last few years, the contractor has slowly increased the dosage of sodium sulfite oxygen scavenger in order to treat the boiler water. This increased dosage grew substantially higher over time. During a recent monthly visit, the water treatment contractor tested the boiler water and found that the facility had switched back to the lower chemical treatment level it had been using a several years ago. No sodium sulfite was present in the boiler water; all of it had been changed by the oxygen into sodium sulfate. No residual sodium sulfite was present (generally levels of residual sodium sulfite are from 30 to 60 ppm). Without a residual test reading, there was no way of knowing if the boiler had been treated sufficiently. The water treatment contractor asked the boiler operator why treatment levels were reduced. The boiler operator said that this is the way the boiler operators had treated the boilers for the previous twenty years, and they didn't have any problems back then. The operator then accused the water treatment contractor of wasting chemicals to make more money. Shortly thereafter, the contractor was told his services were no longer needed. A new water treatment contractor had already given the facility an attractive price quote for their service, and so they switched contractors.

The boiler operator was misinformed, and showed a deep misunderstanding of how the treatment chemicals should be applied. True, the amount of chemicals used did work for twenty years, or at least the previous boiler operator thought so, but the situation had changed. The boiler internal piping was now very old, steam traps were leaking, and numerous water leaks were present in the piping. The amount of makeup city water, which is rich in oxygen, required by the boiler had risen sharply. Without proper oxygen-scavenging treatment, O_2 immediately started corroding the bare metal surfaces of the boiler. The color of the boiler water eventually began turning red with rust. Realizing the mistake, the FM ordered the boiler operator to increase the chemical dosage, and the facility went back to the original treatment contractor.

9.11 CONDENSING BOILERS

All the boilers discussed in this chapter have been noncondensing boilers. Typical thermal efficiencies for noncondensing boilers are 70 to 80 percent. Condensing boilers (see Figure 9.17) can achieve efficiencies of 90 percent, with efficiencies as high as 97 percent under ideal conditions. Condensing boilers capture additional heat that would normally be lost to the stack. The capture of additional heat is accomplished by running the return water and/or boiler makeup water through a special condensing coil. Condensing happens when the water vapor in the boiler exhaust gives up its latent heat to the return water and condenses on the coil. This condensation occurs at a return water temperature of about 130 degrees Fahrenheit or less for a natural gas–fired boiler (see Figure 9.18). At this temperature, an efficiency of 87 percent can generally be achieved. If return water temperature is lowered to 80 degrees Fahrenheit, an efficiency of 93 percent or greater may be reached at full load output. With such low operating temperatures, a biocide should be included in boiler water treatment as a preventive measure against the formation of the disease-causing legionella bacteria. Bacteria exposure could be a problem for personnel when the system is opened for service or repairs. If water containing legionella bacteria should spray, it could become aerosolized and breathed into the lungs, where it may cause a type of pneumonia called Legionnaires' disease.

The condensed water in the flue gas is mildly acidic because of combustion products in flue gas. This acidity in the water necessitates the use of corrosion-resistant materials, such as aluminum or stainless steel, for the condensing section.

The operational difficulty is in ensuring that return water temperatures are low enough for condensing to occur. If temperatures are too high, the gain in efficiency is much less and may be the same as that of conventional boilers. One way to achieve lower return temperatures is to include floor slab radiant heating tubes, which can operate at lower temperatures and still provide heat to the space. Designers

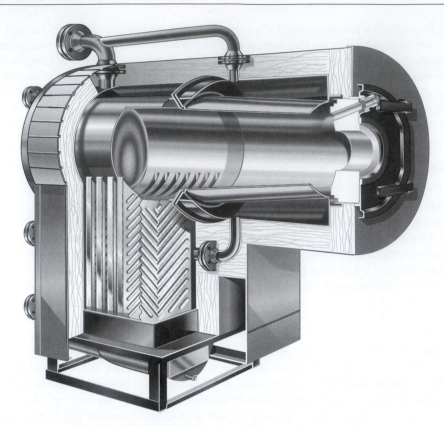

FIGURE 9.17 Cutaway of a high-efficiency condensing boiler: the Vitocrossal 300
Source: Image courtesy of Viessmann Manufacturing Company Inc.

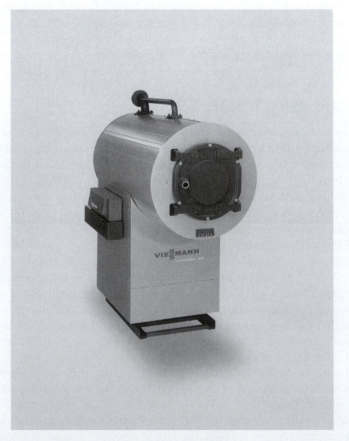

FIGURE 9.18 Vitocrossal 300 high-efficiency condensing boiler
Source: Image courtesy of Viessmann Manufacturing Company Inc.

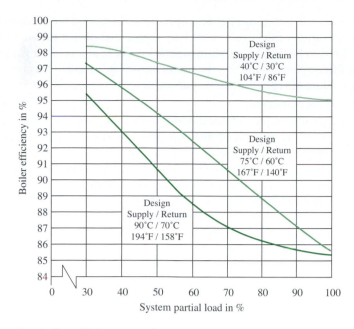

FIGURE 9.19 Condensing boiler efficiency graph. Note the effect that water supply, return, and boiler load has on boiler efficiency.

Source: Image courtesy of Viessmann Manufacturing Company Inc.

may also include oversized heating convectors to extract more heat from the water, thus providing lower return temperatures. Boilers may also be operated at a lower load so boiler outlet water temperatures are less. Boilers need to have an appropriate **turndown ratio** so that they can operate a low load without a drop in efficiency. A boiler with a turndown ratio of 3 can operate at 33 percent of full-load output. A boiler with a turndown ratio of 5 can operate at 20 percent of full load. A good turndown ratio allows the boiler to match a low-load condition without modulating the burner to cycle off. Every time a boiler modulates off, the furnace must be purged to remove any remaining combustible gases, a process called **post purging**. Post purging is accomplished by having the boiler's stack fan run long enough to achieve eight changes of air in the furnace. When the boiler starts up again, just prior to ignition, a pre-purge cycle occurs. Flushing air through the boiler furnace can prevent a boiler explosion, but it also removes valuable heat inside the furnace. A boiler that can turn down to a low level can avoid losing this energy by remaining on during light load conditions.

Boiler System Efficiency

The graph in Figure 9.19 show that a high-efficiency condensing boiler does not operate at maximum efficiency under all conditions. First, the entire heating system must be designed to be able to extract the proper amount of heat from the supply water so that it returns to the boiler at a low enough temperature to ensure proper heat recovery by condensing moisture in exhaust gases. Many factors influence return temperature: the size of the heating convectors, fan speed on fan coil units, R-value of the building insulation, outdoor air temperature, and boiler load. If properly designed a condensing boiler system can operate at a very high efficiency for most of the heating season.

9.12 PLANT SUSTAINABILITY

Due to the complexity and vast array of equipment in a facility, it is difficult to develop plans to keep a facility's power plant sustainable. The specifics of multiyear plans will likely need to be modified due to the constant advances in equipment, but the overall intent and goal of any plan is to provide a sustainable plant. The following steps are important in ensuring the sustainability of an existing boiler plant:

1. Evaluate the useful life remaining on current equipment.

   The useful life remaining on current equipment is sometimes difficult to determine. A well-maintained boiler can operate decades beyond its expected lifespan. A few ships on the Great Lakes have sailed for a nearly a century with their original boilers and engines. Many power plants across the United States are still operating after fifty or more years. Useful life needs to be evaluated based on the current condition of the equipment. For commercial or institutional boilers, useful life is generally considered to be about forty years. whereas residential forced-air gas furnaces have a useful life of about fifteen years. Evaluating the useful life of equipment is necessary to make informed decisions on upgrading or replacing plant equipment.

2. Determine if the plant is outdated.

   Audit the energy use and the physical condition of the existing equipment. A heating plant should have records of the amount of fuel consumed by the boiler plant. The efficiency of the boiler can be determined through an analysis of the stack gas. This analysis can be done by a technician with a handheld stack gas analyzer. Modern gas analyzers can instantly calculate boiler efficiency, which can be compared to the efficiency of a new boiler. The fuel records and the projected efficiency of new equipment can be used to determine the amount of fuel that a new boiler with a higher efficiency will burn. This information is important when doing a cost-benefit analysis. If a new boiler is too costly, installing a new high-efficiency burner and controls on the existing boiler can also be evaluated. In any equipment evaluation, it is important to consider if the building load requirements produce conditions that allow new equipment to operate at its designed maximum efficiency.

3. Evaluate boiler operational efficiency.

   Ensure that control equipment is working properly and the all heating surfaces are clean. The plant should then undergo a thorough retro-commissioning whereby all systems are evaluated on their ability to operate as designed prior to making any plant upgrade decisions. When a plant consists of a battery of boilers, the number of boilers in operation should always meet the building's heating demand in the most efficient manner. Single boiler plants should have the capability of operating efficiently over a wide range of load conditions (i.e., they have a high turndown ratio).

4. Verify boiler chemistry.

   Verify that the proper boiler total dissolved solids (TDSs) reading of boiler water has been maintained at the manufacturer's recommended level. For steam boilers, bottom blowdown frequency and amount of time the blowdown valve is open should be based on TDS readings to save energy. In the past, bottom blowdown was performed daily at a specific time during the day. The blowdown valve was opened for approximately the same amount of time every day, and the actual water chemistry readings and energy lost down the sewer were not considered.

5. Evaluate nonchemical boiler water treatment methods.

   Pretreatment of water coming from the city's main can significantly reduce the amount of chemicals required for water treatment. Boiler water treatment chemicals eventually end up in the sewer system; thus, limiting the amount of chemicals and searching for earth-friendly chemicals are steps toward sustainability.

6. Use waste heat more effectively.

   Heat that normally would be released to the atmosphere should be used for domestic water heating. This and other uses save energy and provide a payback to the facility. Heat can also be recaptured and sent back into the boiler system. Recaptured heat creates a more efficient boiler plant. Another common way to recapture heat is the installation of devices called boiler stack economizers (a series of tubes through which feedwater flows, picking up heat before it enters the boiler). This heat would otherwise have simply left the boiler through the stack. Blowdown water also has a lot of heat. In fact, blowdown water must be cooled prior to discharge down a sewer or the sewer pipe may be damaged by thermal shock. To increase efficiency, the cooling water used is boiler feedwater. The feedwater picks up heat while simultaneously cooling the blowdown water.

7. Ensure safe boiler operations and provide personnel with safety training.

   From a humanistic, a practical, and a sustainability standpoint, personnel injuries are the worst that can happen. Injuries can eliminate much of the cost advantage that is counted on through the operation of a plant. Safety training of personnel with regard to the operations they perform is essential, as is the proper types of *personal protective equipment (PPE)* and the fixed safety equipment necessary to operate a safe plant.

REVIEW QUESTIONS

1. What is a swing chiller and why is it used?
2. Absorption chillers use a(n) _____ instead of a compressor.
3. How do fire tube and water tube boilers differ?
4. Why are condensing boilers more efficient than noncondensing boilers?
5. What is turndown ratio in a boiler?
6. Why is boiler water treated?
7. Why is it important to test boiler water?
8. What is a steam trap and how is it used?
9. Describe latent heat and sensible heat as they apply to a steam boiler.
10. Describe cycles of concentration with regard to cooling towers.
11. How is chemical use reduced in steam boilers?

Essay Questions

12. What are the issues and decisions to be made when replacing a heating boiler for an office building?
13. When is it advantageous to use an absorption chiller?
14. What are the important issues to consider when adopting a chemical or nonchemical water treatment program?

ENERGY MANAGEMENT AND RENEWABLE ENERGY

INSIDE THIS CHAPTER

Can Sustainable Facilities Continue Their Reliance on Energy from Traditional Providers?

This chapter discusses energy, energy efficiency of equipment, power generation, rate structures, and one of the major uses of energy—lighting. It also discusses electrical distribution, the importance of power factor and its effect on a building's carbon footprint, and strategies to help prevent demand charges being assessed by the power utility. Retro-commissioning is introduced as a tool to reduce consumption by improving plant efficiency, as is facilities saving money by utilizing the interruptible rate for natural gas.

Renewable energy devices are becoming more and more necessary in building construction, and the chapter describes geothermal heat pumps, heat pump water heaters, solar water heating, wind power, hydroelectric power, and photovoltaic power. The chapter goes further to discuss microgrids for power distribution and storage, and moves on to fuel cell technology, which is used at the One World Trade Center and three surrounding towers at the World Trade Center site.

10.1 SUSTAINABLE ENERGY MANAGEMENT

Energy management is a broad term that encompasses a multitude of techniques to control the amount of energy consumed and to mitigate the effects of the ever-increasing cost of energy used at a facility. Sustainable energy management (SEM) requires knowledge of engineering, power, efficiency, accounting, and utility use monitoring strategies. Energy can be in the form of natural gas, liquefied petroleum gas, oil, coal, or electricity. The local utility may also provide steam or chilled water which can be used directly for heating and air conditioning.

One of the goals of sustainable energy management for commercial organizations is saving money. It is fortunate that in the process of saving money, the added benefits are carbon dioxide emission reduction and environmental stewardship. Saving money isn't a new goal, but SEM considers much more: the effects of energy decisions on future energy availability, and availability at a reasonable cost. An ultimate goal of SEM is to achieve a net-zero energy building. **Net-zero energy** is a condition whereby the energy used by the building is obtained from its surrounding environment (wind, solar, etc.). This has been possible in a few instances when climatic conditions were generally mild, and the building was able to utilize renewable energy sources and store energy without excessively large amounts of equipment or expense.

An energy management technique is to continuously investigate the efficiency of equipment and operations and continuously look for improvements. Energy management looks at efficiency in three areas:

1. Efficiency of individual electrical fixtures, machines, and components
2. Overall system efficiency: the equipment working together as a whole
3. Operational efficiency: equipment operational strategies and equipment maintenance

Another technique is to look at the operations contained within the facility and seek efficiencies while simultaneously improving the operation as a result of implementation of these changes. For example, an energy manager develops a strategy to shut down desktop computers to save energy when the computers are not in use. The benefit to the information technology (IT) security operation is that unattended computers are inaccessible, which makes the computer system more secure. This technique thus produces a dual benefit of safety and energy savings.

Electric Machinery Efficiency Ratings

Different types of efficiency ratings are used for different types of machinery. For boilers and furnaces, the annual fuel utilization efficiency (AFUE) is used (see Chapter 9). This chapter will discuss machinery that consumes electricity to provide cooling and or heating. For this equipment, electrical efficiency rating (EER), kW per ton, coefficient of performance (COP), seasonal energy efficiency rating (SEER), and integrated energy efficiency rating (IEER) are used. It is important to note that the actual efficiency for a particular piece of equipment depends mostly on how the equipment is operated (mostly part load operation, mostly full load, or a combination of the two), which is known as the equipment's **load profile**. Load profile has a large impact on the actual efficiency obtained in the field due to varying equipment efficiencies, which depend on operating load. The following sections discuss the efficiency rating and which one is most appropriate to use.

Electrical Efficiency Rating   Every item that consumes energy has an efficiency rating associated with it. The electrical efficiency rating (EER) for cooling equipment is equal to BTUs out/watts of input energy. One ton of refrigeration is equivalent to removing 12,000 BTU/hour of heat from the building. For example, a two-ton rooftop air-conditioning unit is serving an office. The rooftop unit consumes 2400 watts of power.

$$10 \text{ EER} = 24,000 \text{ BTU}/2,400 \text{ watts}$$

A rating of 10 EER is a fairly good efficiency rating for an air-cooling unit.

kW per Ton   The efficiency of the two-ton rooftop air-conditioning unit can also be described in kilowatts per ton of refrigeration. This method is common when describing the efficiency of a large electric centrifugal chiller. In this case:

$$2,400 \text{ watts} = 2.4 \text{ kW}$$
$$2.4 \text{ kW}/2 \text{ tons} = 1.2 \text{ kW/ton}$$

This rating calculation provides a very broad, generalized measurement that is useful only in determining which equipment might be considered for installation.

Coefficient of Performance   Coefficient of performance (COP) is used to describe the efficiency of heat pumps.

$$\text{COP} = \text{amount of BTU output/kW input} \times 3413 \text{ (conversion factor)}$$

A heat pump in the cooling mode delivers 72,000 (BTU/hour) while consuming 10 kW of electric power.

$$\text{COP} = 72,000 \text{ BTU/h} \div (3413 \times 10 \text{ kW}) = 2.1$$

Seasonal Energy Efficiency Rating   The U.S. Department of Energy regulates the efficiency of air-conditioning equipment. The seasonal energy efficiency rating (SEER) that is placed on air conditioners is a number that reflects the amount of energy consumed in watt-hours over a cooling season compared to the amount of cooling in BTU provided over the cooling season.

$$\text{SEER} = \text{total cooling output during the cooling season/total energy input during the cooling season}$$

SEER ratings of 13 are the standard today. Many types of cooling equipment have an SEER of 17 or 18.

Integrated Energy Efficiency Rating   The integrated energy efficiency rating (IEER) is a newer rating that is part of Air-Conditioning, Heating, and Refrigeration Institute's (AHRI's) standard 340/360. This new standard more accurately represents part load efficiency of air-conditioning units. It is based on a weighted average of performance levels at air-conditioning output capacities of 100 percent, 75 percent, 50 percent and 25 percent.

The importance of SEER and IEER ratings is that they provide basic guidance for the facility manager (FM) or heating, ventilation, and air conditioning (HVAC) designer during the equipment selection process. More efficient equipment may provide a payback in energy savings that will cover any additional equipment costs over the life of the equipment.

10.2 POWER GENERATION AND DELIVERY

A seemingly contradictory aspect of energy management is created by the real-world engineering aspects of producing electricity. Because electric rates charged by power utilities vary by time of use, achieving savings in energy dollars spent may not be the same as lowering the total amount of energy consumed. The *time of use dollar savings* is a result of overall electric utility efficiency gains achieved by evening out the load and reducing the peak demand for power. This power company strategy ultimately saves energy.

First, consider electric power generation. The world has physical, technological, and mechanical constraints for which the electric utility companies and other energy providers must provide the solution. Two of the biggest constraints are (1) the megawatt-generating capacity of a power plant and (2) the amount of power its transmission lines and substation equipment can carry to the customer at any one time. The major problem is that an oversized generating plant running at a low power output is not a very efficient or economical plant to run. Providing the proper mix of generating plant size/number of generators and properly sized transmission lines and equipment is a complex task that requires multi-year planning by the power company.

Providing plant capacity for an estimated electric load is especially difficult in areas of the world that have four seasons with extremes of temperature punctuated with spring and fall months that are relatively mild. It can be further complicated by providing power to industries that have high intermittent power requirements. The electric load for a service area also varies according to time of day. Most people work 8 a.m. to 5 p.m. and require power to do their jobs. Afternoons tend to be the period of greatest demand for electric power, especially during the summer air-conditioning season, when every building's cooling equipment is running. The period of time when demand is highest is called the *on-peak hours*. For billing purposes, this time period may run from 8 a.m. to 8 p.m., but power companies may have differing on-peak hour periods. (One company gave customers a choice of on-peak hours from 8 a.m. to 8 p.m. or from 10 a.m. to 10 p.m.) During on-peak hours, the power company charges the highest rate per kilowatt hour to their commercial customers. A kilowatt is 1,000 watts. One kilowatt hour is 1,000 watts consumed over one hour.

Commercial Power Point of Delivery

To reduce line losses and infrastructure costs, the power company sends alternating current across power lines at several thousand volts. This voltage may be as high as 300,000 volts. High voltages have the benefit of requiring less current to be carried for the same amount of transmitted power (power = voltage × current). These dangerously high voltages never come anywhere near a facility. High voltages go to the power company's transformer substation, which steps down the voltage to a few thousand volts. From the substation, the power travels once again by wire to the various customers. A facility may receive three-phase, 13,200 volts to the transformer at the facility. The transformer steps the voltage down to 480 volts. The 480-volt power is then sent through solid copper buss bars supported by insulators that run almost the entire height of the building through fireproof electrical closets. Bus bars are an efficient way to run the main electric power through the building because their electrical resistance is low, and it is easy to connect cable to the bars for distribution on each floor.

The 480-volt system provides 277 volts for overhead lighting, 208-volt three-phase power for small motors and special equipment needs (e.g., cooling equipment in the data processing center for after-hours cooling when the building's air conditioning is secured for the day), and 120 volts for wall outlets. Some facilities, such as industrial plants and hospitals, may use high-voltage power of 4,160 volts to provide power to electric motors that drive major pieces of equipment such as centrifugal chillers. This high voltage means that motor windings, power leads, and wiring can be a thinner diameter and still carry the same amount of power as lower voltage equipment. Current determines the thickness of the wire (greater current requiring thicker wire), whereas voltage determines the thickness or type of insulation required to contain the voltage in the wire and prevent it from arcing or grounding.

Power Utility Company Strategy

A power company may use a time of use, power demand, and power factor strategy to make their customers change their power usage habits. As stated previously, utilities charge a higher rate for electricity when overall demand for power is high during on-peak hours. But power companies also provide commercial customers with economic incentives to transfer their electric usage to night and early morning periods, when the demand for power is low. Power companies refer to the period of time when power demand is low as the *off-peak hours*. Check with the power company for their pricing because it varies among power companies. If customers can shift their need for power from on-peak to off-peak hours, they will benefit from the lower rate and save money.

Demand Charge   Power used can be broken down into two basic elements: consumption and demand. *Consumption* is the total amount of energy used over a one-month period. *Demand* is the average highest amount of energy or peak demand a facility registers on its meter over a fifteen-minute period. A power company must size its power-generating plant and distribution equipment based on meeting this peak demand, so they charge a fee based on demand. *Demand charge* is the additional fee the power company charges. Power companies handle demand charges in several different ways. Demand charge can vary due to season (summer or winter), and total amount of power consumed (one rate per kW for the first 300 kW used and a higher rate for any amount over 300 kW).

Example: Demand Charge

One power company took the highest demand that occurred during the current monthly billing period and charged that amount for the next eleven months, even though that high load was never reached again. In fact, if the demand should go higher in a one-month period, the demand charge was reset to the higher amount for the next eleven months. It didn't matter whether the highest demand occurred during on-peak or off-peak hours. If a facility ran all its cooling equipment on an extremely hot day, this could result in a high demand charge for the next year. In this type of situation, it is extremely important to keep track of the facility's demand for power, anticipate it, and plan ways to limit it.

Example: How a Demand Charge Affects Billing

Two buildings, Building A and Building B, are exactly the same. Both buildings use electricity from the same power utility. As shown in Table 10.1, Building A uses 20,000 kWh during the month, and Building B uses 18,000 kWh of electricity, which is 10 percent less than Building A.

     Buildings A and B are charged the same for electricity, even though Building B uses 10 percent less electricity. The maximum demand for power in Building A was 80 kW; the maximum demand for power in Building B was 26 kW higher. The difference is the demand charge. If Building B can reduce its maximum demand to the level of Building A (80 kW), then the charge for electricity would be $2,220, or about 10 percent less than Building A. Building A has compensated for its increased load through its equipment operation (shedding load, staggering startup times, free cooling, optimum start and run times, etc.), which keeps the maximum demand low.

TABLE 10.1 How demand charge can affect billing

	Building A		Building B	
	Amount	**Cost**	**Amount**	**Cost**
Total consumption for the month	20,000 kWh	Off-peak + on-peak	18,000 kWh	Off-peak + on-peak
On-peak consumption ($0.10/kWh)	12,000 kWh	$1,200.00	11,000 kWh	$1,000.00
Off-peak consumption ($0.06/kWh)	8,000 kWh	$480.00	7,000 kWh	$420.00
Demand charge ($10.00/kW)	80 kW	$800.00	106 kW	$1,060.00
Total charge		$2,480.00		$2,480.00

Power Factor

Power utilities also charge their large commercial customers for low power factor. *Power factor* can be thought of as the way that a facility uses power, and the resulting power system efficiency. This is not the same as equipment efficiency. When equipment efficiency is improved, it reduces the amount of energy consumed. Power factor is improved by reducing the amount of inductive loads or adding capacitors to the electrical system. In electrical terms, power factor is the ratio of real power in kilowatts (kW) to apparent power in kilovolt-amperes (kVA). Real power is the energy that does work. Apparent power is the total of real power (in kW) plus volt-amps reactive (VARs) power, also known as magnetizing power. In electric motors, magnetizing power is necessary to create the rotating magnetic field so the motor can turn.

$$\text{Power factor} = \text{real power} \div \text{apparent power} = \text{kW} \div \text{kVA}$$

The facility is most efficient when the power factor is at 1.00, which is also known as the *unity power factor*. If a facility has only resistive loads, such as incandescent lights and resistance heaters, its power factor will be 1.00. But if a facility has electric motors, generators, magnetic ballasts, and/or transformers, the load is inductive. Inductive loads reduce power factor. Power factor is reduced further when electric motors operate at less than full load. At less than full load, they are doing less work but still require magnetizing power. Electric motors frequently operate at reduced load because they are sized for the maximum load. If the facility has a low power factor, it is, in effect, demanding more power than it is actually using to do useful work. Therefore, a low power factor can increase a facility's carbon footprint.

Power factor is calculated by using an electric meter that can register both kilowatt hours and kilovolt-ampere reactive hours. From this reading, a facility's power usage can be corrected. For example, a utility may correct demand if the power factor falls below 0.95.

$$\text{Corrected power usage} = \text{kW demand} \times ([.95 - \text{PF}] + 1)$$

If a facility's power factor is 0.85 and the demand is 1,000 kW, the corrected demand is:

$$\text{Corrected power usage} = 1,000 \text{ kW} \times ([.95 - .85] + 1) = 1,100 \text{ kW}$$

The charge for electricity will be based on 1,100 kW rather than the 1,000 kW that was registered by the electric meter. An electric utility may also charge an additional penalty if the power factor falls below a set level.

The rate a facility pays for power can be seen in the following example. The example shows a charge for a small amount of power under the primary rate.

Example: Office Lighting Energy Cost

An office has 100 light fixtures in the ceiling. The lighting circuit consists of 105-watt, three-lamp, fluorescent light fixtures that burn for 12 hours a day during the peak period of 8 a.m. to 8 p.m. What is the total power required to power these fixtures, and how much power is consumed per day? If electricity costs 7 cents per kWh, how much does it cost to run the lights per day?

Total power in kW (kilowatts)

105 watts × 100 light fixtures = total power = 10,500 W (watts), or 10.05 kW (kilowatts)

Power consumed measured in kWh (kilowatt hours)

10.05 kW × 12 hours = 120.60 kWh

Cost of power consumed (kWh × cost)

120.60 kWh × \$0.07 = \$8.44/day

Charge for a thirty-day month = \$253.20 month

The charge for the power per day is \$8.44, and it is the direct charge that a customer has to pay for energy. Other charges may also be assessed by the power company. Two of these charges are a result of a facility's power factor and a possible low power factor penalty.

The power factor (PF charges and penalty) consists of a usage correction amount and a possible additional penalty charge if the power factor is extremely low. The charge for a low power factor is a charge that essentially raises the power level consumed during the monthly billing cycle.

$$\text{Corrected power use} = \text{kW demand} \times ([\text{acceptable PF level} - \text{actual PF of the facility}] + 1)$$

In our example, demand is 120.6 kWh/day × 30 days/month = 3,618 kWh. The acceptable PF level is 0.97, and the facility's PF is 0.73.

$$\text{Corrected power use} = 3{,}618 \times ([.97 - .73] + 1)$$
$$\text{Corrected use} = 3{,}618 \times 1.24$$
$$\text{Corrected demand} = 4{,}486.22$$
$$\text{Corrected amount charged for power} = 4{,}486.22 \times 0.07 = \$314.03$$
$$\text{Increase due to low power factor: } \$314.03 - \$253.20 = \$60.83$$

An old power factor analogy is a mug of root beer, as shown in Figure 10.1. Just like root beer pouring into a mug, the power company meters electricity. The cashier charges for the liquid root beer, and the power company charges for kilowatts or real power. The reactive power (magnetizing power VARS) is necessary for the electric motors to run, just like foam is necessary for taste. But it would be rather useless to have a mug filled mostly with foam (might taste good but won't satisfy thirst). The power company's infrastructure is like the mug. It must hold both the root beer and the foam. Too much foam (or too low a power factor), and the power-generating equipment and transmission equipment will have to be larger (require a bigger mug) than would have been necessary if the mug was at unity power factor, or full to the top with beer. Capacitors added to the building's electrical system act like the counterperson filling the mug who brushes the foam off the top to allow more root beer to fill the mug. Capacitor banks installed in the building's electrical system raise the power factor.

10.3 FACILITY STRATEGIES FOR MINIMIZING POWER USE AND COSTS

A strategy for achieving energy savings and a reduction in facility utility costs are important factors in achieving a sustainable facility operation. As shown in the next subsection, saving energy does not always equate to having a money savings outcome. Rather a facility manager must balance all the factors

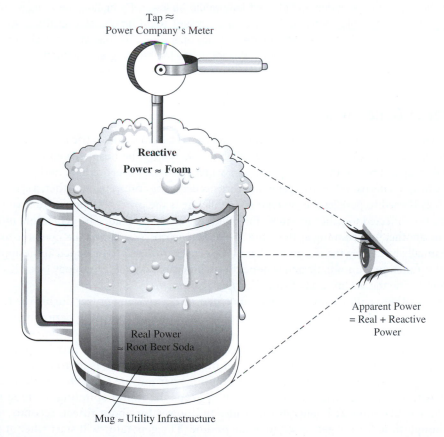

FIGURE 10.1 Power factor and mug of root beer analogy

that go into the delivery and use of electricity to place the facility in the most favorable position related to power.

Facility Strategy to Shift Demand to Off-Peak Hours

One of the strategies facilities use to shift demand to off-peak hours is called ice storage. During off-peak hours, the facility makes ice. The ice is made at night and stored for use in cooling the building during the on-peak hours. But as far as kilowatts consumed are concerned, making ice isn't as energy efficient as running a chiller directly to cool a building. Some of the cooling capacity from the ice that was made is lost during the storage period due to heat gain from the surroundings. We also have to run some of the same auxiliary equipment (e.g., chilled water pumps) when we made the ice and use them again to deliver the cooling effects of the ice. This equipment runs twice as long as it would without ice storage. On the plus side, during the evening hours when the ice is made, outside air temperatures are generally lower, making cooling towers and chillers more efficient. Making ice still *increases* the total amount of electrical energy consumed by the facility. At the same time, however, ice storage saves money by shifting the high electric load items (chiller compressor, cooling tower, and tower pumps) to the period when the rate charged for electricity is lowest.

Example: Ice Storage

An ice storage system was approved, built, and tested for a high school. When the high school experienced very hot summer days, the ice was completely used up by 1 p.m. in the afternoon. The chiller was then started and run at full capacity to keep the building cool. The electrical demand had been kept low until that point, but now the facility would be forced to pay for the higher demand for this billing period and the next eleven months.

What the FM didn't understand was that the way the building was operated in May wouldn't work in July. In May, the chiller was operated at night to make ice, and the ice was used to cool the building during the school day. But for high cooling demand days, the ice system was designed to be operated in conjunction with the electric centrifugal chiller. If the chiller had been operated with the ice storage system on excessively hot days, the peak demand of electricity would be lower. By making ice at night, the small chiller (smaller than what would be needed without ice storage) could be operated at the most efficient load setting during ice production. The same high-efficiency operation was true when the chiller operated during the day in conjunction with the ice storage system to provide building air conditioning.

Facility Power Generation

Self-power generation is another way facilities can avoid charges for a high electric load. Some hospitals have large emergency diesel generators that are capable of handling the hospital's entire electric load. These hospitals may opt to use their own generators when electrical costs are high and provide all the power that the hospital needs. When two generators are needed to be connected together, a synchroscope (see Figure 10.2) is used to parallel the generators. Paralleling is the process where more than one generator's power is connected together to satisfy the building's electrical load. The generators must be in phase with one another and running at the same speed. A synchroscope allows the operator to parallel generators manually by closing the circuit breaker of the generator that is to be added at the appropriate time. Today this function is automatic and is done through master control technology (see Figure 10.3), along with monitoring the on-line and off-line electrical equipment.

Unlike many other facilities, hospitals are concerned with providing power during the transition period between power failure and emergency generator startup. This transition period may only be a few seconds long, but it is long enough to cause problems to mission critical equipment. Some hospitals utilize an uninterruptable power supply (UPS) battery backup system to provide power during the transition period. Another way to provide power during the transition is to utilize flywheel technology. A flywheel is driven at several thousand revolutions per minute (rpm) by an electric motor. When power is lost to the electric motor, the flywheel keeps the motor spinning, turning it into a generator. The flywheel-powered generator provides full power for about fifteen seconds, which is more than enough time for the generator to come on line: The generator will start when it senses a low voltage condition; it will not wait until power is completely lost. A flywheel is a green solution to

FIGURE 10.2 Synchroscope

Source: Photo Courtesy Oconomowoc Memorial Hospital.

the transition problem because it eliminates the need for chemical batteries and is more efficient than a battery UPS.

A facility generating its own power must consider the increased costs of generator maintenance, generator operating personnel costs, air pollution regulations, noise, vibration, and fuel costs. But after a favorable analysis of all the considerations, a facility may find additional benefit in turning a large capital investment in emergency generators into a tool that can be used in the fight to contain energy costs. The fuel used may also reduce the carbon footprint of the facility if the facility's generating system is efficient and the fuel contains less carbon than what is used by the local utility. An example is a facility that uses natural gas to generate power while the power utility uses coal; the facility will have lower carbon dioxide and nitrogen oxide emissions. A facility generating its own power will generate waste heat, which may be captured in a combined heat and power (CHP) system. Efficient local heat recapture can boost efficiency above centralized power generation by a utility. A power company is located a distance away from its customers; heat recapture and delivery of heat

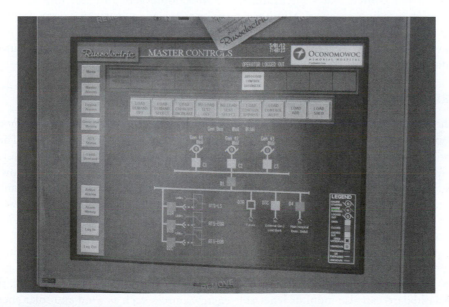

FIGURE 10.3 Master controls for controlling generators

Source: Photo Courtesy Oconomowoc Memorial Hospital.

recapture to its customers (and receiving return flow back to the plant) is more difficult. Also because of the distance away from the consumers, a power utility will have greater electric transmission losses.

Instead of providing power, the hospital can also perform what is known as peak shaving, whereby facility-generated power is provided to only select areas of the hospital. This lowers the peak demand that the power company's meters register because the power company provides only part of the total demand. Peak shaving keeps the peak demand from rising above the normal operating level. To accomplish this in the safest manner, power is supplied by the generator to equipment that is not critical to patients. Any minor disruption to the flow of power when switching between power sources won't cause any problems for the patients.

Temporary Electric/Cooling Load Solution: A Poor Practice

In the past, some facility personnel simply made a guess based on their prior experience on the actions that needed to be taken to reduce a building's electric load. They knew upper management was unhappy when excessive load resulted in a high electric bill, but few pieces of equipment actually told the operator the actual amount of power that was being consumed.

One exception was the electric centrifugal air-conditioning chiller. Chillers had an analog ampere meter mounted on the back of the motor control box. The operator saw the rising ampere draw and took action to reduce it. Frequently, the operator adjusted chilled water temperature to a higher level, which reduced compressor load. This action decreased the amount of humidity that condensed on the cooling coils as the building's supply airstream passed over the coils. This increased humidity levels inside the building, making the occupants less comfortable.

Sometimes operators opened the fill valve on the tower sump and overfilled it with 55-degree city water. This cool water, when sent to the chiller's condenser, reduced the compressor head pressure by more effective cooling of the hot refrigerant gas. This improved centrifugal compressor efficiency and resulted in a drop in the ampere draw by the compressor's motor. Lower power consumption was the good news. Unfortunately, this practice resulted in a loss of tower chemicals, scaling of condenser tubes (the extent of which depended on the hardness of the water used), and an increased charge for the water usage. If city water was used frequently for tower water cooling, the associated water costs for this wasteful practice and the compressor efficiency loss due to scale buildup on heat transfer surfaces, in the long run, greatly exceeded the temporary electrical savings.

10.4 MONITORING AND BENCHMARKING ENERGY

Overall success in energy management dictates a strategy that must include timely monitoring and comparison of best energy practices through benchmarking. Benchmarking can be performed through use of the United States Environmental Protection Agency's (U.S. EPA's) database and through research into best practices. Monitoring can be achieved by using an energy dashboard. Due to the daily demands, an FM can easily forget about monitoring energy, especially when no immediate issues exist. A dashboard makes monitoring energy a simple task, and the software automatically collects data for future use.

Dashboards: An Energy Tool for Today

Through the use of the facility's building management computer, building operators can monitor electric load continuously and make changes to equipment operation that affect electric load. The operator is alerted when load is rising and can take proactive measures to reduce load that won't measurably affect indoor air quality (IAQ) or waste precious water resources. The operator can choose to allow temperatures to rise to a higher level in the entire building or selectively choose spaces within the building for a temporary temperature increase, thereby reducing cooling load. The operator can also prevent a high electric load from occurring by turning off noncritical equipment or reducing services, such as the number of elevators in operation.

To make this aspect of energy management somewhat easier, all the pertinent energy data on energy use is put on one handy screen. This gathering place for energy information is called a dashboard (see Figure 10.4). The **dashboard** provides information on the use of natural gas, electricity, oil, steam, and water. The dashboard can also monitor renewable energy sources such as wind and solar energy. It can gather information, store the information, provide graphs and trending data, and make predictions on the amount of energy that will be used.

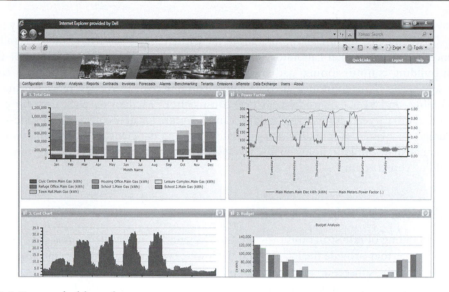

FIGURE 10.4 Energy dashboard

Source: Material courtesy of Trane.

Benchmarking Consumption

Benchmarking consumption is a process of comparing the energy consumption of a facility to other facilities of the same type, use, and age. This comparison allows the facility manager to decide if improvements are needed in the operation of the building to remain competitive. The U.S. EPA's Energy Star Program provides the FM with a web-based computer program and database that compares the facility's energy and water use information with thousands of buildings of the same type across the United States. The program, called Portfolio Manager, is used as a way for the building to achieve the ENERGY LABEL label (see Chapter 1). Building managers from across the United States input data to track their buildings' energy and water consumption. This information allows a facility manager to benchmark his or her building against similar buildings. The program then compares the building's information to the EPA's database and rates the building on a scale of 1 to 100. The system accounts for weather-related factors for the various areas of the country that would otherwise skew the results. A ranking of 60 means that the building performs better than 60 percent of similar buildings.

Energy Star's Portfolio Manager also provides the facility manager with information on the amount of carbon dioxide that the building's operations admit to the atmosphere through energy consumption. The amount of carbon dioxide is adjusted according to the building's location because electric utilities serving a region consume different types of fuel, and burning the different fuels produces different amounts of carbon dioxide for the same amount of electricity generated. For example, a building in the Northwest that is served by hydroelectric power plants emits half the amount of carbon dioxide as the same building in the Midwest that is served by coal-fired electric plants.

Building Commissioning: A Tool for Improving Efficiency

After large commercial buildings are constructed, the building systems must go through a commissioning process to ensure that the equipment is operating as designed. Just because a piece of equipment is mounted correctly, has a high efficiency rating, and is up and running doesn't mean that it meets the total system design specifications. During commissioning, engineering consultants ensure that the water flows, air-conditioning tonnages, energy consumption, airflows, energy consumption, controls, and equipment operation are satisfactory.

Retro-commissioning is done later on in the life of a building to ensure that it continues to operate as efficiently as originally designed. Retro-commissioning is frequently the first step before attempting to upgrade an existing building's HVAC system. It provides a baseline to work from, and it can be used to decide where improvements need to be made, where the equipment must be kept, and which equipment should be eliminated or upgraded.

A study on building commissioning, "Building Commissioning: A Golden Opportunity for Reducing Energy Costs and Greenhouse Gas Emissions in the United States" by Evan Mills, was published by the Lawrence Berkeley National Laboratory on February 12, 2011. The report is said to be the largest analysis of building commissioning, gathering the results of 643 commercial buildings comprising 99 million square feet. The report indicates that commissioning costs in 2009 were $0.30/square foot for existing buildings and $1.62/square foot for new construction. Median payback time for performing commissioning was 1.1 years for existing buildings and 4.2 years for new construction. One-third of the projects studied uncovered over 10,000 energy-related building deficiencies. Correcting these deficiencies resulted in median whole-building energy savings of 16 percent for existing buildings and 13 percent for new building construction.

10.5 RATE STRUCTURES

A basic understanding of rate structures provides the FM with an insight that can lead to equipment or operational changes that provide the facility with the lowest cost for the power it consumes. The information in the following subsections should be supplemented with specific details of the actual rate structure provided by the electric utility serving the facility. Rate amounts and methods for calculating charges can vary among utilities.

Commercial Rate Structures

The primary rate generally consists of monthly charges for the following: the electric service charge, demand charge, charge for electricity used based on time and/or season of use, and an adjustment to the metered amount of electricity through a power factor charge and a possible penalty charge for low power factor. To receive the primary rate (lowest rate per kilowatt hour from the power company), the facility must own the step-down transformer that changes the voltage to a lower and thus more useable level. A facility that operates at this lower level may be billed extra for having a low power factor. Larger facilities, such as high-rise office buildings, are offered the primary rate by the power utility because of the large amount of power they consume and their ownership of the transformer.

If the facility doesn't own the transformer that steps down the incoming voltage, the power company may bill based on the secondary rate. The *secondary rate* is a higher rate per kWh consumed that does not consider power factor in customer billing or levy an additional charge should the power factor be low. But the secondary rate may include incentives for using power during off-peak times through lower off-peak rates and may add a charge for the customer's highest kilowatt demand. This demand charge can be significant, especially because it will be levied the month that it occurred and sometimes for the next eleven months, even if the highest demand for those eleven months is lower.

Interruptible Rate for Natural Gas

Many utility companies that supply natural gas give customers what is sometimes called an interruptible rate. The demand for natural gas is sometimes so high that it is difficult for the gas utility to provide service to all its customers. Utilities pay for the right to order a customer receiving a lower interruptible rate to stop using natural gas until the high usage subsides. Most boiler operators switch to No.2 fuel oil, frequently called furnace oil, during these periods to keep their plant in operation.

The advantage of the interruptible rate is a reduction in a facility's gas bill. In some locations, customers are rarely asked to stop using gas. so it is worth some research to find out how frequently other customers in your area are asked to stop using natural gas before making a decision to go on an interruptible rate.

The disadvantage of the interruptible rate is that many boilers do not operate as cleanly as they do when they are forced to switch to burning oil. Switching to burning oil could create a lot more maintenance involved in the cleaning of the boiler's internal heating surfaces. Maintenance costs are also incurred in maintaining and operating fuel oil storage tanks.

The No. 2 fuel oil has its own maintenance requirements and disadvantages. The fuel must be treated periodically so that it doesn't degrade and so that bacteria don't start growing in the fuel. These bacteria can cause boiler fuel strainers and fuel filters on diesel engines to plug. A biocide is added to the fuel to prevent this condition.

10.6 LIGHTING

For many years, installing energy-efficient lighting has been an easy way for facility manages to save energy and provide the facility with a short payback period. Now lighting choices have increased, and many facilities have already performed energy-efficient lighting retrofits. Facilities today are integrating day lighting, glare control automatic blinds, and occupancy sensors to reduce the electric load created by lighting and provide for a more habitable and productive environment.

The majority of interior lighting today is from fluorescent light fixtures. Exterior lighting employs a greater array of lighting options, such as high-pressure sodium, metal halide, fluorescent induction, and LED. Exterior areas might not require accurate color rendition. For good color rendition, exterior lights are either metal halide, halogen, or incandescent. Otherwise, the lights may be fluorescent, fluorescent induction, LED, or high-pressure sodium. Also the light may not have to be pleasing to the viewer; in many cases, it only has to illuminate the surroundings.

Light Quality

Light quality is as important as the energy consumed by the lamp. Frequently, managers compensate for poor light quality by adding more light, with the resulting waste of energy. One measure of light quality is its color rendering index (CRI). The maximum CRI is 100 produced by incandescent lamps and ranges as low as zero. The *color rendering index (CRI)* describes how accurately colors appear under a particular light source. A CRI of 62 is produced by a T-12 standard cool white fluorescent lamp.

The *color temperature* of a lamp is described in degrees of Kelvin. Zero degrees Kelvin is equivalent to absolute 0 or −273.15 degrees centigrade. You can think of color temperature as the light produced as a piece of metal is heated. At 2,700 degrees Kelvin, the metal gives off a yellowish/orange white light. Heating the metal further, to about 3,000 degrees Kelvin, eliminates the yellowish/orange tinge from the white light. Additional heating, to 5,000 to 6,500 degrees Kelvin, produces a bluish white light, and this lamp is referred to as a daylight lamp. A cool white lamp has a color temperature of about 4,100 degrees Kelvin, whereas a candle has a color temperature of 1,700 and a CRI of 100. A happy medium of 3,500 degrees Kelvin is a popular choice.

An incandescent lamp has a CRI of 100 and a color temperature of 2,700. A high-pressure sodium lamp has a CRI of 25 and a color temperature of 2,100 degrees Kelvin.

Fluorescent Lighting

A 23-watt fluorescent lamp (see Figure 10.5) can provide the same light output as a 100-watt incandescent. It is easy to see the benefits of switching from the incandescent lamps invented by Edison to fluorescent. Fluorescent lighting provides a vast increase in both lamp life and energy efficiency over incandescent. Fluorescents made further headway when light socket screw-in lamps with integral ballasts, called compact fluorescents (CFLs), were developed, thus making conversion easy. CFLs last between 7,500 and 12,000 hours. There were some tradeoffs in the quality of lighting and reduced power factor for the facility's electrical system. But today fluorescent light fixtures provide color temperatures that more accurately mimic daylight, and old magnetic ballasts have given way to electronic ballasts that are more efficient and no longer contain toxic polychlorinated biphenyls (PCBs). Magnetic ballasted fluorescent lamps operating at 60 Hertz flickered 120 times per second because alternating current of 60 Hz (cycles per second) is a zero voltage twice each cycle. Although it should be imperceptible, many people report headaches and eyestrain when working under this light for long periods.

Fluorescent lighting today works by using an electronic ballast to supply power at a very high frequency (20 to 60 kHz) to a tungsten filament inside the glass tube. The tube contains inert gas and mercury vapor, and has a phosphor coating on the inside. The tubes are denoted with a T and a number. Each number is ⅛" of tube diameter. Therefore a T-8 fluorescent lamp has a diameter of 1" and a T-5 has a diameter of ⅝". The lamp's filament excites the molecules of the mercury vapor when current is run through it. This causes ultraviolet light to be produced that strikes the phosphor coating and is converted to visible light. The high frequency eliminates flicker because the phosphor glows longer than the period of zero voltage. The color temperature of the light is influenced by the type or combinations of phosphor coatings used. Depending on color temperature, the light is referred to as cool white, daylight, or warm white. As the lamp warms, its resistance decreases and current flow increases to a dangerous level, except that the ballast controls the flow of current.

FIGURE 10.5 Three-way fluorescent CFL lamp

Source: Osram Sylvania Inc.

Advances in Fluorescent Ballasted Lamps

A new T-8 fluorescent lamp with an electronic ballast may have a life of 24,000 hours to 30,000 hours based on a twelve-hour on/off switching cycle, and a CRI of 82 to 86. Lamp life depends on the switching cycle or on/off times. The greater the number of on/off cycles, the shorter the lamp life. This can be very problematic for fixtures that are connected to occupancy sensors, which turn on and off frequently. A solution is to use a warm-start electronic ballast that preheats the fluorescent lamp for about 1½ seconds prior to full ignition of the lamp. This warming improves lamp life. Ballast themselves last for up to 20 years compared to the ten-year life span of old magnetic ballasts.

T-5 lamps are now known for their high output, and they are frequently used as an energy-efficient replacement for high-bay metal halide lamps. Inside office buildings, these fixtures can provide effective indirect lighting by shining the lamps upward toward the ceiling. After 20,000 hours, a T-5 on a three-hour cycle has an 83 percent survival rate and still produces 90 percent of its lumen output as it did when new.

Disadvantages of Ballasted Fluorescent Lamps

Ballasted fluorescent lamps have some disadvantages. Fluorescent lamps contain mercury and require additional costs to recycle the spent lamps. Fluorescent

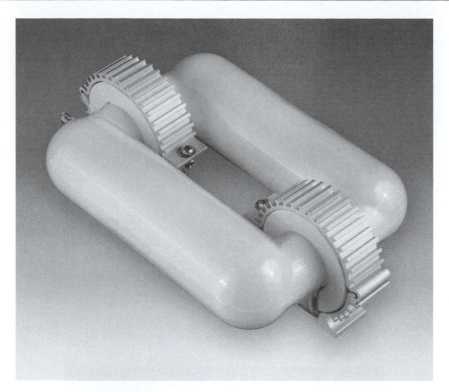

FIGURE 10.6 Fluorescent induction lamp

Source: Osram Sylvania Inc.

lamps have a tungsten filament that deteriorates as it is used, and the end seals that contain the mercury vapor and inert gas may fail. Over time, the mercury is depleted and the light output decreases.

Fluorescent Induction Lighting

Induction lighting (see Figure 10.6) removes the filaments and the end seals of ballasted fluorescent lamps. The lamp is a continuous loop. Rather than a filament, induction coils wrapped around the outside of the lamp excite the mercury inside the lamp. The lamp still requires ballasting, which regulates both frequency and current to the induction coil. The lamp takes one to two minutes to reach full light output. Induction lamps frequently last between 80,000 and 100,000 hours and are suitable for high output lighting in industrial buildings. Induction lamps are frequently more expensive than T-8 or T-5 fluorescent fixtures, but with their long lamp life, they are often used in high or difficult to get to places, and they are frequently more cost-efficient over the long term as a result.

Halogen Lamps

Halogen lamps (see Figure 10.7) are a type of incandescent lamp with a tungsten filament that is surrounded by a halogen gas of bromine or iodine. It has a life span of only about 2,000 hours. This lamp burns very hot, but because it is a type of incandescent lamp, it has a CRI of 100. Halogen lamps provide 20 percent more light than an incandescent lamp. They emit light at roughly 20 lumens per watt, but some lamps used in projectors can put out as much as 35 lumens per watt.

Light-Emitting Diode Lighting

White light-emitting diode (LED) lamps are the most energy-efficient forms of lighting, with a life span of between 30,000 and 55,000 hours (see Figure 10.8). They deliver approximately 60 to 80 lumens per watt and have a CRI of 80 to 85. The life span of a lamp is frequently described as the time it takes for a lamp's light output to degrade to 50 or 60 percent of what it was when the lamps were new. The life span of an LED is affected by the ambient temperature around the lamp. At ambient temperatures above design, the life of an LED will decrease. LED lamps have the unique advantage of not containing hazardous materials.

FIGURE 10.7 Halogen lamp

Source: Osram Sylvania Inc.

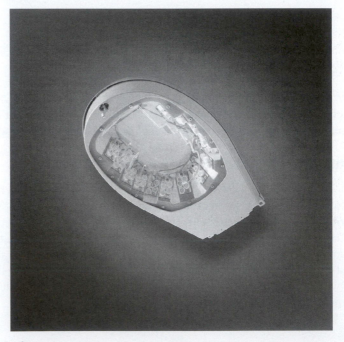

FIGURE 10.8 LED light fixture

Source: Osram Sylvania Inc.

Many mistakenly refer to LED lighting as maintenance-free. Although extremely long-lived, LED fixtures still require the normal cleaning and inspections that any lighting requires. LED lighting isn't a mature science and new discoveries, applications, and improvements still occur all the time.

High-Pressure Sodium

High-pressure sodium (HPS) fixtures (see Figure 10.9) are generally used to save energy for exterior lighting. They can be used where a low CRI of 22 isn't important. Some of the higher wattage (400 watts or more) HPS lamps have an initial light output of 125 lumens per watt, for a total output of 50,000 lumens. This light output degrades somewhat over time, and the mean lumen output at 50 percent may drop to 45,000 lumens (which is still quite high). HPS lamps have a life span of approximately 24,000 hours. As the lamp nears the end of its life, it will begin turning on and off due to a lack of sodium in the arc tube. Some lamps are equipped with dual arc tubes for fast light-up, so lamps restore lighting quickly after being turned off.

FIGURE 10.9 Sodium lamp
Source: Osram Sylvania Inc.

Metal Halide

Metal halide lamps (see Figure 10.10) are used for lighting in parking lots and elevated parking struc-tures, warehouses, decorative pole lighting, retail lighting in shopping malls, and gas station canopies, to name a few. Some higher wattage lamps (400 watts) have a good lumen output of about 80 mean lumens per watt, while lower (100-watt) lamps have about 62 lumens per watt. CRI ranges from 62 to 75 for the

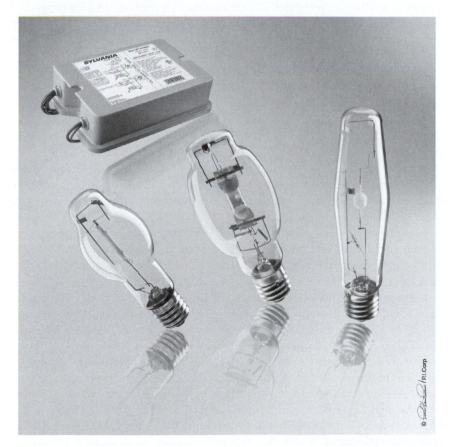

FIGURE 10.10 Outdoor metal halide lamps and ballast
Source: P.I. Corp.

larger lamps, and smaller lamps have a CRI ranging from 80 to 83. Some specialty lamps have CRIs as high as 90. The life of a metal halide lamp depends not only on the switching cycle, but also on lamp orientation, which is either horizontal or vertical. A horizontally installed lamp may last for 15,000 hours, whereas a vertically installed lamp lasts only 10,000 hours.

A common mistake that maintenance personnel make is installing a lamp that has the wrong wattage in a metal halide fixture, which can happen easily when different types of decorative metal halide lighting are used at a particular location. For example, at a residential shopping center, wall sconces might require 50-watt lamps and decorative light poles may require 75-watt lamps. The two lamps may appear very similar. If the 50-watt lamp goes in a 75-watt ballasted fixture, the lamp will burn out in a few weeks because it will be overdriven. If a 75-watt lamp is installed in a 50-watt ballasted fixture, the result will harder to detect because the lamp will put out about 50 watts worth of light output. If your computerized maintenance management system (CMMS) system indicates frequent lamp changes for a particular metal halide fixture, improper lamp wattage may be the culprit.

Managing Lighting

Managing lighting has taken on new dimensions for facility managers. In the past, the most difficult task was to know when to replace lamps or ballast light fixtures to get the most life out of the fixture before too much loss of light output occurred. Then the hazards of mercury and PCBs became better understood, and recycling laws went into effect. Today, building information modeling (BIM) managers can click on a light fixture showing on a computer screen and get the fixture information, including the age of the lamp and ballast in service. This software should cut down on waste with more effective management of each and every fixture.

Lamp and Ballast Replacement

In the past, managers were taught to undertake a program called group relamping. When 60 to 80 percent of one lamp's projected life was reached, all the lamps were changed within a space. Fixtures that may have had their lamps fail prematurely and now had fairly new lamps were also changed. Group relamping basically meant that the facility could now start at "day one" of lamp life in all the fixtures.

Another common practice was a fixture overhaul, where the light fixtures within a new tenant's space were cleaned, relamped and reballasted. The process of relamping involves a light cleaning of the fixture and replacing all the lamps. Reballasting may also be done with relamping, depending on the age of the ballasts; in this case, the ballasts that power the lamp are replaced. This relamping and reballasting effort was generally the result of lease negotiations that required it as a condition of leasing the space. Many tenants would negotiate this replacement so that they could be confident that little disruption and additional costs would be incurred with regard to overhead lighting. Frequently, leases call for additional fees to be paid by the tenant when lighting within their space requires repair or even just a simple change of lamps. New tenants figure that replacing the most important components greatly decreases the need for service during their tenancy.

Both relamping and reballasting have some major flaws. The practice of group relamping, can be extremely wasteful. If we can identify the fixtures that have fairly new lamps, we can avoid having to recycle them before their time. We also need to know the on/off cycling of the fixtures so their life can be properly estimated. If a three-hour cycle gives a 20,000-hour lamp life, then 80 percent of useful life is at 16,000 hours, whereas a twelve-hour cycle might yield a 25,000 hour lamp life, with 80 percent of useful life at 20,000 hours. It would be quite wasteful to dispose of lamps with 4,000 hours of life left in them. New lamps do not degrade as much in light output as the old T-12s did, so lamp replacement at 60 percent of life is much too conservative.

Fixture overhaul may result in lamps and ballasts with thousands of hours of useful life left in them being sent to the recycle bin. A way to avoid this is through good record keeping. For example, let's say that a previous tenant occupied a space for three years that a new tenant is moving in. The lights within the space might have been used for 3,000 hours every year, for 9,000 hours of use during the tenancy. If the lamps are good for 20,000 hours, there is more than enough life for another three-year tenancy. With an accurate estimate of fixture and lamp life remaining, an amicable solution can be negotiated. The FM must be careful that the useful life estimate is accurate because, in general, it is significantly more expensive to relamp an occupied space than an unoccupied space.

Space Junk

An unpleasant surprise that managers may discover when performing a lighting fixture replacement project may be something called space junk: Lighted canopies over sidewalks at strip shopping centers and building entrances might contain a variety of old light fixtures trapped inside. This is more common when recessed can fixtures are used because the fixture is installed in the ceiling from the outside. It is easy for a lighting contractor to disconnect the wires to the old light fixture and push the fixture into the void inside the canopy, leaving the manager with space junk to deal with later. This space junk might be old outdated mercury vapor fixtures that require the recycling of both lamp and ballast as hazardous waste.

10.7 ALTERNATING CURRENT MOTORS

Some of the largest consumers of electricity at a facility are the various electric motors that drive the fans, pumps, compressors, and other types of machinery throughout the building. It is important to have a basic understanding of these rotating machines in order to make wise purchases and save as much energy as possible.

AC Motor Basics

The three major aspects of alternating current (AC) motors are: motor speed, motor power, and motor starting. The speed of an induction motor is controlled by the frequency of the alternating current or the number of magnetized motor poles. In the United States, electric current is supplied at a frequency of 60 cycles per second, also known as 60 Hertz (Hz). The frequency of the supplied current controls the speed of the motor. Motor speed can also be controlled by the number of motor polls in use. A motor with a high, medium, and low speed may employ a switching mechanism to control the number of motor polls the motor is using. The disadvantage of this system is having only a few motor speeds to meet the need, whereas a variable frequency drive (VFD) can provide for almost infinite and instantaneous speed adjustments.

The amount of power consumed by a motor is in direct proportion to the cube of the speed of the motor. In other words, power = (motor speed)^3. Here is an example of speed reduction versus power consumed:

If the motor speed is reduced to 80 percent, the power = $(.80)^3$
$$= .512, \text{ or } 51.2 \text{ percent. The motor then consumes 48.2 percent less power.}$$

Small AC motors of 1 to 3 horsepower are generally single-phase motors that may employ a capacitor attached to the outside of the motor housing. The capacitor is used to apply a higher voltage for more starting power and to create a rotating magnetic field. AC motors over 3 HP are generally three-phase motors. These motors are supplied by three power leads carrying current that is 120 degrees out of phase with each other. As a result of the phase difference, three-phase motors are self-starting (they do not require capacitors for starting), and there is less voltage variation, which allows them to operate more efficiently than single-phase motors.

AC motors in a clean environment require very little maintenance. Most three-phase motors today have sealed bearings that are lubricated for the life of the motor. Three-phase motors have the additional advantage of being easily reversed by switching any two of the three lines that provide power to the motor.

Variable Frequency Drive

A highly successful application of equipment for energy savings that is now extremely common at most facilities is the installation of variable frequency drives (VFDs) for the speed control of alternating current motors (see Figure 10.11). If the AC motor is rated for inverter duty (meaning that the motor can tolerate pulsed DC voltage), it can be connected to a VFD. VFDs are also known as adjustable frequency drives (AFDs). VFDs are used for applications where it is desirable to operate equipment at various speeds that match the load requirements. Unlike the across-the-line starter, where full-line voltage is instantly applied to the motor when starting, the VFD can provide a slow, even starting of equipment that overcomes the equipment's initial mechanical load without drawing excessive current. VFDs do this by simulating an AC waveform with pulsed DC voltage.

FIGURE 10.11 Variable frequency drive providing control of a boiler forced-draft fan

Source: Photo Courtesy Oconomowoc Memorial Hospital.

VFDs are used at facilities for the starting and running of air-handling units, pumps, compressors, and elevator winch motors. By taking input data from control equipment and processing it, the VFD then operates the motor at the most efficient speed for the conditions. A common form of air-handler fan speed control is achieved by monitoring static pressure in the air duct. An electronic static pressure sensor sends a 4- to 20-milliamp signal to the VFD that relates to the amount of static pressure in the duct. As static pressure drops in the duct due to variable air volume (VAV) boxes opening (in turn due to increased cooling, heating, or ventilation demand), the static pressure sensor sends a signal to the VFD to speed up the motor, thereby increasing static pressure.

In the past, motor starters employed soft starters, star-delta-configured motor windings, autotransformers, or resistors that applied increasing voltage levels in steps to the motor during starting. These machines started the motor without the motor drawing a massive amount of current all at once. The VFD eliminates the need for these early forms of starting equipment.

10.8   WATER-SOURCED HEAT PUMPS

Water-sourced heat pumps for commercial buildings are used for heating and cooling (see Figure 10.12). They are refrigeration units that have a water-cooled condenser and an air-cooled evaporator. In a heat pump arrangement that is not geothermal, a water-sourced heat pump condenser is cooled with water from a cooling tower. In cooling mode, the heat of compression is removed in the refrigerant condenser section. Cooling water for the condenser generally comes from a cooling tower located outside the building. The cool water removes the heat of compression and turns the refrigerant to a liquid. The liquid refrigerant is sent to the evaporator through an expansion device, which drops the temperature of the refrigerant. A fan blows air over the cold evaporator coil, and the cold air flows into the space.

In the heating mode, the cooling tower is secured. The heat pump reverses the flow of hot refrigerant. Instead of sending this hot compressor refrigerant gas to the condenser, a reversing valve changes the direction of the gas flowing out of the compressor and sends the hot gas to the evaporator coil. A fan blowing over the coil provides warm air to heat the space. The hot gas that is cooled in the evaporator changes to a liquid. The liquid flows into the condenser. The water in the condenser now needs to be warm to cause the liquid refrigerant to change back into a gas so it can be compressed once again in the compressor. The problem is that cool liquid refrigerant has entered the condenser. The refrigerant is now changing to a gas in the condenser and, in the process, removes heat from the condenser water and might freeze it. To compensate for this, a small boiler or electric water heater is installed in the system to keep the water temperature in the condenser above freezing and to warm the water enough to change the liquid refrigerant back to a gas.

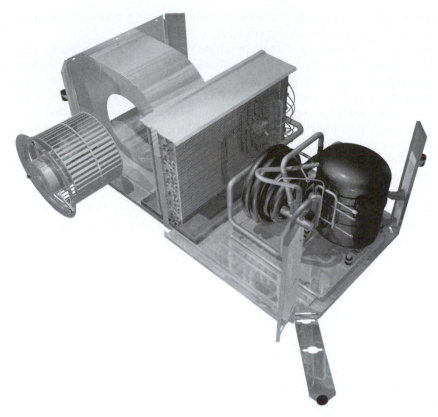

FIGURE 10.12 Heat pump components

Source: Material courtesy of Trane.

Advantages and Disadvantages of Using Multiple Heat Pumps

One advantage of using multiple heat pumps in multiple zones is that they can serve the heating and cooling needs of each zone more precisely, which ultimately saves energy. This precision is important, for instance, when there is a vacancy in a building. Units serving unoccupied areas can be easily secured or the room temperature can be set to meet minimum space temperature requirements. Heat pump installations frequently have only a few tons of cooling capacity, and they usually serve only one zone in the building.

A ton of cooling is equivalent to removing 12,000 BTU/hour. The expression originated in the 1800s when ice-packing houses would store 2,000-pound blocks of ice (one short ton) set in sawdust inside insulated ice houses. The heat of fusion of water to ice is 144 BTU/pound, so 2,000 lbs × 144 BTU is 288,000 BTU. This is the amount of heat that will be removed to melt a block of ice. Because there are twenty-four hours in a day, 288,000/24 = 12,000, or 12,000 BTU/hour. Some commercial buildings have over 100 heat pumps to meet the cooling and heating demand.

A disadvantage of using multiple heat pumps is that the installation requires extensive water piping throughout the building to supply water to the condenser, and to remove condensate water from the evaporator coil. Having multiple pieces of mechanical equipment throughout the building's tenanted spaces is a maintenance concern. Instead of maintaining one or two main units, the maintenance team may have to deal with dozens and dozens of heat pump units. Maintaining so much small tonnage equipment, especially when it is nearing the end of useful life, can be costly and disruptive to the tenants. Mechanical noise from compressors, flowing water in pipes, and electrical hum present within the space can be issues.

Heat Pump Water Heaters

Let's compare an older tank type of electric water heater to a heat pump water heater. Electric water heaters have two submerged heating elements, an upper and a lower. Electric water heaters convert electricity directly to water heat because the elements are in direct contact with the water. Some plumbing textbooks state that electric resistance element water heaters are 100 percent efficient. Heat pump water heaters claim to be about *twice* as efficient as electric water heaters! So does this mean that heat pump water heaters are an impossible 200 percent efficient?

Heat pump water heaters consist of a heat pump sitting on top of a tank of water. They use electricity indirectly to move existing heat from the area surrounding the water heater into the water in the tank. Electric- or gas-fired water heaters must create the heat energy to heat the tank water. Heat pump water heaters use energy to move heat into the water. The hot refrigerant gas coming out of the compressor heats the water, which turns the refrigerant gas into liquid. The liquid refrigerant then goes to an evaporator coil. The warm air surrounding the heat pump is pulled across the outside of the evaporator coil, where the warm space air is moved into the refrigerant, changing it back to a gas. The warm space air that heats the liquid refrigerant is free energy that goes directly into the water tank after the refrigerant gas is compressed. Electricity is used only to compress the refrigerant gas and to run the evaporator fan. The air leaving the evaporator is cool and dehumidified.

This system has an added advantage for a home basement installation that has excess heat and humidity. But it is also a disadvantage for a space that does not want to lose heat because, as heat is pulled from the space and put into the water, the space will be further reduced. A scheme for exhausting cold evaporator air to the outside is employed under this scenario in order to prevent overcooling the space.

10.9 RENEWABLE ENERGY

FMs have taken advantage, either directly or indirectly, of the following forms of renewable energy to meet building needs: geothermal, solar electric, solar hot water, wind turbine power, and hydroelectric power. Renewable energy can be utilized singularly or in combination with other renewable forms. Renewable energy may satisfy the needs of a facility, provide a way to reduce the cost of energy, or be used in concert with conventional forms of energy. Renewable forms of power may be generated by the facility or provided by the power utility company. The goal is to reduce the amount of nonrenewable energy consumed so that the facility's carbon footprint is reduced.

Geothermal Heat Pumps

Geothermal heat pumps are practically the same in all respects except that the condenser is not cooled by a cooling tower or by a fan, as is the case with an air-cooled condenser. Geothermal heat pumps utilize a ***geothermal loop***, which is a closed water loop that may consist of thousands of feet of high-density polyethylene (HDPE) piping installed in the earth at depths of 10 to 300 feet. The earth itself acts as a big heat sink that takes away the heat of refrigerant compression in the condenser. Soil temperatures at these depths are about 55 to 60 degrees Fahrenheit year-round. Pumping water through the loop causes the water to be heated (or cooled) to about 55 or 60 degrees. Water near earth temperature is more than adequate to cool the hot gas in the condenser when the heat pump is acting as an air conditioner. The water temperature is also sufficient to keep the condenser from freezing when the heat pump is acting as a heater.

The geothermal heat pump uses the earth itself as a renewable source. Heat is actually moved into or taken out of the earth depending on the whether the heat pump is in the heating or cooling mode.

Geothermal Loop Configuration   Geothermal loops are configured in a variety of ways. One configuration requires that large coils be placed in a pit that is excavated to a depth of approximately ten feet (see Figure 10.13). As an alternative, this same type of looping may be placed in a lake or pond of sufficient depth to provide the appropriate temperatures required. The most frequently used design is a well-type installation. Wells may be installed both horizontally in (see Figure 10.14) and vertically (see Figure 10.15) through the earth. Frequently, over 100 wells are installed to provide sufficient cooling for a facility.

Advantages of Geothermal Pumps Over Conventional Cooling and Heating Systems   According to the U.S. Department of Energy, the biggest advantage to geothermal loops is that they use 25 to 50 percent less electricity than conventional heating and cooling systems. A geothermal heat pump uses *one unit* of electricity to move *three units* of heat out of the earth. Therefore, geothermal loops can reduce air emissions up to 72 percent compared with a building using electric resistance heat and standard air-conditioning equipment, or up to 44 percent when compared with

FIGURE 10.13 Spiral installation of a geothermal loop

Source: Material courtesy of Trane.

air-sourced heat pumps. The underground piping often carries warranties of twenty-five to fifty years, and the heat pumps themselves have a useful life of fifteen years but can often last twenty years or more.

Advantages of Geothermal Pumps Over Conventional Water-Sourced Heat Pumps

Geothermal systems supply water to the heat pumps at a temperature of approximately 60 degrees Fahrenheit, thereby eliminating the need for a cooling tower to cool refrigerant gas in the condenser when the heat pump is operating in the cooling mode. No boiler or other supplemental heating is required to keep the water in the condenser from freezing in the heating mode. A geothermal heat pump runs with an energy efficiency rating (EER) near 20.

FIGURE 10.14 Horizontal installation of a geothermal loop

Source: Material courtesy of Trane.

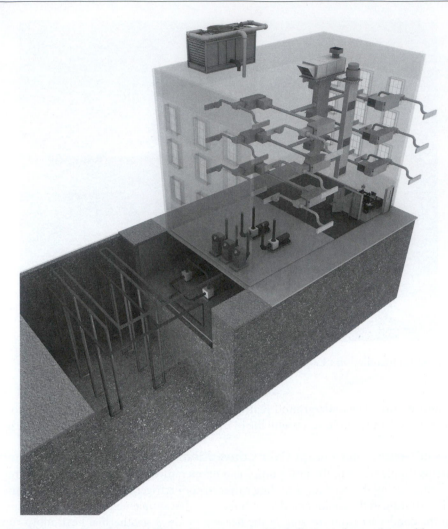

FIGURE 10.15 Heat rejection/heat supply: well-type vertical installation of a geothermal loop

Source: Material courtesy of Trane.

Some residential geothermal units also use waste heat from the condenser to heat domestic water. These units use a type of heat exchanger called a desuperheater. Superheat is the heat added to the refrigerant gas, which is additional heat above the gas saturation pressure. The **desuperheater** is a heat exchanger located at the outlet of the compressor; it cools the refrigerant gas to the saturation temperature without reducing refrigerant pressure. This is accomplished by circulating water from the domestic water heater through the desuperheater. The heat exchanger can remove 10 to 30 percent of the heat that otherwise would be rejected. The heat is transferred to a storage tank for domestic water heater preheating. More heat could easily be removed from the refrigerant, but doing so would result in several mechanical equipment problems. For example, if heat is removed so that the refrigerant temperature drops below the saturation temperature, the refrigerant pressure drops and the refrigerant turns to a liquid. This would result in low compressor head pressure, possible expansion valve malfunction, and liquid refrigerant slugging back to the compressor, with compressor damage the likely outcome. All these conditions can be avoided by the proper use of a desuperheater.

According to Alabama Power, a Southern Company, the amount of heat recovery achievable from a hydrochlorofluorocarbon-22 (HCFC-22) air-conditioning system is approximately 3,000 BTU per ton-hour of host system output. The amount of heat recovery is 2,200 BTU per ton-hour from a hydrofluorocarbon-134a (HFC-134a) air-conditioning system.

Disadvantages of Geothermal Pumps   Geothermal pumps have several disadvantages. They are not suitable for retrofit applications in existing urban buildings because of the lack of accessible open areas for well installation. Extensive internal piping and duct modifications and pipe insulation are necessary. Fireproofing of demising wall penetrations is required, which raises the cost of retrofitting.

Multiple heat pumps inside a space can create noise problems. Heat pumps require significant amounts of maintenance, especially when they near their end of useful life.

Photovoltaic Electric Power Generation

Solar cells are manufactured from a semiconductor material such as silicon with added impurities. Silicon, in a pure state, is a poor conductor of electricity. Boron and phosphorous added to silicon make a more suitable semiconductor. The impurities allow electrons to be released when the silicon is exposed to photons of light, and they cause the resulting dc current to flow in the desired direction.

Photovoltaic cells are wired in a series/parallel arrangement. Each cell, no matter its size, produces about 0.5 volts dc, but larger cells produce more current. Series wiring of several cells provides an additive effect and boosts the overall voltage produced. Parallel wiring is done to produce the necessary current. This arrangement of multiple cells is known as a *solar array*. All solar cells in an array must be exposed to light in order for the array to function properly. A shadow across one of the cells causes the cell to stop producing current. In fact, a shaded cell actually draws a disproportionally greater current from the adjoining cells in the string and becomes hot. To help mitigate the effects of shading, bypass diodes between strings of cells or other type of controllers are installed. Diodes are devices that allow current to flow in only one direction. These diodes result in power being lost only in the shaded portion of the array.

Some solar arrays are designed to track the movement of the sun throughout the day as the earth rotates around the sun (see Figure 10.16). This tracking boosts the output of the array, but it also adds a significant amount of complexity to the installation. In the northern hemisphere, fixed arrays are positioned facing true south (as opposed to the magnetic south direction on a compass); in the southern hemisphere, fixed arrays are positioned facing true north. Fixed arrays have an adjustment on their legs to change the angle of the array and thus accommodate the earth's angular position relative to the sun in the summer or winter months. Other arrays are flat-mounted directly to a pitched roof.

The power generated by a solar array (see Figure 10.17) is direct current. We can use this power directly to charge storage batteries, which are also always dc. Otherwise, we can send the dc power to an inverter and change the power to alternating current. If the alternating current is sent back to the electric power grid, it in effect spins the electric meter backward. This occurs only if the solar array is producing more power than what is consumed by the building. Otherwise, it has a net effect of reducing the amount the electric meter is registering.

Using 1,000 watts/meter squared as the amount of solar radiation power coming from the sun at the equator, a 12 percent efficient solar cell that is one meter square can capture a peak value of 120 watts of power. At the equator, the earth receives the most solar energy and forms a basis for solar power calculations. A reduction factor is used to compensate for the actual amount of sunlight for the area of the country where the array is located. Some solar panel manufacturers today claim an efficiency of about 20 percent.

FIGURE 10.16 Solar arrays with tracking equipment

Source: Photo courtesy of Waukesha County Technical College.

FIGURE 10.17 Solar array

Source: Photo courtesy of Waukesha County Technical College.

Solar technology is changing quickly, with advances happening all the time. Project development, design, and installation of solar technology benefit greatly when a solar consultant is a member of the team. Having a person with the proper credentials may be a requirement when contracting with an electric utility to sell power back to that utility or when receiving government-sponsored rebates for installation costs.

Solar Water Heating

Solar heat can be used to heat swimming pools and domestic hot water, and to provide a portion of the necessary building interior heat. The most common commercial use for solar heat is for domestic hot water. When water heating is the goal, it is inefficient to use photovoltaic cells as the power source because of losses in power conversion and battery storage. Solar water heaters are much less affected by dirt on their surface or by shading. Heat output is reduced in direct proportion to the amount of light blockage, and adjoining units that are still clean or unshaded are not affected. Some solar water-heating systems have heat sensors that shut off circulation pumps when there is not enough solar radiation present to generate the proper amount of heat. For solar water heating, the solar fluid is sent through a coil inside a water heater tank (see Figure 10.18). Sometimes a second water-heating coil fed by a boiler is employed inside the tank to boost water temperature when solar heat alone is insufficient to meet demand. The two most commonly used solar water-heating devices are the plate-type solar heater and the vacuum tube solar heater.

Plate-Type Solar Collector The plate-type solar collector (see Figure 10.19) consists of an aluminum frame box that houses a copper tube containing the solar fluid. Solar fluid is a heat transfer fluid. The fluid absorbs the heat generated by the solar collector, and it is pumped to the hot water tank. A submerged coil inside the tank transfers the heat to the hot water. Control devices sense the temperature in the solar collector and the tank, and operate the pump, as required, to achieve the proper hot water temperature in the tank. The solar fluid is a nontoxic, food-grade fluid that is resistant to freezing down to −35 degrees centigrade and is stable at high temperatures up to 170 degrees centigrade.(On top of the copper tube inside the collector box are a solar heat collector plate and a glass plate. In warm climates, the box is not insulated; in cold climates, it is insulated to help prevent heat loss at the bottom and sides. An advantage of having some heat loss out the top of the collector is that it provides for ice and snow melting.

Vacuum Tube Solar Collector The vacuum tube solar collector (see Figure 10.20) has a clear glass vacuum tube with a solar energy absorbing collector plate inside. It has an advantage over plate-type solar collectors because its vacuum tubes provide insulation akin to a Thermos vacuum bottle. Even on

FIGURE 10.18 Solar collector, hot water storage tank, pump, temperature controls, and backup heating unit
Source: Image courtesy of Viessmann Manufacturing Company Inc.

FIGURE 10.19 Cutaway of a Vitosol 200-F solar water heater
Source: Image courtesy of Viessmann Manufacturing Company Inc.

the coldest days, the tubes can operate with minimal heat loss at the point of collection. Tubes can be rotated so that the collector plate is at the most efficient angle for collecting heat. Vacuum tube solar heat collectors come in two types:

1. Direct flow vacuum tubes
2. Heat pipe vacuum tubes

FIGURE 10.20 Heat pipe vacuum tubes

Source: Image courtesy of Viessmann Manufacturing Company Inc.

FIGURE 10.21 Vitosol 200-T vacuum tube solar water heater

Source: Image courtesy of Viessmann Manufacturing Company Inc.

Direct-Flow Vacuum Tubes   In direct-flow vacuum tubes (see Figure 10.21), heat is absorbed by the collector plate that is attached directly to a tube within a tube, an arrangement that separates the cool incoming solar fluid streams and the hot outgoing streams of solar fluid. Heat is transferred by the collector plate directly to the outgoing stream.

Heat Pipe Vacuum Tubes   In heat pipe vacuum tubes (see Figure 10.22), the captured heat evaporates water in a single sealed tube. The evaporated water droplets rise to the top end of the sealed tube, where heat is transferred to a condenser tube header that has solar fluid running through it. The advantage of this arrangement is that the vacuum tubes are self-contained and thus are easy to replace. A replacement tube can be snapped into place should a tube failure occur.

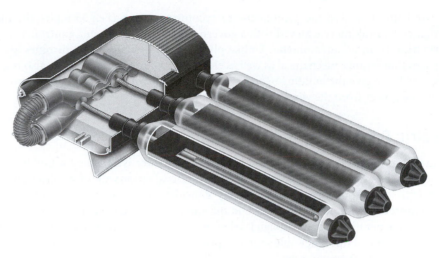

FIGURE 10.22 Vitosol 300-T heat pipe vacuum tube solar water heater, with double pipe heat exchanger. No direct connection exists between the heat pipe and the solar fluid.

Source: Image courtesy of Viessmann Manufacturing Company Inc.

Wind Power

A wind turbine (see Figure 10.23) is essentially a propeller or other wind-catching device that is used to rotate a generator to generate electricity. Small wind turbines often generate direct current that will be stored in batteries and converted to alternating current at the proper line voltage. Large power company turbines generate alternating current using synchronous generators. Upwind turbines need to have their blades facing into the wind; a yaw drive meets this purpose. A downwind turbine does not require a yaw drive because the blades face away from the wind. The wind itself blows the rotor downwind.

FIGURE 10.23 Ten-kilowatt wind turbine at Waukesha County Technical College

Source: Photo courtesy of Waukesha County Technical College.

When wind speeds exceed the manufacturer's recommendation (i.e., 55 mph), the turbine must brake automatically. Braking is achieved with a governor device or through a controller that gets wind speed information from an anemometer. Unlike photovoltaic systems, wind turbines are electromechanical devices that require mechanical as well as electrical maintenance. They are also responsible for bird strikes, which can give the building owner a negative image. But wind turbines do provide a source of free renewable power from the wind.

Hydroelectric Power

In an extremely limited number of situations, hydroelectric power is suitable for building use. Hydroelectric power requires extensive government permits and evaluations about its effect on the environment, along with numerous structural and safety considerations, before any project can begin. If damming is required, the number of permits and the amount of government regulation increase. Once built, hydropower generation requires a high level of maintenance. To avoid these disadvantages, a facility can easily participate in this renewable source of power by purchasing it from the hydroelectric utility through their renewable energy purchase programs. These programs allow facilities to purchase all or part of their energy from a renewable source and pay a rate based on that source of power.

10.10   POWER STORAGE AND MICROGRIDS

Power stability can be achieved with the construction of microgrids (see Figure 10.24) and battery storage systems. Microgrids are electric power grids separate from the electric utility's primary power grid. Microgrids distribute and regulate the flow of power to the facility's electric loads.

During electric utility power outages, microgrids can protect facility operation by isolating the facility electrically from incoming utility lines. Power can then be provided by storage batteries that have accumulated and stored power from wind power, photovoltaic power, or hydropower. Batteries are necessary because of the large variances in power production levels from sun and wind power sources. Electric power can also be provided by diesel generators fueled by biofuels and microturbines fueled by landfill gas. These sources are necessary when the length of the power outage exceeds battery storage capacity. Smart-grid technology gives the microgrid the ability to resynchronize automatically and then connect to the power company's primary grid when power is restored.

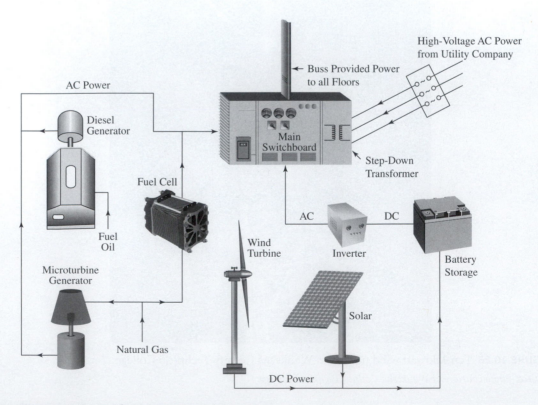

FIGURE 10.24 Microgrid energy management system

10.11   FUEL CELLS

Fuel cells are electrochemical devices that convert hydrogen contained in a fuel source into electrical energy, silently and without moving parts. Fuel cells are both clean and efficient, emitting much lower amounts of harmful exhaust than most other electricity-generating processes. In fact, the major emission is pure water vapor. Five types of fuel cells have been in practical usage, some for well over twenty-five years. Fuel cells are used in the propulsion systems in submarines of the German and Italian navies. They are also used in ships and buses, and to power buildings. The different types of fuel cells have their own particular operating characteristics and therefore are suitable for different uses. One of the main differences in fuel cells is the electrolyte they use, which defines the type of fuel cell.

United Technologies Corporation (UTC) Power has provided fuel cells to NASA for the space program since the 1960s. When fuel cells are used in spacecraft, backup battery power is not needed. Fuel cells have the added advantage of producing potable water as a byproduct—an extremely valuable commodity for astronauts traveling through space.

According to the U.S. Department of Energy, UTC has already installed over 75 megawatts of phosphoric acid fuel cell (PAFC) systems to power buildings. These fuel cells have operated for over 9 million hours in nineteen different countries around the globe. PAFCs can be fueled by methanol; they convert the methanol to electricity, and then use it as a power source to recharge batteries that propel a city bus such as the 100-kW fuel cells that power four electric buses used at Georgetown University.

UTC's new model of a fuel cell, called the PureCell 400, is being installed at the time of this writing at the Freedom Tower (see Figure 10.25), known officially as One World Trade Center, in New York City

FIGURE 10.25 Fuel cell being installed at One World Trade Center

Source: Photo courtesy of UTC Power.

FIGURE 10.26 Small footprint of an installed fuel cell.

Source: Photo courtesy of UTC Power.

and in three other towers at the save site. Each PureCell 400 produces 400 kW of 480-volt, three-phase alternating current. A total of twelve fuel cells will be installed (three at each tower), for a combined output of 4.8 megawatts. The PureCell 400 can be installed in this manner because it occupies a very small area in relation to the amount of power it can deliver (see Figure 10.26). One World Trade Center and the surrounding three towers will be one of the largest fuel cell installations in the world.

Phosphoric Acid Fuel Cell Operation

Phosphoric acid fuel cells operate very quietly, with no moving parts (see Figure 10.27). They are comprised of an anode and cathode plates, with the phosphoric acid (H PO) electrolyte sandwiched between them. Hydrogen and air (oxygen) are combined in an electrochemical process that produces direct current power, pure water, and heat. The gases emitted by a PureCell® Model 400 are nitrogen oxides (NO_x) of 0.02 lb/MWh, carbon monoxide (CO) of 0.02 lb/MWh, carbon dioxide (CO_2) of 1,050 lb/MWh (without heat recovery), and sulfur oxides (SO_x) of 487 lb/MWh (with full heat recovery). Particulates are negligible. Volatile organic compounds (VOCs) are of 0.02 lb/MWh.

The production of electricity and heat has the following five steps:

1. A fuel containing hydrogen is sent to a fuel processor (reformer) that reforms the fuel into hydrogen. The hydrogen is sent to the anode plate of the fuel cell. For commercial building power, this fuel is frequently natural gas.

2. The hydrogen flows to the anode, which has a catalyst layer that separates the hydrogen atoms into protons (also known as hydrogen ions and electrons).

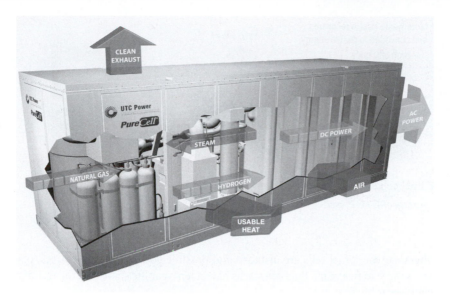

FIGURE 10.27 Fuel cell operation

Source: Image courtesy of UTC Power.

3. The phosphoric acid electrolyte stops the flow of electrons and allows only protons to pass through to the cathode.

4. Electrons are sent via an external circuit around the electrolyte. This power is direct current and is at a low voltage and current. Several cell circuits are connected in series and parallel arrangements, called a fuel cell stack, to achieve the needed amperage and voltage. The output is converted to alternating current for use throughout the microgrid.

5. Air (oxygen) is sent to the cathode. A catalyst layer on the cathode helps oxygen, protons, and electrons combine to produce pure water and heat.

Phosphoric Acid Fuel Cell Efficiency

The electrical efficiency, the process of converting the energy contained in the fuel to electricity, is approximately 42 percent for phosphoric acid fuel cells. This compares favorably with the efficiency of a traditional central power plant operated by a utility company (see Figure 10.28) which, after factoring in transmission and other losses, may only be 30 percent efficient. The overall efficiency of a fuel cell can be greatly increased by capturing the heat generated at the cathode. In a combined cooling heat and power (CCHP) system, efficiency can be raised to as high as 90 percent (see Figure 10.29). The heat generated by this fuel cell can be captured and used in an absorption cooling system. Each PureCell 400 can then produce about 50 tons of air conditioning. The heat, which is 1.7 MMBTU/hour (or about 500 kWh), can also be used for heating the interior of a building or for heating domestic hot water. Depending on the cost of

FIGURE 10.28 Traditional central power plant efficiency

Source: Image courtesy of UTC Power.

More waste heat is recovered
and converted to usable energy.
- High efficiency
- Ultra-low emissions

Waste Heat

10–20%

Natural
Gas

100%

80–90%

FIGURE 10.29 Fuel cell efficiency CHP

Source: Image courtesy of UTC Power.

electricity in the area where fuel cells are installed and possible government incentives, payback periods may be as short as three to five years. Fuel cell stacks have a ten-year life span, and the rest of the cell equipment has a twenty-year life span.

REVIEW QUESTIONS

1. The period of time when power demand is highest is called _____ for utility company billing purposes.

2. The U.S. EPA provides a free computerized tool, known as _____, to benchmark a building's energy usage.

3. A VFD connected to a supply fan motor of an AHU saves electric power by _____.

4. Cooling towers cool condenser water through a(n) _____ process.

5. Group relamping is _____.

6. One of the longest lasting, most energy-efficient forms of lighting is the _____ _____.

7. A type of lamp that has a CRI of 100 and a color temperature of 2,700 degrees Kelvin is _____ _____.

8. An effective way to benchmark an office building's electric energy consumption is to _____ _____.

9. A handy computer program that allows the user to get energy consumption information quickly from a computer screen is known as a(n) _____.

10. EER rating is equal to _____.

11. What are some advantages that geothermal heat pumps have over water-sourced heat pumps using a cooling tower?

12. What happens when one of the photovoltaic cells in a solar array becomes shaded?

13. List the types of renewable energy sources.

14. Why are heat pump water heaters more energy efficient than electric resistance water heaters?

Essay Questions

15. Discuss the importance of having the appropriate electric rate schedule.

16. Discuss which type of lighting is best for each of the following: indoor applications of overhead and task lighting, remote location lighting, and exterior building and parking lot lighting.

17. Discuss the advantages of using fuel cells for electric power in a multibuilding industrial campus. Investigate how system efficiency can be improved.

BUILDING SITE INTERIOR AND PERSONNEL MANAGEMENT

INSIDE THIS CHAPTER

Is There an Ascetic Element to Sustainability Provided by Building Interiors, Exteriors, and Landscaping?

This chapter discusses the building site, including landscaping, drives and walks, and athletic fields. Sustainable landscaping is achieved by using a combination of old tried-and-true methods and new methods such as xeriscaping and the use of recycled materials in place of purchasing new soil additives. Turf maintenance is discussed, with a comparison between artificial turf and natural grass for athletic fields. The chapter then discusses common paving materials and the new porous paving options. New methods of storing and naturally treating storm water runoff are examined in detail. The section ends with a discussion on consumables used in facilities, recycling, waste management, and restroom water conservation.

The chapter then changes direction and goes inside the building for a discussion on housekeeping and facility maintenance methods and management. The chapter considers safety and safe work practices for housekeepers.

11.1 SITE MANAGEMENT

Site management is the maintenance management of all areas external to the building(s) at a property or facility. The areas are generally a blend of constructed areas, landscaped areas, and natural areas. At a college or university, many of the duties of site management fall to the landscaping supervisor. Although overall supervision, planning, and design for new areas are the responsibility of the director of facilities, the director of facilities works with the required specialists (i.e., architects, landscape architects, paving consultants, and site lighting consultants) to obtain their knowledge for effective management.

Constructed areas include:

- Parking lots
- Sidewalks
- Roadways
- Ramps and loading docks
- Piers and boat docks

Landscaped and natural areas include:

- Wooded or grassy areas
- Lawns
- Planting beds and bermed (raised) areas
- Groves consisting of hedges and ornamental trees
- Aquatic areas such as streams, natural ponds and lakes, retention ponds, and storm water treatment and storage

The rather short sections on landscaping in this text are only intended to give some of the important basic points on the subject for site management. They are not intended as a complete landscape maintenance guide.

11.2   SUSTAINABLE LANDSCAPING

The goal of sustainable landscaping is to provide the highest quality natural environment through the use of sustainable practices. Sustainable landscaping can be implemented through design and construction of new or renovated landscaped areas, or by working with the existing landscaping and phasing in improvements. The landscape process can also provide opportunities for other functional areas within an organization to become more sustainable. One example is the recycling of organic materials such as food wastes into compost. This makes the food service operation more sustainable while simultaneously providing needed nutrients and organic matter for the landscape operation. Another example is the use of gypsum from drywall demolition as a soil amendment to make construction activities more sustainable.

Landscape Maintenance

Commercial properties generally seek contracted services for landscaping and snow removal, whereas large facilities such as universities frequently do the work in-house. Landscaping duties are generally thought of as the maintenance of the lawns, flowers, shrubs, and trees, and providing snow removal services. The landscaping department at an educational facility may have several other site maintenance duties. Landscape equipment maintenance, road patching, striping parking lots, repairing concrete curbs and sidewalks, maintaining ponds, installing road signs, eliminating or removing nuisance pests, laying concrete pavers, sweeping parking lots, picking up litter, pressure washing sidewalks, and building retaining walls are some of the additional landscape duties. Sustainable practices in maintenance are to limit water and chemical use, prevent water runoff from paved surfaces, and reduce emissions from combustion engine–driven landscape equipment.

Landscape maintenance techniques, some of which have been used for hundreds of years, are being improved, limited, and/or emphasized to improve sustainability of landscaping. Some landscape maintenance techniques include the following:

- Composting
- Mulching
- Irrigation and xeriscaping
- Fertilization
- Landscape chemicals

Composting   Composting is the process of taking organic matter (leaves, lawn clippings, fruits, and vegetables) and allowing it to decompose by microbial activity, thus producing a nutrient-rich material that can be worked into the soil. Composting material should be shredded in order to provide greater surface area for effective decomposition. Compost provides food for the plants and shrubs, and it gives the soil the ability to retain more moisture, which benefits the plants. Plant and vegetable matter is generally preferred for composting at most facilities. Composting food wastes and other animal matter outdoors in piles produces intense odors and attracts all kinds of vermin and insects. Worm castings (worm manure) are highly prized as compost material. Worm castings can be used to make *compost teas* whereby the material is mixed with water and sprayed around the plants. Compost teas provide nutrients that are quickly taken up by the plants.

Compost should consist of a variety of plant material and not just tree leaves. Fruits, vegetables, and grass clippings added to the compost pile provide the needed nitrogen for microbial activity and keep the PH of the compost closer to 7.0, or neutral. Oak leaves and pine needles are acidic and produce acidic compost, which is favored by acid-loving plants. Acidic compost can be neutralized by the addition of lime during application. Working compost into the soil increases organic content and reduces the need for supplemental irrigation.

Mulching   Mulch is a covering over the soil that is placed around plantings to retain moisture and prevent weeds from sprouting. Mulch may consist of shredded hardwood or various sizes of pine bark nuggets, nut shells, stone, or rubber. Wood mulch may be dyed to prevent it from turning gray. Trees and shrubs naturally mulch themselves by dropping leaves, twigs, and branches on the soil. Groundcovers are living mulch; because they grow quickly and cover the soil, they provide many of the benefits of other mulches. In new planting beds, a landscape fabric of porous synthetic fabric mesh may first be laid down and then staked to the ground. Staking the mesh is necessary to prevent perennial weeds from pushing the fabric up. The fabric allows water to enter the soil but hinders weed root growth. A layer of 2 to 4 inches of mulch is applied over the mesh. Wood mulch decays naturally. A fresh layer of 1 to 2 inches of mulch needs to be applied every year.

Irrigation and Xeriscaping   Many plant types require additional irrigation to supplement moisture provided by normal rainfall. Supplemental moisture is needed when non-native plants are introduced that require more moisture, and is it also needed for all plants when drought conditions exist. Irrigation is now becoming more predictive maintenance–based, whereby the amount of irrigation provided depends on the amount of moisture in the soil rather than the sprinklers being activated solely by a time clock. Rather than irrigating simply by zones, more efficient irrigation directs the water in varying amounts and different spray patterns to the plantings requiring irrigation by using a variety of different styles of sprinkler heads. Drip irrigation, whereby small tubes with nozzles or perforations are placed in the root zone at the base of trees and shrubs, provide water directly to the roots. When a drip irrigation system is used, a much higher percentage of water is then taken up by the plant, and less water evaporates out of the surrounding soil.

Xeriscaping is a method that greatly reduces the need for irrigation. *Xeriscaping* is the use of native plants that, under normal conditions, do not require any supplemental irrigation. It also involves the placement of plants so that plants requiring more moisture are placed closer to a source of moisture. Xeriscaping must also consider soil conditions and the amount of light the plants will receive.

A rain garden is an example of xeriscaping whereby the plants that require the most moisture are placed closest to the downspout that feeds the garden. Another example of plant placement is installing the plants that require the least amount of water at the top of a hill and the plants that require the most at the bottom. Xeriscaping requires a detailed landscape plan showing not only the plants, but all the features of the landscape. Applying xeriscaping to a lawn involves the use of native grasses and allowing them to go dormant during periods of low moisture (late summer) and to revive again in the fall and spring.

Fertilization   Fertilizers are designated by their N-P-K percentage numbers for the content of nitrogen (N), phosphorous (P), and potassium (K). A fertilizer with 18-6-12 contains compounds in the ratio of 18 percent nitrogen, 6 percent phosphorous, and 12 percent potassium. Nitrogen is necessary for plant growth and greening of lawns. Phosphorous is necessary to promote initial root growth. Potassium is necessary for overall plant health. Fertilizers can be water-soluble types that are quickly taken up by the plant, or water-insoluble types that release their chemicals and nutrients slowly into the soil. A bag of chemical fertilizer generally contains a combination of soluble and insoluble chemicals.

Phosphorous can be a problem because, if it enters waterways, ponds, or lakes, it acts as an aquatic fertilizer and promotes algae growth. It has been said that one pound of phosphorous can promote the growth of 500 pounds of algae. The death of the algae depletes the water of oxygen, thereby killing aquatic animals and degrading water quality. Several states have banned the use of phosphorous in general-use fertilizers but may permit their use as starter or winterizing fertilizers. It appears that phosphorous has much less benefit in already established plants. The ban may apply only to nonagricultural fertilizers. Aeration of the soil prior to fertilizer application can reduce the amount of fertilizer runoff.

Landscape Chemicals   Landscape chemicals fall under two main categories: herbicides and insecticides. Herbicides are used to control or kill unwanted vegetation, and insecticides are used to kill insects. They frequently require a state-issued applicator's license in addition to the appropriate personal protective equipment (PPE) to be worn by the applicator.

Integrated pest management (IPM) reduces insect pests through the use of biological pest management techniques. The management process involves a constant surveillance of the pests on a property to identify then and monitor their numbers to decide if the level of pest activity warrants that action should be taken.

Pest control may involve trapping or the use of pheromones to interfere with insect mating activities. Biological controls, such as the use of milky spore to control Japanese beetles when they are in the grub state, may also be used. Pesticides should be used only as a last resort, and then they must be targeted to the type and size of the threat to minimize the impact on people and the environment.

11.3 TURF MAINTENANCE

Turf is one of the most important and challenging features of any landscape for management professionals because of its size, utility, and needs. Turf has the greatest impact on site sustainability: It generally consumes the most water and requires the greatest amount of labor, chemicals, and fertilizers for successful maintenance. On a more positive note, turf also provides building sites with a cooler soil environment through evapotranspiration cooling, thereby helping to mitigate the heat island effect. Turf management is a balancing act dedicated to achieving the best results for the current environmental conditions by using sustainable methods.

Turf management and turf research are two areas to which many professionals across the United States and around the world have devoted their lives, and many areas of turf specialization exist among professionals in both. Some of the areas of specialization are plant pathology, biocontrol of weeds, genetics, plant breeding, agronomy, microbiology and cell structure science, soil and water science, sports turf, chemistry, turf grass insects, soil and water science, and irrigation systems. Landscape workers themselves must have specialized training in landscaping techniques and the proper licensing for chemical application.

Landscape management enters an entirely new dimension when the landscaping crew must maintain athletic fields. Athletic fields, such as football fields, must be maintained to a level specified by organizations that govern the sport played on the field. The field must have proper density, have proper grass height, be cut evenly, have level ground, and provide appropriate drainage so that ponding of rainwater does not damage the turf or interfere with play. Athletic fields in poor condition due to lack of maintenance jeopardize the safety of the players. Lower maintenance is a reason why synthetic turfs have become more popular.

Mowing height depends on the type of play to be performed on the field. A multipurpose field without an irrigation sprinkler system should be mowed to a height of approximately 2½ inches or 3 inches during summer stress periods. An irrigated soccer field that has a lot of ball-to-turf contact should be mowed at 1 to 1½ inches. Mowing affects the density of the turf. The lower the height to which the grass is mowed, the denser the turf becomes above ground. Lower height of grass also means less root structure, which can cause stress on the plant when water is less available during the summer season. Stress shows up as the plant going dormant and turning brown. The use of a mulching mower blade to return finely chopped grass back to the turf can reduce the amount of fertilizer required to maintain the quality and color of turf by 20 to 25 percent without increasing thatch (**thatch** is an undecomposed layer of living and dead stems, leaves, and roots on top of the soil), according to the Ohio State University Extension.

Therefore, the facility manager (FM) must exercise caution when implementing new practices for the maintenance of athletic fields. For example, in an attempt to reduce watering requirements of the sports turf, the supervisor of landscaping might lower the density of turf grass by lessening the amount of overseeding applied to the established turf. **Overseeding** can be accomplished by first aerating the soil, seeding, and topdressing with soil, or through the use of a device called a slit seeder. **Aeration** is a process where ½-inch diameter, 3-inch long plugs of soil and turf are removed, allowing water, fertilizer, and oxygen to reach the turf's roots. **Top dressing** is the spreading of a thin layer of soil that falls between the blades of grass. Top dressing is also used as a cultural method to remove thatch by causing the thatch to decompose naturally rather than by using a dethatching machine. Thatch can be as thick as ½ inch, and this heavy thatch interferes with the spreading outward of established grass plants and the germination of new seed. Applying insecticides to turf has the unwanted effect of also killing the microbes and worms that aid in the decomposition of thatch.

Slit seeders are mechanical devices that cut into the soil and deposit grass seeds for proper seed-to-soil contact, resulting in a high rate of seed germination. Reducing the turf density of an athletic field may diminish the turf's ability to dampen impact when players are tackled or simply fall on the turf while running. Bare spots might also show up in the turf without overseeding. Overseeding in cool season grass regions is a process whereby additional cool season grass seed is inserted into the soil of an established turf to increase the turf density. In transitional zones where warm season grass is used, over-seeding with cool season grasses extends the season of the turf, making the turf green and vibrant in spring and fall. An old trick of many landscaping supervisors is to overseed just prior to a game. The cleats on the shoes of the players increase seed-to-soil contact and provide for higher germination rates.

For common lawn areas, reducing the turf density and allowing the length of grass to increase cuts down on the amount of watering required. For lawns, the turf density should be kept high enough to prevent the establishment of weeds naturally. Otherwise, the gain achieved by the lawn needing less watering could be offset by increased herbicide used to kill weeds.

TABLE 11.1 Comparison of artificial turf to natural turf

	Artificial Turf	Natural Turf
Maintenance	Lower cost	Higher cost
Installation	Much higher cost	Significantly lower cost
Useful life	Approximately 10 years, materials can be recycled	Unlimited life
Grass temperature	Up to 60°F higher than ambient temperatures, increase in heat island effect	About 10°F to 14°F lower than ambient temperature, decrease in heat island effect
Watering	Requires watering before play to reduce temperature	Requires regular watering for plant maintenance
Chemical hazards	Leaching of chemicals from rubber, adhesives, and plastics into the soil, chemical off-gassing into the air	Application of chemicals: fertilizers, insecticides, and herbicides
Safety	Some artificial turfs may compact over time, high grass temperatures and rug type friction burns and cuts	Require regular maintenance to ensure proper height, density and coverage of soil areas. Soil aeration can eliminate soil compaction problems
Environmental	Increased storm water run off	Decreased storm water runoff reduces carbon dioxide provides environment for insects and animal
Humanistic	Unnatural	Mental benefits for interaction with a natural environment

Synthetic Turf

A variety of manufacturing processes are used in the making of synthetic turf. Conflicting claims have been made that synthetic turf is both sustainable and unsustainable (see Table 11.1). Specific situations may vary, but the table lists some issues to consider when deciding between artificial turf and natural turf.

Field preparation is done by first compacting the existing soil and then installing a gravel aggregate base and drain tile for water removal. A finer leveling layer of gravel is smoothed over the aggregate base, and a shock absorbing pad of polyurethane foam or crumb rubber padding from recycled tires is laid down to protect the athletes. One of artificial turf's basic manufacturing processes uses a rubber base with a sand or rubber-and-sand infill from recycled tires beneath a layer of polyethylene fiber grass blades.

11.4 PAVING

Three types of pavement are commonly used today:

- Asphalt pavement
- Concrete pavement
- Brick or concrete block pavers

These products all direct storm water away from their surface to storm water retention areas or storm sewers. The more these materials are used, the more storm water runoff must be treated by municipal sewers.

Asphalt Pavement

Asphalt is the residual product that is finally left after the fractional distillation of crude oil has removed all the lighter compounds (naphtha, gasoline, diesel oil, heating oil, lubricants, paraffin, and heavy oil). It is mixed with stone aggregates at elevated temperatures to produce asphalt pavement mixes. Ninety percent of asphalt by volume consists of aggregates.

The installation process first involves stabilizing the soil through compaction and the addition of a stone base. Poor quality soil may require the use of fairly large stones several inches in diameter and then layered with ever-smaller stones until small 1-inch or smaller chips are used. For normal applications and when stable soils are present, a stone chip base of at least 4 inches thick is required. Once the base is leveled and compacted, a primer coating of liquid asphalt is sometimes applied. Then the asphaltic concrete layer, which is at least 3 inches thick, is installed. A tack coat is applied to the surface of the concrete layer, then a one-inch thick surface layer is laid down and compacted. A typical specification for parking areas is 4 inches of stone chip base, 3 inches of asphaltic concrete, and a 1-inch surface layer. Roadways and loading areas require heavier applications of asphalt, base, and surface layers.

Asphalt Pavement Maintenance   Maintenance of asphalt pavement involves the repairing of pot holes, crack filling, sealing, and sometimes pavement overlays to strengthen the entire pavement structure.

- Pothole repair is a necessary function for safety and to prevent further damage to the pavement. It requires the use of hot mix asphalt or cold patch asphalt for paving repair. After the pothole is cleaned and the damaged pavement is cut back to show good pavement, hot patch asphalt is applied and compacted. Cold patch asphalt is sold in bags and is used when outside temperatures are too low for hot patch, or when hot patch asphalt is unavailable because asphalt plants may shut down during the winter months. Disastrous results can occur if cold patch asphalt is dumped in a wet pothole in a drive lane. The aggregate mix won't remain compacted and will be thrown out of the hole as vehicles drive over it. If thrown onto walkway areas, the aggregate mix makes an unsightly mess, and the stones can be slick and present a serious slip-and-fall hazard.

- Crack filling is performed by blowing out or pressure-washing cracks to remove any dirt in the asphalt, then pouring rubberized liquid asphalt into the crack. Crack filling is necessary to prevent water from getting below the asphalt and weakening it.

- Seal coating is another practice whereby the entire surface is coated with an asphalt emulsion or a coal tar–based seal coating. Coal tar–based seal coats give off unpleasant fumes called polycyclic aromatic hydrocarbons (PHAs) while they are drying. Some municipalities have banned the use of coal tar–based seal coats for health and environmental reasons. Seal coating is touted by manufacturers as a way to keep the sun from drying out the asphalt and making it brittle, thereby extending its life. The motivation for seal coating in property management and real estate circles is frequently for cosmetic reasons: It is an inexpensive way to make an old parking lot look new.

Concrete

It is unfortunate that, in common speech, the word *cement* is used interchangeably with *concrete*. Cement is only one of the materials that make up concrete. Mistakes have been made when cement is used in a project instead of concrete because cement alone has almost no strength.

Concrete is a mixture of Portland cement, aggregates (both coarse and fine), water, and possibly admixtures. Admixtures are chemicals used to provide a variety of desirable concrete properties. Some chemical admixtures are set retarders and accelerators, superplasticizers, and colorants.

An air-entrainment admixture makes tiny pores in concrete that provide spaces for water to expand when it freezes. The pores help the concrete resist cracking from freeze/thaw cycles. Another admixture is a set retarder that offsets the effect of high ambient temperatures, which decrease curing time. Concrete achieves most of its full strength after twenty-eight days. Concrete that cures too fast has a lower compressive strength. Accelerating admixtures help concrete set during cold temperatures. Calcium chloride is generally used to accelerate the setting of concrete. Superplasticizers are admixtures that make concrete flow more readily. Superplasticizers are used for building construction, especially when concrete forms are constructed in shapes that are difficult to fill with unmodified concrete. Superplasticizers added to concrete result in a concrete with a compressive strength of over 14,000 pounds per square inch (psi); therefore, it is also used in bridge and highway construction. Colorants added to the concrete mix are used to beautify concrete, especially when concrete drives and walks are stamped with forms to simulate stone or pavers.

A negative property of concrete is that it has a low tensile strength (about 10 percent of its compressive strength). Tensile strength is achieved through the use of reinforcing steel rods (rebar), or 6×6 wire mesh. Pretensioning or post-tensioning steel reinforcing wires or metal rods are used in building construction.

Sidewalks are generally 4 to 6 inches thick and are poured over a 4-inch compacted aggregate base of stone and gravel, or limestone chips. Reinforcement is usually not required because the slab is fully supported by the base, but wire mesh reinforcement is recommended or required for walks along drive-way entrances, and the concrete should be at least 6 inches thick. Welded-wire mesh helps to control cracks in concrete. It also keeps the faces of the crack in position, which prevents an upward or downward shifting of the halves. Once concrete cracks, the wire mesh provides tensile strength.

Fiber mesh reinforcement consists of small pieces of hair-like synthetic polypropylene fibers that are mixed with the concrete and set in a random pattern throughout the full depth of the pour. Fiber mesh works by blocking the growth of small cracks that occur as the concrete shrinks during curing. Fiber mesh increases tensile strength. It has the additional advantage over other reinforcement of not having to be supported at a specific height within the concrete pour.

Installers debate about whether fiber mesh can be used as primary reinforcement or only as a secondary reinforcement in conjunction with rebar and welded wire. The ultimate compressive strength measured in pounds per square inch (psi) of the concrete relates to the amount of Portland cement in the mix. A six-bag mix that has a compressive strength of 3,500 to 4,000 psi is generally used for driveways. That mix contains six 94-pound bags of Portland cement per cubic yard of concrete.

To help concrete reach its designed strength, a cure-and-seal product (instead of a constant water spray) may be applied to the concrete's surface while it is still wet. This cure-and-seal product slows the drying of the concrete and allows it to reach full strength. Once the concrete is dry, another coating of sealant can be applied to make the concrete more resistant to stains and the effects of deicing chemicals.

Concrete Maintenance　　Well-constructed concrete walks and drives generally require little maintenance. Concrete walkways next to a building might require caulking of the gap between the sidewalk and the building to prevent water from being directed toward the foundation. A pourable self-leveling caulk is available for this application. Pourable caulk can also be used to fill cracks in concrete. A drive that has been sealed should be periodically resealed in accordance with the manufacturer's recommendations.

Should freeze/thaw cycles on a poorly constructed base affect the height of adjoining concrete slabs, mudjacking can be used to level the slabs. Mudjacking is a process whereby holes are drilled in the concrete slab, and a cement mixture called slurry is pressure-pumped under the slab to hydraulically raise and level it.

Concrete or Brick Pavers

Pavers are an alternative to concrete used for patios, walks, and drives. They are also laid over a compacted aggregate base. Sand is used over the base to provide a level surface for setting the pavers and to achieve a slope for drainage. Pavers sometimes have nubs on their sides to provide the correct gap between pavers. After the pavers are laid, fine sand is then brushed into the gap between pavers. The sand may contain a polymer to act as a glue that is activated when the pavers are wetted down in the last step of installation. This type of sand prevents grass and weeds from growing between pavers and ants from digging out the sand.

Pavers have an advantage over all others types of pavement. When utilities that lie below the pavers need to be accessed for repair, pavers can be removed, put aside, and then later replaced once the repair has been made. Unlike asphalt or concrete, replaced or repaired concrete and brick pavers have no noticeable patches.

11.5　PERVIOUS AND POROUS PAVING

Four main types of water infiltration paving are in use today:

- Pervious concrete
- Porous asphalt
- Permeable concrete pavers
- Porous concrete pavers

These types of pavement allow water to flow through them either to recharge the soil areas below the pavement directly with moisture or to collect water to be used in areas that do not require potable water.

Pervious Concrete

Pervious concrete is a concrete mixture that is missing most of the sand in the mix. When properly applied, this structural concrete can have voids that equal from 15 to 35 percent of its volume, allowing for a water flow rate of from 3 to 5 gallons per square foot per minute through the concrete. Then the water can percolate through the soil or reside in the aggregate base area under the concrete. A typical installation may be comprised of 6 inches of pervious concrete on top of a 4-inch aggregate (generally limestone chip) base. The base allows water to drain through the concrete and into the soil below.

Porous Asphalt

Only a very tiny fraction of roads are made from porous asphalt. Porous asphalt is mostly used in parking lot areas. Although more expensive than standard asphalt pavements, porous asphalt has several benefits. One of the benefits is that the pavement is somewhat cooler, thus reducing the heat island effect.

Porous asphalt has tiny holes that allow rainwater to go through the asphalt and into a recharge bed, which consists of a stone chip bed 18 to 36 inches thick. The water then infiltrates through the subsoil. These systems are used to recharge the groundwater rather than sending the water to a storm sewer. Using porous asphalt may also eliminate the need for water retention basins to contain storm water that would run off normal asphalt pavement.

Permeable Concrete Pavers

Permeable concrete pavers (see Figure 11.1) allow water to penetrate through the spaces between the pavers that are filled with sand. Permeable concrete pavers have spacing nubs that provide a larger-than-normal space between pavers. These pavers can be used as a filtering conduit to transfer water from the surface of the pavement to underground storage or to a stone chip base to promote water transfer into the subsoil. A great example of the paver's effectiveness can be seen when a fountain is installed without a catchment (a means of collecting water), and the fountain water that cascades onto the pavers seemingly disappears into the subsoil (see Figure 11.2).

Porous Concrete Pavers

Porous concrete pavers, also known as grass pavers, allow water to travel through them because they have openings that are filled with soil and/or aggregate. The pavers are composed of blocks with an open-cell waffle pattern that allows the grass to grow right through them, or they are a synthetic plastic grid system that comes in rolls that are laid on the ground. Porous pavers provide additional green space when they replace solid pavement or traditional brick pavers for walkways and roads.

Some of the reasons for utilizing permeable and porous paving materials are to:

- lower the amount of rainwater that enters the storm sewer
- recharge underground aquifers
- lessen the burden on municipal and private sewage treatment facilities
- provide a conduit to direct water to capture and storage equipment for landscaping and other uses

FIGURE 11.1 Permeable concrete pavers used to create a patio

Source: Photo courtesy of Aquascape.

FIGURE 11.2 Permeable concrete pavers transferring water from a fountain to storage underneath

Source: Photo courtesy of Aquascape.

11.6 COLLECTING, STORING, AND PURIFYING RAINWATER

A system that collects, stores, and purifies rainwater is the RainXchange™ system, which utilizes the Aquablox® D-Raintanks® tank module for the catchment system (see Figure 11.3). A highly successful system was installed by Aquascape™ at the Boerner Botanical Garden in Hales Corners, Wisconsin, in 2009. The project was funded by the Milwaukee Metropolitan Sewerage District (MMSD), and was designed to be a functional test site. Note that this project is not in an area of the United States that lacks in freshwater resources. This area of Wisconsin has access to the vast water warehouse known as the Great Lakes, which contain approximately 10 percent of the fresh water on earth. MMSD is forced by the vast quantities of storm water runoff it receives after a major rain event to perform the unsustainable practice of dumping partially treated storm water runoff into Lake Michigan. Billions more gallons of water were dumped in the past, but today dumping has been greatly reduced by MMSD's deep tunnel storm water containment project. By constructing several deep tunnels bored through limestone approximately

FIGURE 11.3 Fountain permeable pavers, and water collection and storage system

Source: Courtesy of Aquascape.

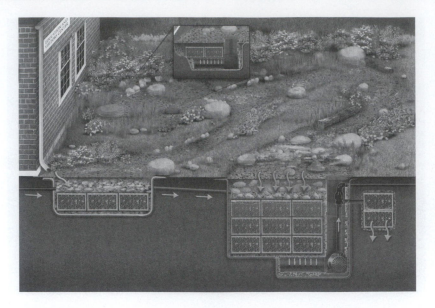

FIGURE 11.4 Containment pools and AquaBlox® rainwater collection system

Source: Courtesy of Aquascape.

200 to 300 feet underground, the tunnel system can hold vast quantities of water for later treatment. The Boerner Botanical Gardens project shows that a system can be installed to collect rainwater runoff, purify, and reuse it. Thereby prevent vast quantities of rain from entering the storm water system.

The project section of the website for Aquascape™ points out that the Boerner Botanical Garden project gathers the water from the roof of the visitor center, roadways, and parking lot. This amounts to 140,000 square feet of water-gathering area, which generates over 2.5 million gallons of water in a year with average rainfall. The entire rain-gathering system can hold 35,000 gallons.

Rainwater for a containment system can be gathered in different ways. It can be collected from roofs through downspout filters, or from elevated concrete parking decks or paved parking lots. Water from parking lots needs to be cleaned through a treatment process, but this process doesn't have to be simply utilitarian. Water from the underground storage (see Figure 11.4) is pumped to a series of aboveground containment pools. The first pool provides a sediment trap and wetland filter with aquatic plants. Water overflow from the pool and enters a stream that is lined with various stone grit sizes to trap tiny organic debris, where it is biologically treated. The water is then aerated through a series of cascading waterfalls as the water flows downstream (see Figure 11.5). The entire stream then filters though several inches of

FIGURE 11.5 Waterfall water purification system

Source: Photo Courtesy of Aquascape.

river stones of various sizes, disappearing completely from view as it flows straight down into the storage block system below.

Buildings, especially those in urban areas, may also opt to install a tank inside the building to collect rainwater from the roof. The weight of the collected water can be quite heavy, and a tank holding 1,000 gallons of water can weigh in excess of 8,500 pounds. Structural reinforcement of the building is required. This collected water can be used to flush toilets and for irrigation purposes.

11.7 HOUSEKEEPING SAFETY ISSUES

Site-specific safe work practices are designed for the hazards that would be considered likely to be encountered at a facility. A list of site-specific safe work practices for a facility maintenance technician includes the proper procedures and operational information for ladder use and tool use, electrical safety, plumbing safety, boiler safety, emergency procedures, confined space entry procedures, hazard communication, hearing protection, hazardous spill procedures, roof work, working on scaffolds and elevated platforms, proper lifting techniques, housekeeping procedures, fire safety and PPE, and licensing and job knowledge requirements.

Safe practices for a housekeeping operation include the following:

- When cleaning areas such as restrooms, where there is a likelihood of blood-borne pathogens and other potentially infectious material (OPIM), employees must refrain from applying lip balm, eating or drinking, applying cosmetics, or handling contact lenses.

- When engaged in cleaning, employees should be mindful not to drip and splash potentially contaminated liquids, which could spread the hazard rather than contain it.

- Contaminated work surfaces and tools should be cleaned as soon as possible after contact with any potentially infectious material. Decontamination is accomplished using a disinfectant registered with the United States Environmental Protection Agency (U.S. EPA) or a mixture containing one part or more of 5.25 percent chlorine bleach to nine parts of clean water. The disinfectant should be allowed to cover the area or item for ten or more minutes prior to cleaning.

- Employees must remove PPE prior to leaving the work area.

- Broken glassware and bottles should not be picked up by hand, even if found outdoors. Outdoors on landscaped areas, broken glass should be picked up with a tool for that purpose; indoors, broken glass should be swept into a dustpan.

- After removing protective gloves or other protective equipment, employees must wash their hands immediately with soap and water.

- When an exposure incident occurs, employees must wash their hands and the affected skin areas immediately. Eyes should be flushed with water.

- Needles should never be picked up by hand. A tool or dustpan should be used to pick up and place them in a labeled sharps container. The tool or dustpan must then be decontaminated.

- Clothing that comes in contact with potentially infectious material should be removed as soon as possible and the skin areas cleaned with soap and water.

- Personnel disconnecting plumbing drain and waste lines should run hot water through the piping, if possible. When disconnecting drain lines and traps, be mindful that needles may have been dropped down the drain. When clearing drains with plungers or snakes, a container or bucket of disinfectant should be available for these tools.

- Employees should know the location of hand-washing or waterless hand cleaner dispensers. Waterless hand cleaner is not a substitute for hand washing with soap and water. It is a temporary solution prior to an employee finding an appropriate location to wash.

- Disposable gloves should never be cleaned and reused. Disposable gloves simply aren't durable enough to be removed and reused, let alone removed, cleaned, and reused. Gloves that are made to be reused must be disinfected after use and tested for air-tightness by closing and rolling the cuff toward the glove's fingers. Gloves that are air-tight but show signs of cracks should be discarded.

- Plastic trash bags containing restroom wastes should be treated as though sharps are present inside them. Bags should never be held against the body; they should be carried only by the closed top.

- Wastes must be disposed of in accordance with all state and federal regulations.

It is the housekeeping manager's responsibility to ensure that safe work practices are followed. Following all of the Occupational Safety and Health Administration's (OSHA's) regulations and guidelines is an essential part of safe work practices. The safe housekeeping practices listed here provide a general guide that covers most situations in the healthcare of office settings, but the FM needs to be mindful of hazards not listed that may be present at her or his facility. The FM should always be looking for ways to improve the work situation and to make the safe ways the best ways to operate.

Blood-Borne Pathogens and Other Potentially Infectious Material

Blood-borne pathogens are infectious viral agents found in the blood. OPIM can be found in bodily fluids and bodily secretions of infected individuals. AIDS and hepatitis B are two diseases that can be spread by contact with bodily fluids and blood. Vomit, feces, urine, sweat, tears, and saliva are not included as OPIM fluids unless they contain blood.

Housekeeping staff members are generally considered as having the potential to come in contact with blood-borne pathogens and OPIM in the pursuit of their normal cleaning duties. Some other facility workers who fall into this risk category are maintenance technicians (especially those working on plumbing), security personnel, and persons likely to provide first aid in an emergency. Facilities are responsible for adhering to the OSHA regulations found in the Code of Federal Regulations, 29CFR1910.1030, which deal with blood-borne pathogens in the workplace.

Exposure Control Plan

Some of the provisions of the exposure control plan found in 29CFR1910.1030 are as follows:

> Employers must have an exposure control plan that lists the steps they are taking to control workers' exposure to blood-borne pathogens to keep the workers safe and keep the facility in compliance with the regulation. Employees must be familiar with the plan and know where it is kept at the facility. Employees are responsible to follow the provisions of the plan. One of the provisions of the plan requires the employer to provide, and the employee every year to complete, training in blood-borne pathogens. OSHA also requires the employer to provide the appropriate safety gear and equipment for the task at hand. Additionally, hepatitis B vaccinations must be offered to employees whose job causes them to be considered at risk for infection. The employee has the option of accepting or declining the vaccination offer. If the worker declines the offer of vaccination, the worker fills out and signs a form acknowledging [his or her] refusal. The employer must keep all records of vaccination and declination of vaccination in the employee's personnel file, and for three years after an employee has left the organization. The employer must have written procedures for employees' handling, cleaning, decontaminating and disposal of hazardous materials. The employer must provide an annual review of their Exposure Control Plan. A sample exposure control plan can be found on the State of Wisconsin Department of Administration website.

Main Elements of an Exposure Control Plan

Worker safety can be broken down into working conditions that can be improved with engineering controls and work procedures that can be improved with work practice controls. These two elements help provide for a safe work environment. *Engineering controls* are physical barriers or devices that prevent employee contact with blood-borne pathogens. Healthcare facilities use items such as:

- self-blunting or retractable needles
- plastic items (rather than glass) for blood-sampling devices
- splash shields
- resuscitation devices
- automatic door openers
- sharps containment receptacles and other hazardous waste containers that are properly labeled and color-coded
- surfaces that smooth and easy to clean and disinfect completely

Work practice controls require that employees use the appropriate personal protective equipment (PPE) for the job. PPE for general housekeeping includes the proper use of:

- disposable and reusable gloves
- protective eyewear with fixed side shields or full face shield

- liquid-proof aprons
- closed-toe shoes and or shoe coverings
- protective cardiopulmonary resuscitation (CPR) devices
- head coverings and hair containment nets
- disposable arm coverings
- blood and bodily fluid spill cleanup kits

Chemical Safety

According to OSHA's Hazard Communication Standard 29CFR1910.1200, employees have the right and the need to know the identity of the chemicals they are exposed to when working. Employees also need to know what protective measures are available to prevent adverse effects from exposure to the chemicals. The OSHA standard also requires facilities to have a hazard communication plan, chemical inventories that are current, and material data safety sheets (MSDSs) for all hazardous chemicals on the premises. Model plans and programs for OSHA Bloodborne Pathogens and Hazard Communication Standards can be accessed from OSHA's website.

Some states already had worker right-to-know legislation in place prior to the federal legislation. If you work in an OSHA-approved state plan state, then the state laws must be followed.

The identity of the chemicals, the hazards they represent, and protective measures can be found in material data safety sheets. MSDSs are required to be provided by chemical companies, chemical importers, or the distributors of hazardous products. When bulk hazardous chemicals are poured into smaller containers or spray bottles, these small containers and bottles must be labeled appropriately as well.

Note that the Hazard Communication Standard (29CFR1910.1200) is changing at the time of this writing. It will be aligning itself with the Globally Harmonized System (GHS), which is an international approach to hazard communication (see Figure 11.6). Three major areas in chemical safety will change:

- Safety data sheets (SDSs) will replace material safety data sheets (MSDSs). The new SDSs will have sixteen specific sections.
- Labels will be in a specific format, which will include a signal word, pictogram designation of the hazard, a precautionary statement, and a hazard statement.
- Manufacturers will provide hazard classification of chemicals, with specific criteria on health and physical hazards, and classification of chemical mixtures.

Important requirements for this new standard are that, by December 1, 2013, employees must be trained on the new labels and SDS format; products with old labels can still be shipped until December 1, 2015; and by June 1, 2016, the labeling and hazard communication plan must be updated, along with employee training on newly identified physical or health hazards. Additional information, as well as the entire standard, can be found at OSHA's website.

11.8 SUSTAINABLE HOUSEKEEPING

It is important to evaluate all products and equipment used for the housekeeping function. Products such as cleaning chemicals should be certified as green, meaning that they have a minimal impact on the indoor environment if used correctly, and that they are made from recycled materials or from renewable sources where appropriate. Although it is difficult to recycle some materials because of concern about contamination and disease transmission, improvements in the collection of wastes for disposal can both help save energy and improve the efficiency of the housekeeping function. Equipment used in restrooms can be improved to conserve water. The FM, through the development of a sustainable housekeeping plan with the in-house or outsourced housekeeping manager, should ensure that all aspects of this essential function are sustainable.

Consumables and the Environment

Several consumable items are used in housekeeping, and other items are restocked in restrooms for building occupant use. All consumables should be monitored as to the amount that is being used, their effectiveness in getting the job done, and the sustainable aspects of each product. For example, paper hand towels are one item that should have a high content of recycled product. The size and utility of

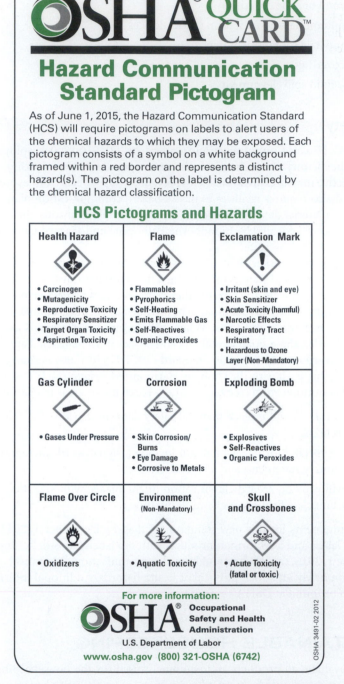

FIGURE 11.6 OSHA's Quick Card™ Hazard Communication Standard Pictogram

Source: Courtesy of the U.S. Department of Labor, Occupational Safety and Health Administration.

towels is important because both relate to generating excess waste through overuse. A hand towel that is too small may encourage a restroom visitor to use multiple towels. Hand towels that are packed too tightly in a dispenser invite visitors to grab multiple towels. Visitors to restrooms may utilize hand towels to cover the door handle when leaving the restroom. Restrooms need to have a waste receptacle near the door, outside the restroom, so that these towels don't end up on the floor. Georgia-Pacific's Safe-T-Guard™ tissue dispenser is mounted near door handles so that people can use them as barriers between washed hands and the door handle. A solution to the paper towel problem is the installation of forced-air hand driers. Driers are available today that almost instantly dry a visitor's hands. It is a nice amenity when both hand towels and hand driers are present because unforeseen circumstances may require hand towel use for other visitor needs.

Another consumable is liquid hand soap. According to the Minnesota Department of Health, plain hand soap is just as good at killing germs outside the healthcare setting as is antibacterial soap. Plain soap is recommended for hand washing outside healthcare settings. Some scientists believe that antibacterial soaps that contain the antibacterial chemical triclosan (TCC) may cause a disruption of a reproductive hormone in fish once the soap is discharged into waterways from sewage treatment plants. Because the chemical is thought to be unnecessary to achieve proper general hygiene, it is an unsustainable practice to put it into general use, even if it is later proven to have no adverse effects.

Recycling and Waste Management

Recycling is an important part of managing the waste stream. Office buildings and institutional facility recycling may consist of separating or single-stream recycling (comingling) white paper, cardboard, plastics, glass (sometimes glass beverage containers are further separated by color when all recyclables are separated according to type), toner cartridges, aluminum, and steel food and beverage containers. Separation of materials has an advantage when recycling costs are reduced by doing so, or money is paid for the materials, thus providing a return on the recycling investment. Separation is most effective when the persons who generate the material are responsible for recycling the material by placing it in the appropriate container. Recycling is also required by law in many states because of its effect on diversion of material from the landfill.

When the waste stream of a facility is defined, recycling and trash removal are most effective from cost, labor, and space perspectives. Over time, waste stream type and amount data provide the FM with the knowledge to design an efficient waste and recycling operation. The most efficient sizes and types of trash and recycling containers and compactors can then be utilized.

Waste collection over an entire site is another issue. Recycling and trash removal becomes more difficult when both commodities and amounts vary, and the geographic area of collection is large. Another problem is the not knowing when a container is ready for pickup, or whether the pickup effort will consist of removing plastic trash bags that are partially full, thus wasting bags, labor, vehicle use, and fuel.

Today, solar-powered, sidewalk-mounted trash and recycling containers are available (see Figure 11.7). These containers compact trash and, through a wireless cell telephone–based network, provide bin level

FIGURE 11.7 Solar-powered trash compactor on a Chicago sidewalk

Source: Courtesy of BigBelly Solar.

information to network computers. According to BigBelly Solar, Inc., a maker of this product, compacting trash in itself can save up to 80 percent on trash pickups. Savings are achieved by eliminating unnecessary pickups, and truck-routing computer software provides the most efficient pickup routes for sanitation workers. Historical bin-level information can provide data for planning purposes. Planning may consist of the manipulation of the number of containers and their location to accommodate varying demand. Demand can vary due to the season (cold weather can reduce the amount of food consumed outdoors) or events that bring a lot of people to the area. Containers can then be moved to the locations where they are needed most.

Restroom Water Conservation

Great success has been achieved in the area of water conservation for commercial buildings. FMs have many choices in this regard. Restrooms and shower rooms have fixtures are required to save water by code, and several devices on the market further increase savings.

Toilet and urinal fixtures have flushometer valves for reduced flows. Some flushometers even provide the user with a choice of two flush volumes that saves even more water. Automatic motion sensor flushometers on the market today are flush valves that can detect when a person is entering or leaving the stall. Metering faucet valves that are activated by pushing the handle downward can be adjusted to limit water flow to a specific number of seconds and shut off automatically. Automatic flushometer valves and faucet valves are electronic and operate automatically by detecting motion. Proper adjustment of the motion sensor is necessary to ensure that they are delivering water at the appropriate time. Waterless urinals use a liquid-filled cartridge that allows wastes to pass to the drain while keeping odors to a minimum. Automatic soap dispensers can also save water by helping to prevent overuse of hand soap that must be then washed off.

Training of maintenance staff in the proper adjustments and maintenance of valves and fixtures is necessary to keep the equipment operating efficiently and to save water. Commercial plumbing distributors can be utilized, as can maintenance information from manufacturers' websites.

11.9 WORKER TRAINING

Worker training is essential to a sustainable facility operation. Not only are the nuts and bolts of facilities changing, along with their operational requirements, but the face of the workforce becoming more diverse. The makeup of tenants and customers as a group is diverse, so a diverse workforce is a strength for an operation. Training programs need to accommodate this diversity. Manuals, test materials, and instructions should be given in the language the trainee is most comfortable speaking. Such translation is not easy because translating technical words is much harder than translating conversational speech. Accommodations should be made for people with disabilities who may require spoken tests, large print, or a sign language interpreter.

Training Budgets and Planning

When the economy is in a recession, one of the first things many organizations cut is the facility budget. This frequently forces the FM to make cuts in worker training because deferred worker training is perceived as no immediate threat to the operation. The reduction in available funds for training is an unfortunate situation, but this challenge can present some opportunities and lead to greater efficiencies. Many managers have embraced the importance of education as a driver for continuous facility improvement. Cutting training is difficult because all relevant training has value, but cutting the budget may have unexpected positive outcomes: It forces the FM to search for value in training and to closely evaluate every potential training opportunity.

Managers should take a proactive approach to workforce training. Rather than waiting for new governmental regulations, , or worse an accident that forces the need for training to prevent future mishaps, managers should allocate the time and money into a training that will serve the facility operation. Forward-thinking managers will take into account the needs of their clients and tenants when designing a training plan. Worker training will provide a better value for the tenants lease payment. Ultimately, a structured training plan for facility personnel provides the workforce with training experiences that benefit the facility through increased productivity and higher quality.

Worker Training Goals

A manager must develop a formal, structured plan for worker training which is presented to upper management. A formalized written training plan that is already part of the budget is more difficult to cut because training is integrated into a greater plan for organizational success, and the loss by not providing training is therefore more obvious. This formal plan should spell out areas for individual worker improvement and a real outcome that the FM expects the training to achieve. The training goal must be realistic. A worker taking a basic electrical course can't be expected to perform extensive troubleshooting of electronic circuits. Workers must also think about their own goals for the future and how training can lead to achieving those goals. Workers who have already received training need to practice their new skills. Job assignments must be modified to give newly trained individuals a way of performing their new skills. Skills not practiced can quickly become skills lost. The FM should expect that additional future training and/or additional educational materials will be needed to keep the workers' knowledge current. Workers also need feedback on how they are performing their new skills. The FM can utilize the services of the local technical college and/or seek training from equipment manufacturers for information about training.

Not everyone who goes through a training program can apply the newly learned skills effectively to their work assignments. In this case, the FM needs to speak to the individual and make her or him aware of this failure, discover the reasons behind it, and develop a plan to prevent future training failures. Facility maintenance is a technical service position. As technology changes, so does the job of facility maintenance. In fact, as economic pressures increase on the facility, facility maintenance technicians must do more to keep the operation economically viable.

11.10　PRODUCTIVITY OF MAINTENANCE PERSONNEL AND FORMAL WORKER EVALUATIONS

At many organizations, evaluating employees is a single annual event. It occurs just prior to the managers making their recommendations for salary increases for their workforce. This type of evaluation schedule is unfortunate because it misses all the value that can be provided by a well-run evaluation process based on continuous and timely feedback to the employee. And it can actually demoralize employees, who quickly realize that management puts little effort in something that directly affects the welfare of the employees and their families. Workers may also sense that their salary increases have more to do with the fortunes of the company (higher profits equaling higher salary increases), rather than their individual contribution to the facility operation.

Worker evaluations should be a continuous process throughout the year, with quarterly evaluations and a summary evaluation once a year. Written assessments should be provided to the worker each time. The assessment should include a plan for improvement of the worker's skills. Any written evaluation, whether it comments positively or negatively on a worker's performance, must be placed in the worker's personnel file. If, at any time, an incident of poor performance occurs, the FM must document the issue. An evaluation should be set up as soon as possible under these circumstances. If a reprimand is needed, a copy of the reprimand is given to the worker to read and then it is signed by the worker. The signed reprimand should go immediately into the worker's personnel file. Exceptional performance also needs to be documented and placed immediately in the employee's file: It is just as important to record the successes. Three tools that are valuable for worker evaluations are:

- Position descriptions
- Qualification standards
- Performance standards

Position descriptions are used to provide information in employment advertisements for the position. Position descriptions describe the most important aspects of the position in the view of the employer. These descriptions list the major job responsibilities. They give new employee a basic sense of what the employer's goal was in hiring them for the position.

Qualification standards list the outside requirements the employee must bring to the position. Qualifications begin with basic physical requirements such as the ability to lift 50 pounds, climb ladders, have excellent hearing, possess 20/20 vision, and be a nonsmoker. Qualification standards may include some humanistic traits such as the ability to work with others on a team, the ability to deal with difficult people, or the ability to meet a deadline and work under pressure. Qualification standards also include the required training, licenses, and certificates that the employee must have to be qualified for the position. Typical requirements for a facility maintenance technician are a heating, ventilation, and air-conditioning (HVAC) diploma or degree, boiler/power plant operator's license, EPA refrigerant handlers certificate, and asbestos operations and maintenance certificate.

Performance standards list the important duties of the job and what constitutes a job well done. Performance standards provide a basic roadmap for employees to follow so they know when their work is satisfactory. Performance standards are not the same as position descriptions because position descriptions, however lengthy and detailed, describe the job but fail to identify when successful performance has been achieved by the worker. Performance standards should state the following:

- The time that a job should take or time standards used for evaluation
- The quality level the work should achieve
- The expected customer satisfaction level with the work
- The resource management requirements (use of tools, equipment, transportation, and supplies)
- Cleanup requirement for the job site
- Record-keeping requirements during and after job completion

ADA and Performance Standards

According to the Equal Employment Opportunity Commission, "an employer should evaluate job performance of an employee with a disability the same way it evaluates any other employee's performance." An employee with a disability can request a reasonable accommodation according to the Americans with Disabilities Act (ADA), to help him or her meet the organization's performance standard for the position. Accurate evaluations of employee performance frequently contribute to employees making a request for such accommodations. If the employee makes a request for an accommodation, the employer must comply unless doing so would cause an undue hardship for the company or incur excessive costs. But simply lowering standards so the employee with a disability can meet them is not considered a reasonable accommodation. Eliminating ancillary or marginal functions that may prevent an employee with a disability from completing work that is part of the performance standard (essential job function) *is* a reasonable accommodation. Providing the employee with special tools, allowing the employee to perform the job in an alternative way, or allowing the employee to telecommute are also be reasonable accommodations.

Facility Maintenance Worker Productivity Measurement

Facility maintenance worker productivity is measured for several reasons:

- In a healthcare setting, measurement provides data to regulatory authorities to show that government-sponsored funds are well spent.
- Measurement data provide ways for facility management to check on management procedures and methods.
- Data to top managers can help them evaluate the performance of facility management personnel.
- Data provided to human resources can justify the creation of new positions or position elimination.
- Measurement data can be examined to evaluate individual worker productivity for their contributions to the success of the organization and to provide the basis for salary increases or bonuses.
- Managers can base their decisions on such data, and not solely on cost, about whether the facility should outsource maintenance personnel or keep maintenance personnel in-house.

Understanding Worker Productivity Level

According to the Bureau of Labor Statistics, U.S. Department of Labor, in their *Occupational Outlook Handbook* (2012–2013), facility maintenance technicians are "general maintenance and repair workers."[*] They repair machines, mechanical equipment, and buildings, and carry out many varied tasks in a single day working both indoors and outdoors at single buildings or at campuses consisting of multiple buildings. This varied work situation and multiple requirements placed on the worker make it more difficult for FMs to evaluate their productivity. To understand productivity evaluation, the FM can compare it to other work situations.

One of the easiest worker categories to evaluate is the industrial worker engaged in producing parts on an assembly line. The worker is required to produce a certain number of parts x with a failure ratio of y. If the worker has met this goal, commonly known as a quota, his or her productivity is judged to be

[*]Bureau of Labor Statistics, U.S. Department of Labor, *Occupational Outlook Handbook*, 2012–2013 Edition, General Maintenance and Repair Workers, http://www.bls.gov/ooh/installation-maintenance-and-repair/general-maintenance-and-repair-workers.htm (accessed January 12, 2013).

satisfactory. An industrial service worker might also be able to "bank his or her productivity" by going over quota on some days while falling short on others. As long as their cumulative numbers fall in the satisfactory range, the worker need not fear any disciplinary actions by management.

Another category is a nontechnical service worker, such as a housekeeper or janitor. Workers in this situation must perform a series of nontechnical tasks such as vacuuming and emptying trash. Such workers in this situation cannot bank their productivity because they must always complete their assigned tasks every day. In other words, housekeepers can't mop the same floor twice to improve their productivity ranking.

Facility maintenance technicians operate in an environment that requires them to perform a variety of repair and maintenance tasks throughout the day. The first step in measuring productivity is to understand what is really being measured by the method employed. Is the method providing data on the individual worker or on the work-order (WO) system.

For individual worker productivity, the reviewing manager examines the actual time it took the worker to complete a task and compares it against a generic time standard for the type of work. This provides a good measure of individual productivity, notwithstanding the mitigating circumstances surrounding completion of the work.

Looking at the entire system's productivity, the manager examines the job from the moment the work order was issued until the time the work order is closed out. The actual job might take only an hour to complete, but it might take two or three days for the work to be properly issued and recorded. This amount of time can be misleading. For day 1 of the three-day work-order closeout scenario, the mechanic is completing work orders compiled the day before. On day 2, new work orders come in and are given to the mechanic to execute, and the completed work is recorded on day 3.

The system examines individual work completion, and it can also examine the maintenance operation on a macrolevel. For example, if a high percentage (90 percent or better) of work orders are being completed with little or no unproductive time, the system is working well. Conversely, if there is 40 percent unproductive time and 100 percent WO completion, the data may tell the manager that the facility is overstaffed. If WO completion is below 65 percent and slack time is 35 percent on a daily basis, the WO system, supply system, outsourcing work, and managerial methods should be thoroughly examined to uncover the reason. Once in a while, all facilities have a bad day when little seems to go right, so conclusions based on only a few days of data shouldn't be made. FMs should look for a trend in the data over several months in order to draw a conclusion.

Measuring Worker Productivity

Facility managers can measure worker productivity in several ways: the "percentage of completed work orders" method, the "time to complete one work order" method, the "days to enter and close out all work orders" method, and the "24/72-hour" method.

"Percentage of Completed Work Orders" Method

The *"percentage of completed work orders" method* examines the percentage of all the work orders completed compared to work orders issued for the particular day. If the number of incomplete work orders is dropping, then it is generally viewed that productivity is increasing. This method can be misleading, however, because of factors such as the amount of outsourcing employed, using outsourcing to perform time-consuming or difficult tasks, deferring maintenance, combining work orders, and seasonal variations. This type of productivity measure includes all trades in a maintenance department and cannot single out the work of individual employees for analysis. An employee's productivity is based on how his or her productivity relates to the percentage of the entire department. For example, if 10 percent of all work orders for the department are incomplete and the worker being reviewed is at 95 percent completion, then the individual's productivity would be rated as being above the norm. This method can easily be implemented without the help of a computer, and that is its main advantage.

"Time to Complete One Work Order" Method

After an analysis of 976 work orders over a two-month period at a hospital in Germany, one study Lennerts and others pointed out in 2005 that 67 percent of work orders were corrective and 33 percent were for preventive maintenance. Since the advent of computerized maintenance management systems (CMMSs), a common measure of productivity has become the *"time to complete one work order" method*. Many CMMS systems provide these calculations automatically. The process examines all work orders, considers them equal in all respects, and displays a number based on the average time (in hours) to complete one work order. Survey results show this number to be in the range of 1.2 to 1.3 hours per work order for hospital mechanics performing corrective work.

The idea behind this method is that, in any industry group (e.g., healthcare, education, heavy industry), the type of work generally performed averages out in complexity and difficulty level over numerous work orders, and provides an average time per work order. This method is convenient for a facility manager to gauge whether the department's productivity is increasing or decreasing. It also provides a way in which facility managers can compare their operation to similar operations within an industry.

To improve accuracy of the "time to complete one work order" method, survey results have indicated that separating preventive maintenance work orders (PMWOs) from corrective maintenance work orders (CMWOs) is vital. Preventive maintenance work reflects more on the skills of the manager in ensuring that the maintenance technician has everything needed to perform the work properly and that the work is scheduled at the most efficient time. Preventive maintenance work orders are predictable, and the planning effort results in them taking less than 0.8 hours per work-order to fill. At first glance this low number may be surprising because preventive maintenance work can be very time consuming. But on further consideration, a PMWO, such as "grease fan bearing on air-handling unit no. 1" can be completed in 5 minutes. These quick work orders compensate for more lengthy jobs such as "replace pump seal on chilled water pump no. 2," which might take 4 hours to complete.

"Days to Enter and Close out All Work Orders" Method

The *"days to enter and close out all work orders" method* is one of the simplest method for measuring worker productivity. It sets the goal that all work orders are completed in a specified number of days. The required number of days is not a universally accepted value; it is different from one workplace to another. The method has more to do with the work-order system then the actual work. The method begins when work orders are entered into a work-order system. The work is then completed, and each work order is closed out. The goal of this method is to make sure that the work is getting done (and the paperwork is in order) and completed within an acceptable time frame.

The 24/72-Hour Method

The *"24/72-hour work order" method* was discovered through a discussion with a plant manager at a Wisconsin hospital. This method differentiates between work orders that should be completed in a day and those that are more difficult and may take up to three days to complete. The rationale behind this method is that it is unfair to penalize the worker or the department for handling difficult and thus more time-consuming tasks. Work-order computer systems are not capable of predicting how extensive a job may be because the extent of the repair is generally unknown for corrective work orders at the time the work order is issued.

Example: The "24/72-Hour Work Order" Method

A work order states, "Room 2304 is too warm." The mechanic may discover that the repair is just a simple pneumatic thermostat recalibration, and the situation is corrected in under an hour. In this case, the repair falls under the 24-hour correction requirement. If the repair instead requires the replacement of an old, rusty, ceiling-mounted heating valve (a much more extensive repair), then it falls under the 72-hour correction completion goal. The work order is then recoded to reflect this change.

The goal of the "24/72-hour work order" method is to achieve a 92 percent completion rate of all twenty-four-hour work orders, and a 95 percent completion rate of all seventy-two-hour work orders. An essential added step is the reclassification of work orders once the necessary extent of a correction is known. Maintenance personnel performing preventive maintenance might also discover that more extensive repairs are required on a piece of equipment. Personnel in this situation can request a "demand work order" and compensate for the additional time required to finish the job completely.

The "24/72-hour work order" method looks at the work as a whole, meaning that the entire amount of work at the hospital facility is being accomplished in a reasonable time for the type of work being performed. The "24/72-hour work order" method works well, as signified by the subject hospital's recent increase in the completion goal of twenty-four-hour work orders from 87 percent to 92 percent. It should be noted that the direct involvement, knowledge, and ability of the plant manager also contributes to these excellent results.

The work-order systems can be abused and, above all else, rely on the integrity and competency of the person(s) managing them. The "24/72-hour work order" method could easily be misused if twenty-four-hour work orders are reclassified (because they are not completed in a day) and moved to a seventy-two-hour completion. In the "time to complete one work order" method, reducing the scope of a work order can decrease the time necessary to complete it and give the appearance of increased productivity. In the "percentage of work orders completed" method, a manager can increase the level of outsourcing or shift outsourcing work to include difficult and time-consuming tasks to give the appearance of improved productivity in-house. In the "days to enter and close out all work orders" method, a work order could be closed out, even if the work is not finished and the employees stretch the time to include the next two- to three-day period to compensate.

The results of all preventive maintenance work-order issuance methods can be improved by checking on parts and equipment availability prior to issuing the work order, and then performing a quality check after the work order is completed. These two important points can improve worker productivity and can help managers to assess productivity more accurately.

The check on parts availability should be accomplished through an automated system that reviews parts requirements against parts inventory to meet the needs of a specific preventive maintenance work order. The computerized system should identify any missing items and alert managers so that they can procure the parts prior to work-order issuance. If managers are physically checking for parts prior to issuing a work order rather than letting automation handle the task, overall department performance will suffer. It could actually result in a performance decrease because of the increase cost of supervisory staff doing it compared to employees. It is more efficient to issue a PMWO that can be completed the same day because all tools, parts, and material are available than to issue a PMWO that forces the supervisor to pull the work order from the maintenance system until parts arrive. When parts do arrive, they must be physically set aside, or they will end up in inventory rather than being installed.

An automated inventory control/work-order issuance system has great benefit for a large facility with a professional staff that handles numerous preventive and corrective maintenance work orders with a significant level of difficulty on a daily basis. This system ensures that the job can be done before it is attempted.

Quality of Work Evaluation

Job quality evaluation applies to both the PMWOs and the CMWOs. A quality check is performed by supervisory staff members, who review the job completed and assign it a quality factor that is applicable to the maintenance situation. When workers perform tasks that exceed the expected quality, the tasks are assigned a number greater than 1, and items below the quality standard are assigned a number less than 1. Quality evaluation is more than checking the physical quality; to be complete, it must review the customers' reactions to the completion of the work. A supervisor then must query the personnel or outside customers.

The same job performed at the same quality level at a different time might not receive the same quality rating. This difference could result from job conditions at the time at which the work was performed. A necessary job performed in a high-traffic area at a peak time that was accomplished without disruption should get a higher rating then the same job performed after hours. Note that multiplying a quality factor that is above 1 times output/input may result in a number exceeding 100 percent productivity. This is a correct assessment. A person who goes the extra mile to ensure customer satisfaction might not be as productive as a person who quickly performs job after job. But if the ultimate goal of the facility is to ensure customer satisfaction, the worker who achieves customer satisfaction at the cost of productivity should not be penalized for ensuring customer satisfaction unless the time required to do so is excessive.

Productivity Quality Factor Matrix

It could be argued that a maintenance supervisor does not have the time to assess every job that a worker completes for the various aspects of quality and professionalism. In response to that dilemma, the supervisor can utilize the worker productivity factor matrix shown in Table 11.2.

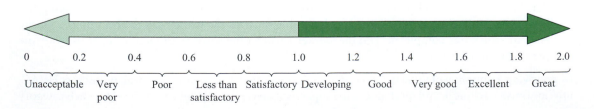

TABLE 11.2 Worker productivity factor matrix

Evaluation Area: (Name Type of Work)	0 to 1.0	1.0	1.0 to 2.0	Total for Service Area
1. Work completed compared with time standard for the work				
2. Quality of the work				
3. Efficient use of parts, supplies, and equipment				
4. Worksite safety				
5. Technical difficulty				
6. Customer service				
7. Work performance did not interfere with others'				
Total for category				
Productivity factor total for category divided by 7				

The productivity factors listed in Table 11.2 can be described as follows:

1. *Work completed compared with time standard for the work:* Supervisory personnel are generally well aware of how long a job should take for the circumstances surrounding it. A job that took 1 hour and 15 minutes to finish but should have taken 1 hour would be given a 0.75 rating because the job took 25 percent longer to complete.

2. *Quality of the work: This* category is an evaluation by the supervisor about the variation of the work quality from 1.0, which is the acceptable work level. A rating of 1.1 would indicate that the job was 10 percent above the norm in quality.

3. *Efficient use of parts, supplies, and equipment: This* category rewards people who diagnose a problem and then make a repair rather than swapping out parts until the defect is found. A supervisor would have to examine the part that was replaced in order to make an accurate determination about whether the part should have been replaced or repaired.

4. *Worksite safety:* This category requires a little investigatory work on the part of the supervisor, who needs to determine whether the work was performed in a safe manner. For example, did the worker use the correct PPE for the job? Did the worker make the work area safe for him- or herself and for others entering the area?

5. *Technical difficulty:* This category requires an understanding of the repair and the processes that went into troubleshooting the equipment. Not all repairs are technically difficult, and a low grade should not be received just because the work was not technically difficult. Rather a low grade for work that wasn't difficult technically is given if the worker underestimates the work (because it is easy) and fails to do a thorough job as a result. Work that is technically difficult is given a high grade when it is performed well. Work that isn't technically difficult but was performed correctly should be given a 1.0 evaluation.

6. *Customer service:* This category requires a discussion with the people who were served by the repair. Questions to ask the people being served include the following: Did the worker properly inform the people being served? Did the worker follow up with the people being served performed? Did the worker arrive on time?

7. *Work performance did not interfere with others':* This category applies to preventive and corrective maintenance work. Did the worker use his or her equipment or place supplies in a way that blocked access and prevented others from being productive? Was the worker speaking too loudly? Was there unnecessary conversation going on? Did the worker fail to clean up the jobsite after the work was completed? Did the worker perform the work at an inconvenient time for the personnel in the work area?

By reviewing a sampling of three to five completed jobs in categories such as electrical, plumbing, HVAC, interior finishes, clinical, and mechanical, the supervisor can assign a productivity level for each category in which the maintenance technician performs. Then an overall productivity rating can be assigned to each worker. Productivity levels can be used for determining salary adjustments and for identifying areas of worker strengths and weaknesses. The supervisor can also see the direct effect that improvement initiatives, such as increased training, have on an individual worker's productivity.

According to the matrix in Table 11.2, a worker who is working at a standard level produces a rating of 1.0. Working at a level below standard produces a decimal rating less than 0.8, all the way down to a 0. A rating of 0 indicates that the work was a failure and had to be completely redone by another maintenance technician, or it indicates that the process of performing the work interfered completely with the work of other trades or departments. Conversely, a rating above standard indicates above-standard performance. A rating of 1.4 to 1.6 is considered very good performance, and 1.8 to 2.0 indicates excellent performance. Excellent performance indicates that the work was done so that it prevented additional work from being necessary and another order from being issued (i.e., the maintenance technician found a hidden malfunction and corrected it), or the worker's performance made a lasting impact on the operation.

After reviewing a worker's productivity, the information should be provided to that worker as a useful means of self-improvement. All technicians can then improve their ratings in substandard areas. Maintenance technicians may also realize that they must start working "smarter" in order to improve their own productivity, and they can see how they will be rewarded for doing so.

By using the matrix in Table 11.2, Supervisors can determine an average score for each category. They can determine an overall average by adding up the average scores and dividing by the number of categories. To determine a worker's productivity level, first find the standard number of work orders per day that each worker is required to complete (average number of work orders per day divided by the number of workers equals the standard number of work orders per day per worker):

Productivity = (actual average number of work orders per day completed by the mechanic/standard number of work orders per day) × (overall productivity level from matrix)

Example: Productivity Factor Matrix

Maintenance Technician No. 1, commonly known as a hospital mechanic or mech., works at a healthcare facility with a facilities department staff of ten workers at a hospital, which generates an average of 120 work orders per day. Ten workers, or ten full-time employees (FTEs), work each day for the period reviewed. Therefore, each worker has a standard responsibility of twelve work orders per day. As represented in Table 11.3, Mechanic Technician No. 1 is assigned a productivity number from a thorough analysis of data drawn from several worker productivity factor matrices representing related functional areas.

TABLE 11.3 Example matrix for a maintenance technician

Evaluation Area: HVAC	0 to 1.0	1.0	1.0 to 2.0	Total for Service Area
1. Work completed compared with time standard for the work			1.2	1.2
2. Quality of the work			1.3	1.3
3. Efficient use of parts, supplies, and equipment		1.0		1.0
4. Worksite safety			1.1	1.1
5. Technical difficulty		1.0		1.0
6. Customer service			1.5	1.5
7. Work performance did not unnecessarily interfere with others'		1.0		1.0
Total for category				8.1
Productivity factor				
HVAC review total: **Category totals divided by 7 = (8.1/7) = 1.17**				

Example: Overall Worker Productivity

From the three simulated results tabulated in Table 11.4, one can see the importance of quality. Quality is also stated as the overall productivity factor in estimating the productivity of a mechanic.

TABLE 11.4 Simulated tabulation of matrix results for overall worker productivity

Skill Area	Review Scores			Average of Three Reviews
Electrical	1.20	1.15	1.15	1.17
Plumbing	1.15	1.15	1.15	1.15
HVAC	**1.17**	1.19	1.21	1.19
Interior finish	1.25	1.10	1.25	1.20
Clinical equipment	1.0	1.2	1.3	1.16
Mechanical	1.5	1.1	1.1	1.23
Total of averages				7.10
Overall worker productivity factor (total) divided by 6) = 1.18				

Case A

Assume that Mechanic No. 1 completes an average of 10 corrective maintenance work orders per day. Apply the following formula:

10 average WOs per day / 12 standard WOs per day = 0.83

0.83 × 1.17 (overall productivity factor) = 0.97, or 97%

Mechanic No. 1 appears to be a highly productive worker.

Case B

Survey results indicate that corrective maintenance work orders take 1.2 to 1.3 hours per work order to complete. Mechanic No. 1 actually completes 6.15 corrective maintenance work orders per day, based on survey results of 1.3 hours per work order. Apply the following formula:

6.15 average WOs per day / 12 standard WOs per day = 0.51

0.51 × 1.17 (overall productivity factor) = 0.60, or 60%

Case C

Mechanic No. 1 completes 6.7 corrective work orders per day, based on survey results of 1.2 hours per work-order completion rate. Apply the following formula:

6.70 average WOs per day / 12 standard WOs per day = 0.56

0.56 × 1.17 (overall productivity factor) = 0.65, or 65%

In actuality, Cases B and C are more accurate representations of actual productivity levels, as confirmed by survey data.

Summary of Productivity Evaluation

Productivity measures that lead to improvement of the individual worker and the maintenance function as a whole should get the most attention. The chosen method for determining these productivity measures must also provide data for analysis so that the effect of changes can be readily seen and evaluated. Evaluation method using the matrix in Table 11.2 provides the necessary data to meet many changing situations.

It is reasonable to hope that the productivity of the maintenance department will always trend upward. The reality, however, is that many factors can change productivity level. For example, worker productivity may change as a result of any of the following:

- An economic downturn in the economy, whereby less funds are available for major maintenance. Workers may have to perform more repairs or more difficult repairs on existing equipment.

- Reduced payroll funds may restrict the issuance of overtime pay to finish jobs. Workers may have to perform the setup work again when the job is completed the next day.

- Handling the challenge of terrorism or a possible future pandemic has the direct effect of increasing cost and reducing productivity. Terrorism and pandemics divert effort to these necessary issues, but they do not contribute to worker productivity.

It is essential that worker productivity must be reviewed on an individual basis and in the context of the overall organization. A worker may be significantly more productive if the technical capabilities of the organization are at the appropriate level to match the worker's ability. Workload may also influence productivity. Too large a staff may encourage workers to work less when work cannot be properly partitioned, and too small a staff might make the workload impossible to handle, leading to worker discontent. Sometimes a worker who is productive when assigned a workload that can be completed in one shift may become unproductive with a higher workload. A 10 percent increase in workload might result in a 20 percent decrease in productivity. One explanation might be the inability of the worker to properly prioritize work orders, which can result in failure to complete the work orders or only partial completion of important tasks. In situations where the facility has expanded in size or has given the workers additional responsibility, the matrix points out failing areas that are not always easy to see without it.

Example: Examining the Matrix to Determine Productivity Issues

A large facility has expanded 10 percent in size, but the staff size remains the same. Even with an increase in workload, however, overall facility work-order completion remains unchanged. If the facility wasn't overstaffed by 10 percent before the expansion, overall facility work-order completion should decrease because more work orders must be completed now.

Something changed in the way work is being performed, and the FM needs to discover the reason. Possible explanations can be found by reviewing matrix data. Data may show that, to keep the output up, the employee's proper parts utilization has decreased. Workers may be throwing away parts that require only minor repairs to make the parts useful again. Facility maintenance technicians may even be throwing away parts that require only minor cleaning before they can be reused. Another problem could be that, in the pursuit of work-order completion, customer satisfaction has dropped. The matrix in Table 11.2 can reveal these important changes.

When a facility has expanded or taken on larger, more demanding tenants, managers must avoid the temptation to rate employees higher to compensate for the increased workload. FMs need to show the true rating numbers of their staff members. If most of the worker ratings are tracking lower after facility expansion, for example, it may provide justification to increase the number of staff members. A manager can easily reduce the performance level that must be met before issuing worker pay increases to reflect the current understaffing condition. In this way, good workers are not punished by the data.

The individual worker's contribution level is the critical element in developing workforce requirements for the facility. Customer satisfaction, as part of the overall quality of the work completed, is an essential element of the output of the facility maintenance technicians. FMs must evaluate worker productivity beyond the simple evaluation of "the amount of time to complete a particular work order type." FMs and trade supervisors need to expend the necessary time to do a quality worker productivity evaluation. An assessment that includes the multiple factors in the matrix is indispensable for reflecting worker contribution to the facility's operation.

REVIEW QUESTIONS

1. Name some advantages of using native plants for landscaping.
2. What is xeriscaping?
3. What benefits can be achieved by aerating turf grass?
4. What is integrated pest management?

5. What is the difference between porous concrete pavers and permeable concrete pavers?

6. The two main elements of an exposure control plan for blood-borne pathogens are _____ and _____.

7. What diseases are transmitted by blood-borne pathogens?

8. The worker right-to-know requires that employers provide information to their employees on _____.

9. Name three site-specific safe work practices.

10. Name some elements needed for sustainable housekeeping.

11. What chemical in some lawn fertilizers can cause problems if it enters our waterways?

12. What is thatch?

13. Name the three types of pavement commonly used today.

14. What are engineering controls in an exposure control plan?

15. Metering faucets are activated by _____.

16. What are the benefits of using superplasticizer additives in a concrete mix?

17. Compared to common concrete, what is missing from a pervious concrete mix?

18. Name three important elements of a sustainable training program.

19. When using a worker productivity matrix like the one shown in Table 11.2, a satisfactory grade for a new employee is _____.

20. Name three ways for measuring the productivity of a maintenance worker.

Essay Questions

21. What steps should be taken in order to establish a sustainable housekeeping program at an office property?

22. Discuss the ways to reduce the need or amount of supplemental irrigation.

23. What are some of the advantages to using a trash container system that can wirelessly report whether it needs to be emptied?

24. Research different methods for making landscaping activities more sustainable.

25. What factors should be considered when deciding which equipment to choose in order to reduce the amount of water used in a restroom?

26. Discuss the factors that should be evaluated when estimating a worker's productivity.

GREEN BUILDING CONSTRUCTION

INSIDE THIS CHAPTER

Can Building Structures Be Sustainable?

This chapter begins with some basic commercial building construction methods. It discusses the equipment used to mitigate damage to buildings due to seismic disturbances and wind loads. It describes the different types of common roofs, their benefits and maintenance requirements. Current technology in these areas; descriptions of green walls, cool roofs, and green roofs; and new methods of water leak detection, are covered. The chapter ends with a discussion of people-moving equipment (elevators and escalators) and current methods for improving the efficiency of this equipment.

12.1   SUSTAINABLE BUILDINGS

Sustainable cities require a sustainable built environment. The cities themselves are or should be products of sustainable urban planning. The goal of such planning is to attract the types of buildings whose operations complement the surrounding services that other buildings provide and create a synergy between them, strengthening the entire city as a whole. City planners must consider the needs of the city's current population, for example, transportation, infrastructure, or livability issues, while planning for the future. Facility managers (FMs) have the task of making their buildings as sustainable as possible.

Green building designers take advantage of the climate, geography, natural surroundings, and specific site characteristics in their designs. Building materials are evaluated for sustainability. Locally available and recycled materials are given priority to reduce the buildings' carbon footprint. A consideration too frequently ignored in the past, but essential in green building design, is the *maintainability* of the building. The ability to maintain the building without major costs or disruption is essential. The design must therefore consider the life-cycle costs associated with the choice of equipment and building materials.

Green building construction also wants to mitigate the damage that the built environment inflicts on the natural surroundings. Paving over large areas of soil surface in urban environments captures and directs water to storm sewers. Building materials absorb or reflect heat into their surroundings. The absorbed heat is given off as temperatures drop in the evening, and reflected heat causes other structures to heat up more than they normally would with just daylight shining on them.

12.2   BUILDING FOUNDATION AND SUPPORT

The aboveground structure of a building rests on a foundation, which supports the massive downward force created by the weight of the building. In small buildings, spread footings provide more surface area to distribute the weight over a larger area. Large buildings use pilings driven into the soil that come in contact with stable bedrock, or friction pilings that support the foundation through the grabbing force of the soil around the piling along its length. In special situations where the soil cannot bear the weight of the building and pilings would have to be excessively long, a raft foundation may be employed. A raft foundation (or continuous footing, as it is sometimes known) extends under the entire building. This foundation causes the building to float atop the soil.

In the design phase, the building, foundation, and the natural forces it will encounter are considered. Great success has been achieved in making buildings capable of withstanding seismic forces generated by earthquakes and resisting the influence of strong winds generated by hurricanes and tornados. Wind exposes the building to a direct lateral force due to the wind's direction, but the building also creates forces due to oscillations and pressure differentials as the wind blows over and around the building. Earthquakes exert up-and-down and side–to-side forces on the building's foundation and supports. Side-to-side motion at ground level creates a shear force that attempts to tear a building from its foundation. Different strategies can be employed to mitigate the effects of these forces on the building. In addition to reinforcing the building structure in key areas, methods to lessen the effect of exterior forces are base isolators, sliding isolators, viscous dampers (see Figure 12.1), and tuned mass dampers. Some of these methods have also been employed on old historic buildings to prevent their destruction in the future.

A seventeen-story housing complex in Tokyo employed metallic roller bearings as a sliding isolator between the building and its foundation. The sliding isolator allows movement between the foundation and the building when lateral motion is exerted by earthquakes. A seventy-story mix-use hotel-and-condominium building in Chicago uses a tuned mass damper which in this case is a 300-ton steel weight supported on four cables. Its purpose is to dampen building sway created by winds. In tall buildings, such as the Taipei 101 tower (see Figure 12.2), the pendulum's mass can be as great as 800 tons. The weight, located near the top of the building, moves like a pendulum but in opposition to the direction of sway. This is accomplished by hydraulic rams that move the weight to dampen the building's motion (see Figure 12.3).

Tuned mass dampers can also dampen some of the building motion caused by earthquakes. Tuned mass dampers stabilize violent motions generated by earthquakes that create harmonic vibrations. They dissipate this energy by using hydraulic (viscous) dampers that convert the energy into heat.

Strategically locating viscous dampers (see Figure 12.4) throughout a building's structure is another way to dampen the effects of an earthquake. Viscous dampers absorb some of the energy generated by the earthquake and convert it to heat. Conversion to heat is accomplished by a hydraulic cylinder filled with a viscous fluid that is forced to flow through an orifice by a plunger. This action dissipates the earthquake energy as heat and can protect the major structural components of a building from failure (see Figure 12.5). Viscous dampers can be horizontally, vertically, or diagonally mounted in the building's structure. It is very important to provide seismic protection for buildings and structures where large numbers of people congregate (see Figure 12.6).

Whatever the foundation type, making a building sustainable requires a stable structure resting on a foundation, but maybe not rigidly mounted to that foundation. The seismic and weather-driven natural forces that a building is subjected to, now and decades into the future, must be considered in sustainable building construction.

FIGURE 12.1 Seismic viscos damper providing base isolation

Source: Taylor Devices, Inc.

FIGURE 12.2 Taipei 101, built in 2004, is a landmark tower that was the world's tallest building for six years. Later undergoing extensive modifications, the tower achieved LEED Platinum Certification in July 2011.

Source: Chang-Chih Liao/Fotolia

The foundation should not be a source of other building problems. One such problem, water infiltration through a building's foundation, can cause mold growth, poor basement indoor air quality (IAQ), and possible loss of usable or rentable basement space. To keep problems from occurring, the FM needs to include inspection of the basement and foundation areas in any maintenance plan for the building's structure.

Upper Building Support Framework

The most common types of support framework for buildings consist of structural steel, reinforced concrete, or a combination of the two. The framework provides support for the floors and roof deck, and a place to mount the wall materials that will comprise the building's exterior. A reinforced poured concrete building framework is constructed on site; pre- or poststress concrete support columns, planks, and beams are premanufactured and brought to the building site.

Steel Construction   A steel-supporting framework is constructed out of various steel shapes for columns, beams, and joists. The most common shape is the steel I-beam. Beams can be bolted or welded

FIGURE 12.3 Tuned mass damper located in the uppermost floors of Taipei 101

Source: Peter Tsai Photography/Alamy

FIGURE 12.4 Building viscous damper mounted diagonally to structural steel

Source: Taylor Devices, Inc.

FIGURE 12.5 Building in Tokyo, Japan, with viscous damper diagonally mounted. The damper braces the major structural building components.

Source: Taylor Devices, Inc.

FIGURE 12.6 Viscous damper protecting the structural components of a stadium with a retractable roof
Source: Taylor Devices, Inc.

together. Hot riveting was done in the past, but this method has been replaced by high-strength bolts in most cases. The steel framework provides the building's structural strength, and once the building's useful life has ended, this material is completely recyclable. The exterior walls, constructed of glass panels in a metal frame, are known as curtain walls. These glass exterior walls provide no real structural strength for the building and act, as the name implies, as a curtain to keep out the elements. A masonry core gives rigidity to a high-rise building. The masonry core also provides a fire rating for elevator shafts, ventilation duct enclosures, and electrical and plumbing chases. The floors of the building are constructed by laying corrugated steel pans and pumping concrete on top of the pan. Cellular floors with voids (also known as floor duct systems) can be constructed to provide access for electrical and data wiring. Masonry fire exit staircase towers are also found within the steel structure to provide safe emergency exits.

All exposed steel in a building must be fireproofed with a spray-on fire-resistant insulation. If any of the spray-on insulation is disturbed during remodeling, it must be patched to ensure that a one-hour fire resistance is maintained. Fire resistance is necessary to allow time for firefighters to battle a blaze and evacuate the building's occupants before the building steel structure starts to buckle and fail because of the heat.

Pre- and Post-Tensioned Concrete   Concrete beams, columns, and decking made of concrete planks may be factory-made and come with reinforcement that is pretensioned or post-tensioned. A pre- or post-tensioned beam, supported at either end with a load in the middle, experiences a tension load near the bottom of the beam. Tendon reinforcement is placed near the bottom part of the beam to resist this tension. The side of the beam is squeezed by the tendons. Compression caused by this stress is offset by the load placed on the beam.

It is very important to monitor for corrosion of tensioning cables or rods. Loss of this reinforcement can have catastrophic consequences from sudden failure. Any damage of this nature must be inspected by a licensed professional structural engineer, who can provide the recommended method of repair.

Reinforced Concrete   Reinforced concrete constructed on site is created by building an interior steel-mesh structure consisting of reinforcing rods that are welded or wire-tied together and embedded in concrete. After the steel-mesh structure is created, a wooden form is built around the mesh, and concrete is poured inside the form. As discussed in Chapter 11, concrete that has a superplasticizer additive (a water-reduction admixture) is used to allow the concrete to flow easily all the way around the form and self-compact while simultaneously producing a high-strength concrete. Superplasticizers cause concrete to flow quickly and fill many forms without having to move the concrete truck or the pumping hose. The compressive strength of concrete with a superplasticizer additive generally ranges from 7,300 to over 14,000 pounds per square inch (psi).

12.3   BUILDING EXTERIOR WALLS

Several types of envelopes seal the exterior of the building. Some of these for building walls are tilt-up concrete construction, masonry construction, exterior insulation finish systems, and glass curtain wall construction. All these exterior wall systems have one thing in common: They must shield the building interior from the weather outside. Masonry and tilt-up construction can be designed as structural walls and can carry some of the weight of the floors and roof structure. Exterior insulation finish systems and glass curtain walls only provide protection for the building's interior and provide no structural support.

Tilt-Up Construction

Tilt-up construction is frequently used for warehouse and light industrial buildings. A method of constructing a tilt-up building is first to pour the building's floor slab and footing around the slab, which will support the wall sections. The floor slab is used as a base for the concrete wall forms, which are pieces of wood attached together and laid on the floor. The wood pieces act as a mold for the wall sections. The wood forms, in addition to simple wall sections, frame out windows and doors. (An alternate method is to pour a separate slab on grade near the building to act a base for the concrete forms.) A release agent is sprayed on the floor and the forms. Reinforcing steel mesh and reinforcing rods are placed in the mold, and then concrete is poured into the mold. A colorant may be added to the concrete to enhance its appearance. Using the floor slab as the base of the forms yields savings on materials used in form construction.

Any mechanical anchoring of the forms to the building's floor must be removed and the floor must be patched once the walls are up. The cured panels are then tilted in position, braced with temporary supports, and bolted together with fastening attachments embedded in the concrete. The gap between wall sections can be caulked at this time. These reinforced concrete wall sections, which are 6 to 8 inches thick, act as bearing walls supporting the roof by open-web bar joists, whose ends sit on top of the wall. This type of construction can save a lot on transportation costs of wall sections because concrete is frequently available locally.

Exterior walls are also built from precast concrete panels that are brought to the building site and anchored to the exterior of the building structure. This method saves a lot of time during construction, but the cost savings of not having to transport these heavy wall sections is lost.

Masonry Block or Brick Construction

In masonry or brick construction, the walls of the building are built on top of the foundation. Masonry blocks are known as concrete masonry units (CMUs) and come in several different styles today to meet various construction situations and to make construction easier. CMU walls can be made stronger by adding reinforcing rod through their openings and grouting with cement. CMU walls can be either bearing or nonbearing walls. Nonbearing walls rely on steel columns and beams for building support of floors and the roof structure. Bearing walls have a roof structure that is frequently comprised of open-web steel bar joists that rest on the top course of block. Bearing walls have either the top course of block filled with concrete, or they are solid masonry units.

Exterior Insulation Finish System

The exterior insulation finish system (EIFS) allows the architect to design complex stucco or stone-like building exteriors without the expense of actually constructing with stone or stucco. The two basic types of EIFS are the barrier type and the drainage type. EIFS construction consists of exterior sheathing, cement based adhesive, expanded polystyrene foam (EPS) ¾ inch to 4 inches thick, ¼-inch fiberglass square mesh fabric, ¼-inch-thick base coat, primer, and $1/16$-inch-thick finish coat.

The first component of an EIFS system, the exterior sheathing of the building, can be either exterior-grade gypsum board, oriented strand board (OSB), particle board, plywood, or concrete block. In a drainage system, a moisture barrier is placed over the sheathing. The insulation can be attached directly to the exterior sheathing of a building, or it can be set by first installing a drainage system. The drainage system stops moisture from becoming trapped between the foam and the sheathing due to either imperfections in the EIFS or failures in caulk joints. Trapped moisture can cause mold growth and damage the sheathing. The drainage system may consists of a rolled-on or wrapped-fabric moisture barrier that has ridges, or a grooved foam board is used to provide a gap to allow the moisture to drain out bottom holes. The foam board is either mechanically fastened or set with a polymer-modified cement adhesive. A plastic-coated fiberglass mesh is placed over the foam's exterior. The mesh provides impact strength and prevents cracking of the base coat at foam board seams. Heavier meshes are used to provide more impact strength in areas where impact is likely. It may be wiser, however, to provide barriers to prevent impact or use materials such as concrete and brick that are more impact-resistant where damage is likely from site activities (such as the use of shopping carts at a supermarket). The base coat and finish coat are troweled on and smoothed. The finish coat is an acrylic co-polymer about $1/16$ of an inch thick.

The advantage of EIFS construction is low cost relative to more traditional methods, and the insulating value of the material. EIFS system components and specific application materials and methods vary per product manufacturer's recommendations.

Glass Curtain Wall Construction

Glass curtain walls are constructed of aluminum-framed glass panels that are attached to the supporting framework of the building. The weight of the panels (dead load) along with wind, snow, ice, seismic activity, and maintenance loads (live loads) are transferred to the framework and floor structure of the building. The building structure expands, contracts, and moves differently than the curtain wall. Allowances must be made so this movement does not damage the building and the entire system can remain weather-tight. Caulking around window frames should be inspected and renewed as necessary because this is a common source of leaks into the building. Moisture that gets past the rubber window seals around the frame drains out through weep holes in the bottom of the window frames. The glass is generally double-pane, insulated, and tempered. Insulated glass whose interior face gets fogged over should be replaced. Fogging indicates that the seal separating the glass panels has failed and that the insulating inert gas contained between the sheets of glass was lost.

Spandrel glass panels are colored panels used in nonvision areas to hide building construction elements or mechanical equipment while maintaining an entirely glass curtain wall structure. The panels hide items such the edges of floors where they meet the building exterior, or the end of acoustical ceilings and the heating, ventilation, and air conditioning (HVAC) ductwork above the ceiling. Clear glass in these areas would be very unsightly. Spandrel glass panels may be laminated glass panels backed with ceramic, with an insulating core. They can a variety of different colors, or they can be designed to match the color tinting or mirroring of the glass plate so they blend in and appear as any other glass panel on the building façade.

Green Walls

Green walls are living vegetation systems. They provide shade to exterior walls and mitigate temperature fluctuations on the wall surface. Evapotranspiration limits the heat rise of the wall to ambient and sometime even slightly cooler than ambient temperatures. Plants also provide additional habitat for insects and birds. Green walls may also be utilized in building interiors to improve the indoor air quality. They come in the following varieties:

- Green façades
- Semi-intensive
- Intensive
- Freestanding
- Indoor

Green Façades   Green façades are composed of plants, such as vines and ivy, that can be installed in the ground and allowed to grow upward along the outside of the building. Buildings constructed of porous materials, such as brick, make good grabbing points for the plants. Some maintenance is necessary to prevent the plants from blocking windows, although sometimes plants are purposely used to shade windows.

Semi-Intensive Green Walls   Semi-intensive green walls are used when the building surface is not appropriate for plant attachment, or plant attachment could cause damage to the building's surface. Stainless steel cables or mesh can be run upward along the building exterior to allow plants to attach themselves.

Intensive Green Walls   Intensive green walls are constructed almost entirely of plants and devices that increase their height range. Plants that grow from the ground level up are limited in the maximum height they can grow to about 50 feet. To increase the coverage of tall buildings, modular panels containing stainless steel planters filled with soil are attached at various intervals to the building's exterior (see Figure 12.7). The planters have a drip irrigation system. Some intensive green wall systems claim to reduce the temperatures behind the green wall system to as low as 25 percent below ambient temperature on a south-facing wall (see Figure 12.8).

Freestanding Green Walls   Freestanding green walls are shrubs and hedges that grow independently of architectural structures and function as walls. Although freestanding green walls seem common, and it sounds rather obvious to mention them, freestanding green walls are much overlooked today. However, neatly trimmed hedges and shrubs were and are an essential part of English gardens. Hedges provide sound insulation, filter the air, and provide homes for small birds.

Indoor Green Walls and Plants   When indoor plants are watered, about 90 percent of the water held by the soil evaporates into the interior space. Adding plants to building offices can be another way of humidifying the area. But controlling the need to water interior plants to match humidity requirement is a challenge. During already humid seasons that require dehumidification, plants inside a building can

FIGURE 12.7 Green wall on One PNC Plaza

Source: PNC Financial Services Group, Inc.

FIGURE 12.8 Close-up of the green wall on One PNC Plaza

Source: PNC Financial Services Group, Inc.

add to elevated humidity levels; however, indoor plants can also absorb harmful volatile organic compounds and carbon dioxide in the air and improve air quality overall.

12.4 ROOF CONSTRUCTION AND MAINTENANCE

Many types of roofs are in use today. The list of available roofing material keeps expanding due to reformulations, new developments, and new regulations. For instance, today some municipalities require that cool roofs or green roofs be installed on newly constructed commercial and municipal buildings. Roofs today can also provide accessible green space for building tenants and visitors. On the other hand, built-up roofs have been in the United States for more than a century, providing both value and long life.

Whichever roof type is chosen, managers must consider the useful life of the roof, the term length of the roof warranty, warranty cost, maintenance requirements, disposal and recycling costs for eventual removal of the roof, and the availability of contractors to provide competitive pricing. Some of the common types of roofs in use today include the following:

- Cool
- Built-up
- Modified bitumen
- Rubber
- Thermoplastic
- Metal
- Green

Cool Roofs

Cool roofs are a new category of roof; they can be composed of many different types of roofing materials. Cool roofs are either made of a light-colored material, or they are covered with a light-colored paint or aluminized coating that reflects solar radiation. For added assurance that the coating performs as designed, the coating should be Energy Star rated by the United States Environmental Protection Agency (US EPA). When a roof reflects solar radiation, the surface temperature of the roof is reduced from what it would have been if it absorbed solar energy.

Cool roofs reduce the amount of mechanical cooling required to reduce a building's interior temperature. Unfortunately, cool roofs may also increase the amount of heating required to raise interior temperatures during the heating season. A cost/benefit analysis should be performed when deciding whether cool roof technology is appropriate. The US Department of Energy's Oak Ridge National

Laboratory has developed two online calculators to estimate cooling and heating savings or penalties for flat roofs with nonblack surfaces. One calculator is designed to be used on small to medium-size facilities that are charged for their electric power on the secondary rate. The other is for large facilities that purchase electricity with a demand charge based on the peak monthly load.

Using the calculators requires knowledge of the R-value of the proposed roof, its solar reflectance (SR), and its infrared emittance (IE). The FM must also know the summertime cost of electricity (the dollar amount divided by kilowatt hour [$/KWh]), the air conditioning efficiency (*coefficient of performance [COP]* provided by the air conditioning equipment manufacturer), the heating energy source and costs (fuel or electric), and the heating system efficiency (boiler efficiency). The result shows cooling savings as a dollar amount per square foot per year ($/square foot/year); the heating penalty is shown as a negative number ($/square foot/year). The calculator also tells the user how much additional R-value in roof insulation will yield the same effect as installing a cool roof. This information is most useful in cold climates, where heating penalties are the highest.

Built-Up Roofs

For many years, built-up roofs (BURs) were considered the best roofing installation. Built-up roof construction begins by mopping a layer of hot asphalt on the roof deck to act as a waterproof cement that holds the sheets of insulation in place. (This process is known as mop 'n' flop in roofing terminology.) More asphalt is applied over the insulation, and then the first layer of roofing felt is rolled out. The next roll overlaps the first by 4 inches. The entire roof is covered in this manner, and the process is repeated an additional three times for a four-ply built-up roof. A flood coating of asphalt is applied over the top layer of felt, and pea gravel is partially embedded in it to provide a surface that can be walked on. With the high cost of petroleum today (asphalt is made from petroleum), built-up roofs are expensive relative to many types of single-ply roofing.

Different grades of asphalt are available, and the use of each grade depends on the slope of the roof. Coal tar pitch can also be used in lieu of asphalt. FMs can tell if coal tar pitch was used instead of asphalt by dropping a small piece in a cup of gasoline. Coal tar pitch does not dissolve in gasoline, but asphalt dissolves easily. Coal tar roofs have an advantage of being somewhat self-healing. Cracks in the roof seal once temperatures get high enough in the summer so that the coal tar becomes liquid. Coal tar is self-leveling and flows together, eliminating any gaps or cracks. Conversely, coal tar roofs can be used only on flat roofs because, on a steeply pitched roof, the coal tar may melt right off the roof. Installing coal tar at elevated temperatures can be a problem because of the pungent, acrid odors that burn the eyes and throat. The fumes are also carcinogenic to a person who is exposed to concentrations in the air that exceed the workplace exposure limits as defined by the Occupational Safety and Health Administration (OSHA). According to the New Jersey Department of Health and Senior Services, many scientists believe that there is no safe level for human exposure to a carcinogen, and they recommend that protective breathing equipment and other personal protective equipment (PPE) be worn whenever there is a possibility of exposure.

Maintenance of BURs involves a thorough survey of the roof to detect problem areas. Roofing consultants use an infrared camera to detect leaks by finding areas of wet insulation, which have an elevated temperature in the cooler months because of their ability to conduct heat easily. BURs frequently suffer from blisters caused by moisture trapped between layers of felt or between the felt and the insulation. The moisture expands when heated, balloons the felt layers, and creates a blister. If an unbroken blister is two feet long or less, it may be left alone, or it may be marked with fluorescent spray paint to prevent workers from inadvertently stepping on the blister and breaking it. Another common maintenance task is to refill pitch pockets with asphalt. Pitch pockets are generally square, open-bottom metal containers mounted on the roof that are used as a way to provide a waterproof seal around pipes and conduits that run through the roof. Over time, the pitch shrinks and actually provides a recess that contains water rather than shedding it. This depression eventually causes the contained water to slowly seep into the building. Pitch pockets need to be refilled to prevent this condition.

As a built-up roof nears the end of its life, consideration should be given to the possibility of putting an additional layer of felt and asphalt (adding a single ply) to extend the life of the roof. The useful life of a built-up roof is generally twenty-five years.

Modified Bitumen Roofs

A modified bitumen roof is a roll-type of roofing that has layers of nonwoven polyester or glass fiber mesh reinforcement impregnated with rubberized modified bitumen. The rolls come 3 feet wide and cover 100 square feet of roof (100 square feet of roof is known as 1 square in roofing terminology). The roof may be laid as a one-, two-, or three-ply installation. First, insulation is placed over the steel roof deck. Next a harder, more durable roof board is placed on top. Then the insulation and the roof board are mechanically fastened to the deck, or the layers are individually attached with adhesives. The first layer of modified bitumen is installed, and this layer can be directly adhered by hot-applied or cold-applied adhesives, torched on, or cold applied with overlapped seams that are torched. Self-adhering modified bitumen that requires no torching or adhesives can also be used. The top layer of bitumen can have a smooth surface or a mineral granule coating.

Modified bitumen roofing should be coated immediately or within five years of installation with a heat-reflective elastomeric coating that is Energy Star rated. The coating prevents the plasticizers from leaching out of the roofing material. The coating also reflects heat, keeps the roof cool, and thus makes this roof a cool roof.

Rubber Roofs

Rubber roofs are made of single-ply ethylene propylene diene monomer (EPDM). The seams of rubber roof membranes are glued together. A seaming tape has been developed that has drastically cut the time it takes to seam a rubber roof. Rolls of EPDM may even come with the tape already applied to one edge. EPDM is frequently secured to the roof by ballasting with smooth river stones. Rubber roofs generally have walkway pads installed to prevent damage that might be caused by someone stepping on a broken stone with sharp edges.

Thermoplastic Roofs

Thermoplastic roofs use polyvinyl chloride (PVC) and thermoplastic olefin (TPO), which are sheet materials reinforced with fiber matting of polyester or glass fibers. These materials can be heat-welded with hot air. The material changes from a solid to a semisolid when heated. Sheets are overlapped and bonded together, resulting in a weld that is stronger than the surrounding sheet. As a result, seam failures are uncommon. Roof maintenance includes washing the roof to remove dirt, or cleaning with a light detergent and bleach to remove any algae buildup.

Thermoplastic sheets can come in a variety of colors, with white or gray being the most common. They are attached to the roof mechanically with screws or by fully adhering or welding plates to the ends of the sheet that are set at intervals. Lightly colored or white thermoplastic roofs are cool roofs that save energy required to air-condition a building by reflecting solar heat energy rather than absorbing it.

Metal Roofs

Metal roofs can be made of copper, aluminum, or steel, with an aluminum/zinc coating and a protective paint coating over that. These roofs have narrow panels approximately 12 inches wide that run vertically up the roof. Metal roofs are commonly called standing seam roofs because the panels are connected together via a raised (standing) interlocking metal seam. Because these roofs are so smooth, they have a one-inch, L-shaped metal bar attached perpendicularly to the metal seams. The metal bar is used to hold snow and ice on the roof to allow it to melt and prevent it from creating a hazard by sliding off the roof.

Metal roofs require a waterproof synthetic underlayment that is first applied to the roof. The underlayment may be a type that is attached with plastic-head roofing nails, or it can be a self-adhesive type very similar to ice and water shields used on private residences when there is a possibility of ice dams forming on the roof. Paint coating on metal roofs can also reflect heat and provide for a cool roof design. Metal roofs have a fifty-year life span.

Green Roofs

A way to reduce the surface temperatures of buildings, provide for a more natural environment, and lessen storm water runoff is through the use of vegetation walls and roofs. In addition to or in lieu of absorbing rainwater in a green roof, some buildings collect rainwater from the roof and use it for nonpotable purposes. Rainwater is also known as gray water. A building may incorporate cool roof technology and rainwater storage to provide water for toilet flushing, cooling tower makeup, and watering green roofs and walls.

Green roofs provide a method of keeping the water they collect from entering the municipal storm water sewers. This is possible through the moisture-holding capacity of the soil on the roof. The green roof assembly (see Figure 12.9) also provides support for the planting bed and a way to store water and

Components of a Garden Roof® Assembly

The range of Hydrotech's Garden Roof Assemblies ensures that the right design can be provided to meet the requirements of any landscape option, including:

- Extensive, Intensive Lawn and Intensive conditions
- Ponds, paved and play areas
- Sloping and flat roofs
- Pedestrian access onto the roof

A typical Garden Roof Assembly consists of...

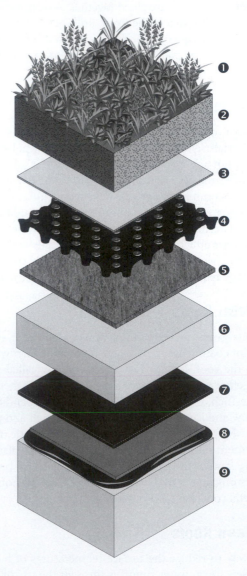

❶ **Carefully selected plants** – extensive plants for low maintenance landscaping including the drought resistant, self-regenerating varieties available from Hydrotech. There are a wide range of plants for intensive landscaping that can be supplied by garden centers and nurseries. GardMat® erosion control blankets (not shown) are also available to aid in plant establishment.

❷ **Engineered lightweight growing media** – must have a well balanced structure and low weight. The pH values, nutrients, degree of porosity and vapor permeability must be suitable. The type and thickness of the growing media ultimately determine the plant choice as well as the structural load imposed on the roof structure. Hydrotech's LiteTop® growing media is specifically blended to meet the requirements of each project.

❸ **SystemFilter** – to prevent fine particles from being washed out of the substrate soil and therefore maintaining the efficiency of the drainage layer.

❹ **Drainage/retention / aeration element** – Hydrotech's Gardendrain™ retains water in the profiled troughs, even on sloping roofs. Excess water drains away through channels between the troughs. Strategically located holes provide the necessary aeration and ensure that moisture below diffuses up into the growing media.

❺ **Moisture Mat (optional, contact Hydrotech for recommendation)** – made of non-rotting fiber to retain moisture and nutrients as well as provide physical protection to the root barrier and waterproofing membrane.
(An air layer is required when placed directly over insulation - not shown)

❻ **Insulation (optional)** – situated above the roof membrane, an extruded polystyrene insulation which exhibits excellent moisture resistance, is dimensionally stable and has a high R-value. Dow Chemical's STYROFOAM® brand insulation is typically utilized.

❼ **Root barrier** – prevents roots from affecting the roof membrane. The type, thickness and method of installation depend on the nature of the landscape planned and the shape and slope of the roof.

❽ **Roofing membrane** – Only the best, Hydrotech's Monolithic Membrane 6125EV-FR® (Environmental Grade) Assembly with protection layer.

❾ **Structural roof deck** – must be designed to support the weight of the green roof as well as any other dead or live loads. Acceptable deck types include cast-in-place concrete, precast concrete, metal deck with cover board and plywood.

FIGURE 12.9 Components of a Garden Roof® assembly. Note the individual components, from the roof deck to the plantings.

Source: Image courtesy of American Hydrotech, Inc.

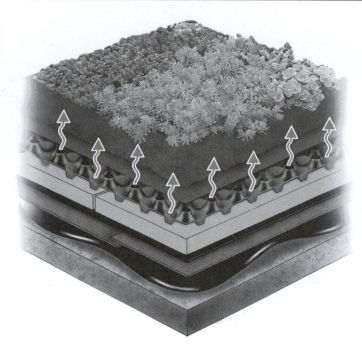

FIGURE 12.10 Rainwater collected on a green roof is used by plants later

Source: Image courtesy of American Hydrotech, Inc.

drain off excess water (see Figure 12.10). Green roofs can also be quite beautiful because they can provide garden-type green roof installations with attractive plants and shrubs. Green roofs can take advantage of roof space that is normally wasted and use it as an amenity for tenants and visitors. Through a process called ***evapotranspiration***, the water evaporates from the soil and transpires from the plants. This form of evaporative cooling can lower roof deck temperatures 50 degrees Fahrenheit. Water that the plants can't use travels into a raised area beneath the soil (see Figure 12.11) and can be sent to a storage system for irrigation purposes or sent to a roof drain.

It is predicted that green roofs installed throughout a city can lower the mean summer temperatures and reduce the heat island effect. Lower roof temperatures reduce the amount of air conditioning required during the hot summer months. Green roofs also increase the amount of heating required during the winter heating season.

FIGURE 12.11 Cutaway of a green roof component showing how excess water flows to a roof drain

Source: Image courtesy of American Hydrotech, Inc.

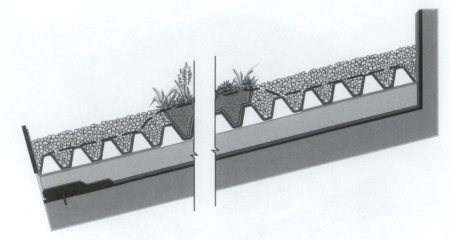

FIGURE 12.12 Extensive green roof with moderate slope

Source: Image courtesy of American Hydrotech, Inc.

Green roofs or vegetation roofs are divided into four main categories:

- Extensive
- Lawn
- Sloped
- Intensive

The use of each type of green or vegetation roof depends on the steepness or pitch of the roof and the desired outcome.

Extensive Green Roofs   Extensive green roofs (see Figure 12.12) provide a depth of 2 to 6 inches of growing media, resulting in a wet weight of 17 to 41 pounds per square foot. With this comparatively small amount of additional roof loading, these installations can cover extensive areas of the roof. Their use is limited to absorbing storm water and reducing roof temperatures. Extensive green roofs can be installed over a standing metal seam roof. Figure 12.13 shows a green roof covering an existing roof, and a metal seam roof without a green covering.

FIGURE 12.13 Two roofs with identical construction. The one on the left is a green roof; the one on the right is metal standing seam.

Source: Photo courtesy of Waukesha County Technical College.

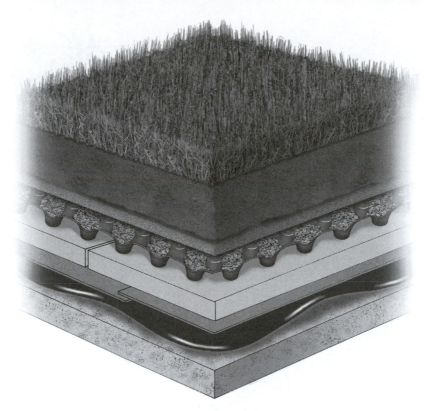

FIGURE 12.14 Cross section of a component part of a lawn roof

Source: Image courtesy of American Hydrotech, Inc.

Lawn Roofs   Lawn roofs (see Figure 12.14) provide a depth of 8 to 12 inches of growing media resulting in a weight of 53 to 78 pounds per square foot when wet. They are intended to be used as any lawn would be used, and they require essentially the same maintenance, including watering, fertilizing, and even cutting in some cases.

Sloped Green Roofs   In roof construction terms, the pitch of the roof is the how high the roof rises for every 12 inches of span. Having no rise, a flat roof would have a 0:12 pitch. A ranch house in the Midwest might have a 5:12 pitch (the roof rises 5 inches for every 12 inches). Another way to define pitch is to picture a triangle where the base side is 12, the height side is 5, and the resulting pitch is the hypotenuse. The higher the height of the triangle, the greater the resulting angle of pitch. Roofs with slopes of up to 12:12 pitch can be covered with a green roof (see Figure 12.15). A 12:12 pitch is a fairly steep roof. The green roof may be extensive or limited intensive. These roofs require a specialized installation to prevent soil and plant materials from sliding off.

Intensive Green Roofs   Intensive green roofs (see Figure 12.16) require the type of landscape maintenance that would be found in any garden. The plant material might even include shrubs and small trees. Soil depth ranges from 8 to 36 inches. The weight of the soil and soil containment requires that roofs be solidly reinforced to support this additional weight.

Green roofs of this type are frequently used to increase the green space and the useful areas of the building, thus providing more livable space (see Figure 12.17). Green roofs can also provide a beautiful addition to a basement extension that would otherwise result in a drab, waterproofed, concrete basement ceiling (see Figure 12.18). The result of this green roof installation in the figure is a beautifully landscaped garden walk.

Managing Green Walls and Roofs

Green roofs, with their multilayered construction, make access to a roof's waterproof membrane rather difficult. Roof leaks are a major issue for green roofs because of the cost and coordination of trades (facilities maintenance technicians, landscapers/irrigation technicians, roofers, leak detection specialists) necessary to find and fix leaks. Leak problems can frequently be mitigated through the

FIGURE 12.15 Gardnet™ assembly for steeply sloping roofs
Source: Image courtesy of American Hydrotech, Inc.

FIGURE 12.16 Intensive green roof assembly
Source: Image courtesy of American Hydrotech, Inc.

FIGURE 12.17 Green roof on a building that provides more livable areas

Source: Image courtesy of American Hydrotech, Inc.

installation of leak-locating equipment prior to the installation of a green roof, or using advanced leak detection tools.

A tool use to find leaks on some vegetation roofs is electric field vector mapping (EFVM). The theory behind EFVM is that two conductive plates are established on the roof. One plate is the grounded roof deck, and the other plate is a film of water on the roof surface. First, a technician places a noninsulated loop of wire around the roof area to be tested, and then the roof is wetted down. Then a 40-volt electric pulse is sent through the wire for one second every three seconds. The electric pulses seek to complete the electric circuit by finding a path to ground. The path that a pulse can follow to ground is through the roof leak to the roof deck. The survey technician enters the loop with two probes connected to a

FIGURE 12.18 Green roof over a basement area that extends out from a hospital building and provides a landscaped garden walk

Source: Photo courtesy of Oconomowoc Memorial Hospital.

potentiometer and analog meter. The probes are placed on the roof deck. The swing of the meter indicates the direction of current flow and leads the technician to the leak. The leak can then be repaired and then tested again.

EFVM won't work if the roof membrane is electrically conductive, which is the case with ethylene propylene diene monomer (EPDM) roofs that use carbon black in their formulation. Roofs that have aluminized coating painted over them might also be inappropriate for testing. It has also been reported that plastic root barrier membranes that sit on top of the roof can act as an insulating layer. When a root barrier is present during leak testing, small incisions are made in the root barrier so that electric contact can be made by the technician's probes. These incisions can be repaired later.

A way to improve the accuracy of EFVM is to have a conductive wire mesh, which is grounded to the building steel, laid under the roof membrane during roof membrane installation. The mesh greatly lessens the distance that water from a leak has to travel to reach a suitable ground path.

Managing green walls and roofs is similar to the management of any landscaped environment. It is often similar to the care and feeding of indoor potted living plants, with all the exterior environmental benefits and hazards placed on them. For example, torrential rains and high winds may wash the soil and nutrients from the elevated planters. These environmental events should be anticipated in the original planning by placing the plants away from building entrances, and afterward by reviewing the condition of not only the landscaping, but the supporting mechanical elements.

12.5 COMMON BUILDING PROBLEMS

Some common building problems arise from water infiltration through the building envelope, mold growth, corrosion, and cracks in masonry. Water infiltration is a major concern of building managers. The obvious problem created when water enters an office building is damage to expensive equipment (computers, copy machines, network equipment, etc.). When water enters a retail shopping center, damage to the items for sale many render them totally unfit for purchase. Water entering an industrial facility can damage equipment and stop production. The damage scenarios are endless. But just as important are the not-so-obvious problems, like mold and corrosion that can occur over time when water enters a building.

Mold

Mold is an immediate concern once evidence of water infiltration from outside the building has been discovered. The addition of water provides one of the two ingredients necessary to start mold growth. The other ingredient is food, and molds are not picky eaters. Mold spores can be thought of as tiny seeds that are ready to hatch. They float in the air and are literally everywhere. A problem with mold is that, once it starts, it is difficult to stop. A mold test kit provides the perfect environment for mold growth. It detects mold if spores land on the media and start growing. The amount of time required to grow mold on the surface of a test kit indicates the amount of mold present: the faster the mold grows, the more serious the problem.

If it is certain that mold is causing indoor air problems, it may be better to find the source of the mold's food and moisture and eradicate both to stop mold growth. Where mold is concentrated on a surface, it must be removed by chemical application. During the process of killing mold for removal, the mold releases spores that increase the chances of mold spreading to other areas, possibly making the current situation worse. If a lot of mold spores are present in the air, chances are much greater that they will find a suitable spot to grow before the mold's food source is removed and/or the area is dried out. The first step in professional mold removal is having the contaminated area enclosed in plastic sheeting, putting it under a negative air pressure, and filtering the exhaust air to capture the spores. This process allows the mold to be killed and removed without allowing mold to spread to other areas of the building.

Corrosion

Water infiltration can cause structural steel reinforcement embedded in concrete to rust and expand. The expansion of corroded steel causes fissures and voids in the surrounding concrete. The rusting and breaking of steel reinforcement reduces the tension strength of the concrete. Damaged reinforcing steel may degrade the strength of concrete spans to the point of failure. Deteriorated concrete support spans require evaluation by a masonry structural engineer in order to determine the seriousness of the damage, specify the repair methodology, and supervise the repair work.

Water can also enter masonry materials such as brick. Freezing temperatures can cause the water to expand and cracks to form. Water can also transport corrosion-accelerating chemicals, specifically chlorides found in deicing chemicals for roadways. Corrosion can wreak havoc on the decks and support beams of elevated parking garages. Performing a simple chain drag can detect hidden voids in corroded, reinforced concrete decks that have not yet caused a noticeable concrete crack or spall at the surface. A *chain drag*, as the name implies, involves dragging one or more chains across the concrete deck. When a void is found, the chain will vibrate differently, and this vibration is felt easily by the person dragging the chain. Spray paint can be used to outline the hidden damage for later jackhammering or hydrodemolition and repair. This maintenance can prevent a disruptive failure from occurring.

Cracks in Masonry Buildings

Cracks can be a sign of a very serious problem. Cracks in masonry buildings can be caused by the settling of the building, impact, contraction or expansion of masonry, or corrosion of steel set behind masonry. To prevent cracking due to temperature change, expansion joints are cut into brick or block walls or designed into the wall construction. An expansion joint is a small gap about 3/8 inch wide running the full height of the wall, depending on the design, that has a foam backer rod recessed into it. The gap is then caulked to ensure that it is watertight.

Cracks in masonry frequently begin at the corners of windows and doors. These 90-degree corners are stress points. Cracks that run between the mortar and the brick or between cement blocks are frequently due to shrinkage of the masonry when it dries out. Cracks that run through masonry may be a result of stress. Another cause of cracking of masonry is when water gets into the wall of a building and corrodes steel that is behind the brick. The rusting causes the steel to expand, pushing against the brick and cracking it.

It is important to remember that cracks need to be monitored to make an effective repair. There is little value in trying to make a repair if the crack is still growing. The source of the stress must be found and corrected first. When in doubt, contact a structural engineering consultant skilled in concrete and masonry evaluation and repairs.

12.6 ACTIVE BUILDING EXTERIOR MANAGEMENT

Building exterior management should always include management by walking around (MWBA). A hands-off approach can allow problems to accumulate until a major failure occurs. This does not mean that the FM must do everything. The FM needs to develop a building exterior maintenance management team. Facility maintenance technicians (FMTs) who know what to look for in order to identify problems can be the first source of information. They can also make minor corrections, such as patching small leaks with plastic roofing cement, applying roof coatings to exposed flashing, picking up debris from the roof and roof drain, noting missing vent caps and missing traps on condensate pans leading out from rooftop units, looking for evidence of water ponding, and noting cracks in masonry. The FMTs can report back if they detect anything unusual on the roof: improper antenna installations, unauthorized material storage, damaged mechanical equipment, natural gas odor, surface pea gravel or ballast stones that were swept away to find a leak but not put back in place, and water hoses run on the roof to help cool fan coils for refrigeration (constant water flow can sometimes damage roofing, and puts a layer of scale on the coils as well). The FMT can also be sent on a flat roof with a set of plans and pencil in any equipment that is present on the roof but not on the original roof plan. This frequently happens in older shopping centers where tenants have come and gone, and the spaces below have been reconfigured in size to meet their needs. If lacking a computerized maintenance management system (CMMS) for the property, the FMT should also record the nameplate data of all the equipment and number or tag equipment in a manner appropriate so it can be identified later. These types of activities help with the day-to-day preservation of the roof and mechanical equipment located on the roof.

The FM also needs professional advice when dealing with roofs and building exteriors. Consultants need to be called in periodically to evaluate conditions and provide a professional maintenance plan that spells out the course of action over several years, with the estimated cost of the proposed maintenance. It is also wise for an FM to hire a consultant to provide plans and specifications, and to function as the owner's representative during roof replacement.

12.7 BUILDING INTERIOR CONSTRUCTION

Interior construction provides the actual working and living spaces. It makes a statement about the organization's cultural values. A building might be beautiful from the outside, but its beauty quickly diminishes if the interior does not live up to the exterior. Great progress toward sustainability has been achieved with interior finishes in use today. Not only are a variety of materials recycled and reused, but existing materials are frequently salvaged, restored, and reinstalled in new construction. The creativity of interior designers comes into play when materials that were used for one application are used in an entirely different way in another. Clever reuse applications occur, for example, when items such as heavy ornate brass and glass doors salvaged from an old theater building, no longer suitable to be used as doors, are reused to construct decorative display cases.

To save on costs, most buildings have instituted a strategy of using **building standard materials** for construction, renovation, and repair. The chosen materials will be used throughout the building. Care should be taken so that the materials reflect the building class, are durable and functional, have a short lead time, and are cost-effective. Building materials such as doors and door frames; door hardware such as locks, door closers, and hinges; light fixtures; plumbing fixtures; bathroom partitions and hardware; heating valves; thermostats; carpet type and face weight; wall base; wall coverings; and paint are some of the typical building standard items. Building standards for materials save on both maintenance time and waste. Standardization means that FMTs can quickly learn how to install and troubleshoot building standard items. Parts inventories can also be kept lower than would be the case with multiple different items that each require their own replacement spares.

Metal-Stud Construction

Steel studs have been used for many years in commercial construction. Steel offers a lightweight, noncombustible alternative to wood. It is manufactured from recycled steel and can gain Leadership in Energy and Environmental Design (LEED) points. Unlike wood, steel studs don't warp and are dimensionally correct all the time. Steel studs have prepunched openings that can be lined up for easy installation of small water pipe and electric conduit. Light-gage metal studs come bundled together, so several studs can be cut to length with one stroke of a chop saw. Cutting the light 25-gage studs individually can be achieved by scoring the width with a utility knife and speed square, nipping the flanged ends, and then snapping the stud by hand. The thickness of studs ranges from 18 mills for a 25-gage stud to 188 mills for a 10-gage stud. Small self-tapping, No. 8 pan head screws about 5/8 inch long are frequently used to attach light-gage steel studs to their bottom and top plates. The bottom plate is secured to the concrete deck by a powder-actuated nail gun. To maintain stud spacing and for additional wall rigidity, prenotched spacer bars and U-channel bars with notches for 16 or 24 inches on center stud spacing can be inserted in the prepunched holes. To use metal studs for headers and other load-bearing applications, heavy-duty, roll-formed, shaped studs are used.

Drywall

The most popular form of office interior wall construction is drywall over steel studs. Dry wall frequently comes in 4 foot by 8 foot or 4 foot by 12 foot sheets. Drywall has a strength axis and therefore should have its long side run perpendicular to the steel studs for increased strength. Gypsum board has a tapered edge along its long side to create a recess and thus make the application of seem tape and compound in the joint less noticeable. Standing gypsum up and running the thinner edge along the stud means that a row of fasteners will be drilled into the weaker tapered edge. The fasteners used to attach drywall to the stud are self-tapping screws with a drill point for a tip.

Gypsum board frequently goes by other names depending on the manufacturer or geographic location. Common names for gypsum board are drywall, wallboard, and Sheetrock (a trademark of USG Corporation.). Drywall is basically gypsum sandwiched between a finished side of paper and a backing side of heavier paper. Drywall comes in several thicknesses: ¼ inch, 3/8 inch (9.5 mm), ½ inch (12.5 mm), 5/8 inch (16 mm), ¾ inch, and 1 inch. One-quarter-inch drywall is used to laminate over old drywall that has been damaged rather than removing it. This sometimes occurs when vinyl wall coverings are removed from drywall that was not primed prior to pasting on the wall covering. Prime painting seals the drywall and is an important installation step, whether the drywall will be finished by painting or covered in wallpaper. It is also bent to form decorative arches in layers for strength.

Drywall is designed to meet a variety of building situations:

- Exterior drywall has additional fire resistance and water-repellent coatings. The board is used as exterior sheathing underneath metal, engineered foam and stucco, wood, or other types of siding.

- Moisture-resistant drywall can be used in shower and bath areas. Another name for this drywall is green board because of its green color. Green board was found to be unsatisfactory for use in showers due to water penetration past the ceramic tile attached to the face of the board, which caused mold growth. Cement board is one of the approved choices for shower areas with direct water-spray contact.

- High-strength or sag-resistant drywall can be used for ceilings.

- Fire-rated type X drywall has fiberglass pieces added to its core for increased fire resistance. Five-eights-inch fire-rated type X drywall has a one-hour fire rating.

Interior Walls

Two types of interior walls are found in buildings that define tenant spaces: partition walls and demising walls. Partition walls are used to separate rooms within a single tenant's space. These walls are frequently not sheathed with drywall all the way up to the floor above. In fact, partition walls might simply end a few inches above the ceiling. This works especially well if the area above the ceiling acts as a plenum for return air. Return air can then flow unimpeded between the uncovered wall studs back to the grill on the return air duct.

Demising walls run from the floor to the bottom of the floor deck above. Wall sheathing runs from the floor to the deck above. Demising walls are fire-rated walls. The level of hazard presented by the tenant's operations governs the number of hours that the wall must be able to contain a fire. Demising walls are used to separate different tenants and to separate tenants from common areas. In many commercial office buildings, they are constructed with 5/8-inch fire-rated drywall over steel studs. The voids between the wall and the deck above must be sealed with a fire-stop material to prevent the transfer of fire and smoke.

Phase Change Materials Used in Construction

Cathedrals and castles built hundreds of years ago were constructed of thick stone walls, which provided a huge thermal mass. Cooler temperatures at night reduced the temperature of the stone, whereby the stone could absorb heat during the day and provide a nonmechanical form of natural cooling. The huge mass was required because heat transfer was through sensible heat. *Sensible heat* is the amount of energy required to raise the temperature of a substance one degree. On the other hand, the amount of energy required to change the phase of a substance from one phase to the next without a change in temperature is known as *latent heat*. Latent heat can be much greater than sensible heat. The heat required to change the phase of liquid water to gaseous steam at 212 degrees Fahrenheit is about 970 British thermal units per pound (BTU/pound) of water at normal atmospheric pressure. The heat required to change the temperature of water by 1 degree Fahrenheit is only 1 Btu/pound. The phase change of some materials is at a very specific temperature. Replacing thermal mass with phase change material inside a building provides essentially free air conditioning that requires little or no maintenance.

BASF makes a phase change material called Micronal® PCM (see Figure 12.19) that can be incorporated in the manufacturing process of several different types of construction materials (i.e., gypsum board, gypsum compound, CMU). The PCM melts at 23 to 26 degrees Celsius, or 73.4 to 78.8 degrees Fahrenheit. The PCM material is based on a paraffin wax that is microencapsulated in an acrylic polymer. These materials can be drilled and cut without the loss of paraffin. As interior building temperature starts to rise outside the comfort zone, the paraffin melts and absorbs heat. If enough PCM is present, it will keep the interior temperature comfortable.

DuPont™ makes a wall panel product called Energain®. Panels made of this product are installed behind gypsum board panels. These panels absorb room heat at about 22 degrees Celsius and store the heat until the temperature drops to about 18 degrees Celsius, releasing the heat back into the room.

The challenge with managing a building with PCM is to operate the building's cooling equipment and ventilation system to take advantage of the beneficial properties. The building systems should be set to allow the PCM wall and ceiling panels to absorb heat during the day and give it off at night, which requires setting cooling equipment higher during the day and providing cool ventilation air at night to allow the material to solidify. It is also important to train workers in any special repair of maintenance issues required.

ThermalCORE™
PCM Panel by National Gypsum

NEW WALL PANEL WITH LATENT HEAT STORAGE CAPACITY
with BASF Micronal® phase change microcapsules

Fiberglass Mat

Enhanced
Mold Resistant
Gypsum Core

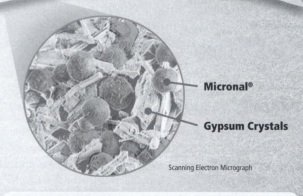

Micronal®

Gypsum Crystals

Scanning Electron Micrograph

ThermalCORE™ panels contain BASF Micronal® phase change material. Micronal* material is a microencapsulated high-purity paraffin wax that changes phase from solid to liquid when it reaches 73° F, absorbing thermal energy to help moderate a room's temperature. When temperatures fall, the wax solidifies and releases heat. This alternating process of melting and solidifying allows ThermalCORE to absorb daytime temperature peaks, ideally providing a more consistent temperature.

FEATURES/BENEFITS:

- Provides added thermal mass not typically found in traditional lightweight construction.

- Moderates indoor climate and provides a more consistent temperature.

- Phase change material is contained within virtually indestructible microscopic acrylic capsules which will not leak.

*Micronal® is a registered trademark of BASF.

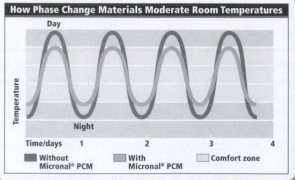

FOR MORE INFORMATION VISIT
www.thermalcore.info

National Gypsum®

Technical Info: 1-800-NATIONAL

FIGURE 12.19 PCM wall panel

Source: Reprinted with permission by National Gypsum.

12.8 FLOORING

Floor covering can be broadly divided into two categories: carpet and hard-surface flooring, Hard-surface flooring includes polished concrete, terrazzo, cork and bamboo, various tile surfaces, and wood laminate. Choosing the appropriate flooring type means evaluating the use of the floor, the likely problems it will encounter, and the life-cycle costs. For example, it would be inappropriate to install carpet where frequent liquid chemical spills are likely. Conversely, carpet would be a great choice in areas where moderate dust and soil can be contained by the carpet until it is vacuumed. Life-cycle costs are based on the estimated life of the carpet for the use and location, the cost to maintain the carpet, and the carpet replacement costs. For example, the best carpet for durability, one with a long useful life, might not be a wise choice for areas that have low traffic and overall light use.

When choosing flooring, a frequently missed point is the need to perform a thorough evaluation of the flooring's protection requirements and associated cost to provide this protection. The seasonal need for a protective cover of walk-off entrance mats can add to costs while simultaneously detracting from the beauty of the flooring. With vinyl composition tile (VCT) flooring, the need to periodically strip and finish the floor adds to costs, whereas removing the VCT and polishing the concrete can provide an attractive low-maintenance flooring solution.

Carpet

One of the most popular types of floor coverings in office buildings is carpet. The part of the carpet that we see is the pile. The pile is secured to backing material. Carpet comes in different pile configurations: cut pile, level loop pile, multi-level loop, shag, plush, Berber, Saxony, frieze, random sheared, and tip sheared. The pile configuration affects the wear characteristics of the carpet and how it will look when it becomes worn. The three types of synthetic carpet are nylon, olefin (also called polypropylene), and polyester (also known as polyethylene terephthalate [PET]). Another type of synthetic fiber carpet, acrylic, is used in bathroom rugs and throw rugs. Acrylic has a velvet feel and looks similar to wool, but it is much less costly. Wool is one of the most common types of natural fiber carpet. Carpet pile may also be produced as a blend of materials in order to take advantage of the positive qualities of the individual materials. Some carpet blends are wool/nylon, acrylic/olefin, and nylon/olefin.

Carpet comes in rolls that are 12 feet, 13 feet 6 inches, or 15 feet wide. Some carpet manufacturers have a 6-foot-wide roll to accommodate getting the roll into elevators, stairs, and other tight spaces inside a high-rise building. Carpet can also come as carpet tiles that are laid on the floor but are not glued down or fastened to the floor in any way. When the carpet starts to wear, maintenance workers can just replace the damaged or worn tiles in traffic areas rather than replacing the entire carpet. By not gluing the carpet tiles in place, off-gassing problems associated with some adhesives is eliminated. Carpet tiles have another advantage: They can be removed and cleaned in a remote location should spills or other problems occur. Temporary replacement carpet tiles can be installed in areas where the tiles were removed by unskilled labor. Carpet tiles have a heavy vinyl, resin, or polyurethane cushioned backing and come in squares that are either 18 inches by 18 inches or 36 inches by 36 inches. Raised floor installations frequently use nylon carpet tile.

Wool carpet is the most expensive carpet. It displays a natural richness that makes it most suitable for conference rooms and the private offices of the corporate hierarchy. Wool has an advantage over other carpets of being self-extinguishing when exposed to fire. Olefin carpet is frequently the least expensive carpet and is suitable for moderate traffic areas. It has an advantage over nylon carpet of being less likely to generate static electricity during periods of low humidity. It is resistant to mold and mildew. It is also used for indoor/outdoor carpet. Nylon carpet is the carpet most frequently used in commercial applications. Nylon carpet makes up about 65 percent of the carpet sold in the United States. It is strong, has good abrasion resistance, resists stains, and is easily cleaned, and its fiber memory retains its twisted yarn configuration. Nylon's drawback is its ability to build up static if the humidity in the space is low, but nylon carpet can be treated with antistatic chemicals.

Polyethylene terephthalate (PET) comes from clear plastic, postconsumer recycled soft drink bottles (see Figure 12.20). This polyester carpet is more resistant to mildew than nylon, and it is bright in color and less expensive than nylon. PET carpet's greatest advantage is that it helps provide a solution to the amount of landfill space taken up by beverage containers (see Figure 12.21). The disadvantages are that it may squeak or gleam, it sheds more than nylon, it attracts oily soil, and currently it is difficult to recycle.

FIGURE 12.20 PET Carpet installation
Source: Courtesy of Shaw Industries.

FIGURE 12.21 Beverage container recycling center
Source: Courtesy of Shaw Industries.

Product and Equipment Certification

Several organizations and government agencies certify carpet products and carpet installation equipment used both commercially and residentially:

- The Carpet and Rug Institute (CRI) is a nonprofit trade organization that has performed third-party testing of many brands of carpet, carpet cleaning products, and floor care equipment. As part of their Green Label Plus rating system, CRI has instituted a testing program to ensure that carpets, carpet pads, and carpet adhesives give off the lowest levels of volatile organic compounds (VOCs). Carpets, pads, and adhesives are tested by an independent laboratory following guidelines and methodology established in cooperation with the US EPA.

  Many parameters are evaluated in order to earn the CRI's Green Label and Green Label Plus designation. Some of the points evaluated for cleaning products are a cleaning chemical's ability to remove the dirt and stain and not damage the carpet or leave residue that attracts dirt. CRI's Seal of Approval/Green Label program for vacuum cleaners rates their efficiency at removing dirt and dust through X-ray fluorescence technology. CRI evaluates a vacuum's ability at keeping the tiny particles of dirt in the vacuum and not spreading them throughout the space, and how a vacuum affects the texture of the carpet over a one-year cleaning period. CRI has a similar program for carpet extractors.

  FMs can consult the CRI website to determine whether a carpet carries a Green Label Plus certificate. This website also has links to companies that provide carpets comprised of a percentage of recycled materials and companies that recycle carpet into other materials.

- The US EPA has a program known as Designation for the Environment (DfE) that promotes partnerships for developing products that are safer for people and the environment. Products labeled with DfE have been tested for various safety and green aspects. A database of companies and their products that have the DfE label can be found at the EPA website.

- Green Seal is an independent, nonprofit organization that provides third-party testing for certification of a multitude of products and services for companies that seek the Green Seal™ logo for their products. Products are evaluated according to their environmental impacts in the areas of product manufacturing, use and disposal, product toxicity, product reusability, energy efficiency, environmentally responsible packaging, recycled content, and biodegradability. Green Seal also checks that products meet their labeling requirements.

A manager needs to evaluate all aspects of equipment and products to ensure that they fulfill their intended mission and are cost-effective, green, and sustainable.

Hard-Surface Flooring

In the past, flooring was always thought of as a walking surface placed on top of a subfloor. Today, with polished concrete, the subfloor is the walking surface. Depending on the location and use, hard-surface flooring can be the flooring that will have the best performance, the least maintenance, and the lowest cost.

Polished Concrete   Polished concrete is a green alternative to installing new flooring. Most decks of commercial buildings are made of concrete. Making an existing concrete deck into a durable and attractive floor saves on floor-covering materials, the landfill waste they produce when they are replaced, and the energy it takes to manufacture the floor covering. Creating a polished concrete floor begins by removing the existing flooring and any mastic from the concrete deck. This is done by grinding the floor. The floor is then grouted with cement, and a clear liquid hardener is spayed over the floor. Then the floor is reground. Later the floor is polished with a series of finer and finer abrasive grits. The new shiny floor is then sealed to prevent stains from seeping into the concrete. Under moderate normal use, polished concrete floors require only mopping to keep them clean. Depending on floor traffic, polished concrete floors may require some periodic resealing.

Terrazzo Flooring   Terrazzo flooring is extremely durable (with a life span of more than forty years) and seamless, and it comes in different types. The two that are most popular today are polyacrylate and epoxy-based terrazzo. Polyacrylate is a cement base that contains acrylic additives. Floors made using this method can be as thin as ½ to ¾ inch thick. The other type is an epoxy-based terrazzo that can be applied at thicknesses ranging from ¼ to 3/8 inch thick. Chips of marble and sometimes postconsumer glass are added to the terrazzo, and then the surface is ground smooth and sealed. Terrazzo floors add three to five pounds per square foot to the weight of the floor. Terrazzo must be cleaned only with a neutral cleaner because acidic cleaners etch the marble and cement.

Cork and Bamboo Flooring   Cork and bamboo flooring are sustainable flooring materials because they can be replaced by growing new trees, as is the case with bamboo, or simply harvesting cork from the outside of the existing oak cork tree (the harvesting of cork does not damage the tree). When it is time to replace this flooring, it can be recycled. Bamboo forests regenerate themselves eight times faster than hardwood forests. Bamboo flooring is made by fusing fibers of bamboo together under extreme pressure. This process results in a flooring material that is more resistant to dents than red oak. Cork flooring is made from 100 percent postindustrial recycled material scraps from the production of cork for wine bottles. Cork trees absorb carbon dioxide to regenerate their cork bark. A cork tree that is harvested removes more carbon dioxide than one that isn't.

Stone Tile Flooring   Stone tile flooring, such as marble, has frequently been used in lobbies to convey richness to the visitors and in entranceways for its durability, as is the case with granite. Several buildings have a slab of granite set under revolving entrance doors to absorb the abuse generated by this extremely high traffic spot. Stone tiles are laid over concrete with a thin-set mortar. A gap, which is filled later with grout, is left between tiles for expansion. When stone tiles are installed, it is important that they are perfectly level with one another because, at some point, the tiles will need to be wet-honed through a diamond abrasive process to remove scratches and embedded dirt. If the tiles vary too much in height, the honing process will be time-consuming and expensive. After honing the tiles, a penetrating sealer is applied and the tiles are buffed to obtain a glasslike shine.

Stone flooring can be considered an option for sustainability if the materials can be obtained locally. No matter where stone tiles are obtained, several spare tiles should be kept in storage for replacement. Stone tiles, even when they are obtained from the same quarry, will vary in color and veining due to the specific location or depth at which the stone was mined.

Vinyl Composition Tile   Vinyl composition tile (VCT) is a product that has been around for many years. VCT is durable and comes in boxes containing 12-inch-square tiles. An adhesive is applied to the floor to secure VCT. A few companies have started recycling VCT, and one manufacturer has developed a product with 25 percent postconsumer recycled content from recycled VCT. VCT's life-cycle costs have come under increased scrutiny because of the necessity to perform the labor-intensive process of periodically stripping and refinishing VCT tile in order to keep its clean and shiny luster. Early tiles were vinyl asbestos tile (VAT), which should be maintained only by trained personnel. Facilities frequently eliminate this potential asbestos exposure problem by removing VAT.

Ceramic Tile   Ceramic tile is mostly used in restroom flooring because of its moisture resistance. In the past, ceramic tile was used as flooring in entranceways and hallways. Ceramic tile is long lasting, but the grout joints in ceramic tile are difficult to clean. Grout joints must be cleaned and sealed periodically. Ceramic tile is difficult to recycle, although some manufacturers use recycled content from other sources in tile production.

Wood Laminate Flooring   Wood laminate flooring was introduced in the 1980s. Laminate floors are floating floors that are not adhered mechanically or glued to the subfloor. Instead wood laminate floors rest on top of a soft pad, or they have padding attached directly to their backs. Laminate floors are also known as floating floors. Laminate floors have a wear layer that is frequently made of melamine, and below that is a craft paper print of the type of wood, stone, or tile the manufacturer is trying to emulate. Below that are the pressed wood layers bonded together with adhesives, which may contain formaldehyde (more ecofriendly laminates contain little or no formaldehyde and are made from recycled wood products). A melamine backing layer helps prevent moisture from entering the wood from below. The pad can be made of a variety of materials depending on the subfloor and the moisture and sound conditions. A pad made of foam is suitable for dry conditions; a foam pad with a vapor barrier is suitable to prevent moisture transmission on a floor that is sitting atop concrete. Cork pads are used for cork's excellent sound-absorbing qualities and because its tight cell structure resists the transfer of moisture.

Laminate floors are easy to install: The planks of the flooring literally snap together. Glues may be used for certain types of laminate flooring, but this is usually not necessary today. Some laminate floor planks have a padding fused to its bottom, eliminating the need to install padding. Laminate flooring requires minimal maintenance: just vacuuming and damp mopping. Green cleaning solutions, such as solutions of vinegar and water, are sufficient to clean laminate floors in most situations.

12.9 PASSENGER CONVEYANCE

Elevators, escalators, and movable sidewalks make up the majority of mechanical conveyance devices intended to transport people. These devices can move thousands of passengers during the course of a day in safety and comfort. They are also becoming smarter by providing their own logic systems to analyze the traffic situation and modify routing or speed. As equipment gets smarter, FMs need to understand how the equipment operates under both normal and emergency situations. Safety is the major concern of FMs managing people-moving equipment.

Elevators

Two basic types of passenger elevators serve the vertical transportation needs of the building:

- Hydraulic elevators
- Traction elevators

Hydraulic elevators are powered by a hydraulic pump that sits on top of a reservoir of hydraulic fluid. The pressurized hydraulic fluid pushes a hydraulic piston upward, causing the elevator to rise. The hydraulic cylinder is buried in the ground, which limits hydraulic elevators to buildings that are about seven stories in height because the cylinder must buried in the earth seven stories deep in order for the piston to travel seven stories up. These elevators most often use hydraulic oil, which would contaminate the soil if a leak should occur. Some buildings use vegetable oil in their hydraulic systems to prevent contamination of the soil and groundwater should leaks occur. Other building may have their hydraulic cylinder encased in a large polyvinyl chloride (PVC) pipe enclosure that has devices inside that monitor for oil leakage.

Hydraulic elevators traveling up require electric energy to power the hydraulic pump. Hydraulic elevators going down require little energy because gravity does most of the work. The speed of the fall is controlled by the hydraulic fluid returning to the storage tank.

Traction elevators have cables and a cable drum that winches the elevator up and down. Traction elevators come in two types: geared and gearless, depending on elevator speed requirements. Gearless traction elevators have the ability to travel much faster, but geared traction elevators utilize smaller electric motors to turn the drum. Traction elevators have no height restriction and thus are used in the tallest buildings in the world.

To make traction elevators more efficient, a counterweight system is employed. As the elevator moves up, a counterweight attached to the other end of the cables that is equal to 40 percent of the weight of a fully loaded elevator moves down. This 40 percent of full-load weight uses gravity to assist the elevator in moving up. An elevator that is filled to 40 percent capacity is most efficient moving either up or down. But an empty elevator cab going down uses more energy than an empty cab going up because the winch has to work against the counterweight in order to move the elevator cab downward.

In the past, motors used to move traction elevators were direct current. Direct current motors had the advantage of their speed being easily controlled by varying voltage. Direct current was provided by using motor generators, whereby an AC motor would drive a DC generator to provide DC power. The elevator control system consisted of dozens of open relays in a metal cabinet. Carbon dust from wearing DC motor brushes could get into the control system and cause problems.

Elevators today employ AC motors whose speed is controlled by variable frequency drives. One manufacturer employs permanent magnet synchronous motors in combination with frequency control for energy savings (see Figure 12.22) and speed control. The new system also eliminates the need for a separate room for the hoisting equipment (see Figure 12.23). All of the hoisting machinery and can now be mounted in the hoistway (also known as the elevator shaft). The elevator system is computer controlled, and the computer constantly evaluates traffic patterns and equipment information. Utilizing voltage and frequency adjustments, the elevator's acceleration and deceleration can be adjusted automatically. The system also employs a regenerative drive that captures the energy generated when the elevator goes down. The elevators also have a sleep mode that shuts off the lights and fan when the elevator is not being used. These systems can save as much as 75 percent over conventional elevator systems.

FIGURE 12.22 Krone EcoSystem MR™ machine room above traction elevator
Source: Krone Inc.

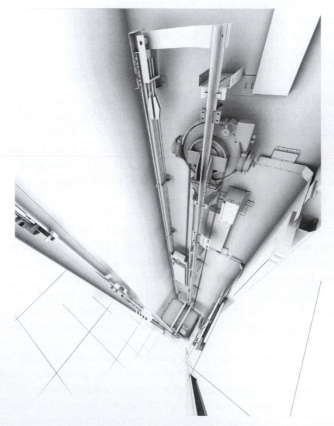

FIGURE 12.23 Krone EcoSpace™ machine room-less traction elevator
Source: Krone Inc.

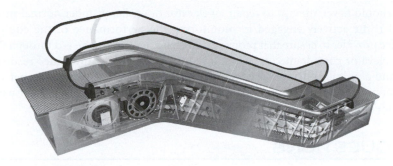

FIGURE 12.24 Krone Travel Master™ 110 escalator

Source: Krone Inc.

Escalators and Moving Walks

Escalator systems (see Figure 12.24) are in constant operation for safety and practicality reasons. A stopped escalator appears to be malfunctioning and therefore won't be used by passengers. Escalators are moving stairs. Moving walkways are basically conveyor belts. Escalators are generally limited to a speed of 100 feet per minute (fpm), and moving walkways can be 100 fpm or 130 fpm. Some escalators can go into a standby mode: When no passengers are detected at the landing through the use of radar or photocell detectors, the escalator slows to 40 percent of normal speed. Escalators moving downward can also generate power the same way that elevators do through regenerative drives that send power back to the building's power grid.

Managing Passenger Conveyance Equipment   Most elevators, escalators, and moving walkways are maintained through a service contract with an elevator/escalator maintenance provider. This service provider may be part of the elevator manufacturer's organization or a separate maintenance company entirely. Due to the specialization of the equipment and the legal liability of operating passenger conveyance operations, few facilities attempt to perform their own elevator or escalator maintenance.

Elevator and escalator maintenance (see Figure 12.25) is a hands-off affair for facility maintenance personnel. Facility personnel who lack the specialized training and certification provided by elevator

FIGURE 12.25 Krone Travel Master™ 110 escalator. Proper maintenance of the disembarking area is critical for passenger safety.

Source: Krone Inc.

manufacturers should never attempt to repair an elevator or escalator under normal situations. Facility managers need to be actively engaged in monitoring the performance of the equipment and the contracted service provider to ensure that proper maintenance levels are achieved through the maintenance agreement. An elevator consultant can be employed to review the maintenance records and contract requirements, and physically operate the equipment to ensure that it is working properly. Decisions can then be made about modifying contract requirements. Consultants can also help with decisions to upgrade elevator equipment and provide the budget information necessary to do so.

REVIEW QUESTIONS

1. Why are cork and bamboo considered to be sustainable flooring?
2. PET carpet is made from _____.
3. What type of carpet is most frequently used in commercial applications? Why?
4. How do intensive and extensive green roofs differ?
5. Name three types of gypsum board that are in common use today.
6. What is the difference between a black roof and a cool roof?
7. What are some of the devices that have improved the efficiency of traction elevators?
8. What methods can be used to protect buildings from excessive movement during high winds or earthquakes?
9. The two types of interior walls are _____ and _____.
10. What is the US EPA's DfE program?

Essay Questions

11. Green roof installation can turn an area of the building from something that is unusable into a space for visitors and tenants. What special precautions should be taken when tenants are given roof access? How can a green roof become a profit center for the building?
12. A facility manager must make a choice between a photovoltaic installation for generating electric power and a vacuum tube water-heating installation. What are the benefits versus the challenges of each system?

STRATEGIC PLANNING AND PROJECT FINANCIAL ANALYSIS

INSIDE THIS CHAPTER

Can Facility Managers (FMs) Plan for Sustainability?

This chapter starts a discussion about beginning strategic planning through team development. It then talks about some of the methods a manager may employ to achieve a strategic plan. But plans are only effective if they meet the financial goals of the organization. The chapter then discusses various tools of financial analysis and master planning. The chapter turns to space planning and churn management, and ends with a discussion of the Council House 2 project in Melbourne, Australia.

13.1   STRATEGIC PLANNING

Goals represent a facility's wants and desires. Goals become more of a reality when an organization has a strategic vision in keeping with its cultural values. To achieve this vision, strategic planning focuses on a central theme that is part of the strategic vision and that provides a framework and functionality for achievement. In other words, strategic planning is the systematic steps for goal attainment.

Strategic planning must include organizational sustainability. For example, a building may have budgeted funds years ago for a building system upgrade as part of an overall strategic plan for a phased modernization of the facility. In the past, *modernization* was the term frequently used for facility upgrades. Basically, it meant removing the old and replacing it with the new using the theory of "bigger and better."

Today, FMs must do much more for their facilities, and the goal is sustainability, not just modernization. A manager can modernize an unsustainable facility with some temporary positive results. A strategic plan, however, guides the manager toward a sustainable facility that is part of a sustainable vision for the organization.

Assemble the Planning Team

Strategic planning is not a one-manager process. The first step involves developing a team consisting of representatives from the departments with a vested interest in the project's outcome. A planning team leader is then chosen. Depending on the scope of the project, specialists from outside the organization and members of local government and community organizations may also need to be part of the planning team.

Strategic planning is most often initiated by upper management, and the FM generally has a lead role in spearheading all facility aspects of the plan. Even if upper management is not in the lead position, including upper management as part of the planning team is important. It makes top managers aware of the aspects of the plan that require funds for development and ultimately of plan execution. Assembling the necessary research (construction documents, energy audits, building commissioning, environmental site assessments, etc.) can be costly, so top managers awareness and consent are essential at the beginning stage of the project.

Strategic Planning Meetings

A series of planning meetings are scheduled with the goal of developing a written strategic plan. This document will be a guiding force for the organization for years to come. Team efforts and involvement have the added benefit of providing team members with motivation during the project implementation phase. Prior to every meeting, agenda suggestions are obtained from the team members, which helps to ensure their participation. From these suggestions and from input from the team leader, an agenda is drawn up and distributed to team members. A time limit needs to be set for the meetings to ensure the continued efficiency of the day-to-day facility operation. If meetings exceed the time limit, individual members associated with the items not covered are directed to meet and report back to the group at the next meeting.

Meeting minutes are recorded, formatted so that items can be referred to easily (generally number coded), and then distributed to the members. Members having comments or issues with certain items can bring these up at the next meeting or via e-mail, instant messaging, or software specifically designed for groups. If planning members are spread out over great distances, Internet programs such as Google Docs™ can be helpful for distributing and collaboratively modifying and storing document information. Teleconferencing can often achieve many of the same tasks as can face-to-face meetings. Members who do not need to be present through the entire meeting should be dismissed by the planning team leader. Ideas from the planning sessions obtained by brainstorming should be tested through a SWOT analysis, a technique that is discussed in the next section.

13.2   SWOT ANALYSIS AND THE ENVIRONMENTAL SCAN

An environmental scan is performed early in strategic planning, during the discovery or research phase. An *environmental scan* is focused research performed on the major elements of the project or the important ideas under consideration. Environmental scans provide the topics, investigation, and backup data for the SWOT analysis. Environmental scans are also a broad look at the global challenges that will be encountered during plan execution. Planners often look only inward and deal only with problems internal to the organization. Environmental scans force the team members to look outside the organization.

A great way to start is to use mind maps and brainstorming techniques to uncover the topics that should be scanned. Newspapers, research papers, historical data, professional society publications, trade journals, financial and real estate publications, and various web-based professional journals should be researched. This research provides information to back up ideas and the proposed courses of action. Research can provide new benefits through the discovery of new information that was previously unknown to the team and that could send the project in a new direction.

The accumulated information is put into a format for evaluation known as a SWOT (strengths, weakness, opportunities, and threats) analysis. The FM needs to address these four major evaluation topics for any project or situation. Strengths and weaknesses are generally thought of as areas that are internal to the organization; opportunities and threats are thought of as being external to the organization. As an evaluation tool, a SWOT analysis is more sophisticated than a simple list of the positive and negative aspects that could occur when attempting to achieve a project goal. SWOT makes the FM consider what could happen and where hidden opportunities lie. A SWOT analysis can also quickly eliminate project alternatives when it is discovered that weaknesses and threats are great, and opportunities and strengths are few. SWOT analysis is also not a one-person job; all the stakeholders in a project need to be part of the team performing the analysis. A SWOT analysis can also be performed for various stages of a project that is part of a strategic plan.

Example: Project One: Building Expansion Considerations

A property operation wishes to expand through adding an addition to an existing office property. The environmental scan has indicated a current need for space, but a possible glut of space in the years to come. The expansion requires the installation of new heating, ventilation, and air conditioning (HVAC) equipment and work will have to be done in the existing tenants' spaces.

Strengths: The building is currently populated by loyal tenants who have occupied space in the building for many years. This, in no small part, is due to the building's staff members, who are a hardworking group of true professionals and who cater to the tenants' needs. The building has also used technology to improve indoor environmental conditions as much as possible with the existing equipment. The building practices sustainable practices, has a reputation for quality, and has a premier location in the city.

Weaknesses: A high percentage of materials from overseas has been specified. Staff size will have to be increased. New equipment needed for the expansion will require some worker training. Current rental rates in the building are higher than many similar buildings nearby. New tenants may not be willing to pay the current rate that existing tenants are charged. Construction may cause a disruption to existing tenants. The success of this property has resulted in management devoting more time to the oversight of other properties in their portfolio. Upper management participation is vital.

Opportunities: The various businesses of new tenants, if chosen carefully, can provide synergies among all the tenants, thus strengthening the customer base for all the tenants in the building. New equipment will reduce energy costs and make the building more competitive overall. The reduced consumption will lead to a reduced carbon footprint for the building.

Threats: According to the environmental scan, market saturation will occur in 2025. The expansion will only make this condition worse. Tenant loyalty will be even more important in the years to come. Low lease rates can affect the ultimate success of the project. Energy costs are projected to rise, which will increase costs and rents, which in turn will have to be adjusted. It is possible that government regulation will become more stringent with regard to a building's carbon production.

Table 13.1 recaps the key points outlined in this example.

TABLE 13.1 SWOT analysis for building expansion

Strengths	Weaknesses
Superior staff capabilities.	Higher rental pricing than competitors.
Solid income stream.	Small staff size means more space for each staff member to maintain.
Customer/tenant loyalty.	Insufficient upper management participation.
Effective use of technology.	High percentage of materials from overseas has been specified.
Reputation for quality.	Disruption to tenants during construction.
Sustainable operation.	Worker training for new equipment.
Established location.	
Opportunities	**Threats**
Expand customer base and to new business areas both locally and globally.	Market saturation was estimated to hit the area sometime in 2025, after which the demand for space will be less than the supply of space.
Provide a superior built environment.	Building and equipment is at an advanced age.
Reduce carbon footprint.	Low lease rates.
Save on energy costs, thus providing competitive advantage.	Unfavorable government regulations.
Could become a trendsetting property.	Rising interest rates.
	Rising energy costs.

Example: Project One: Building Exterior Considerations

The designer of the expansion project is considering a smooth, white-plastered, cement-coated surface for the expanded office building, which will cover both the old and new sections. Such a finish is unique to the area of the country where construction will take place. Using a SWOT analysis for material maintainability, we can see more clearly if this is the best product for the building.

Strengths: The surface is durable and long-lasting, with moderate to high water resistance. It is reflective to light and heat, and can contribute to lower cooling costs. It can be made with recycled materials that are available locally with no lead time (in construction terminology, **_lead time_** is the time required to procure an item) It is visually attractive, and many other prominent government and museum buildings have a bright white appearance, which will put the building in this elite class.

Weaknesses: The exterior surface is difficult to maintain when damaged; rust stains and paint are difficult to remove. The surface could be damaged by high-pressure washing. Staff members are not trained in the maintenance of this surface, and few contractors in the area know how to maintain it. The surface may require more security than what is currently being provided, and additional lighting may be needed to deter vandals.

Opportunities: The surface could provide for highly visible signage that would be attractive to new tenants and therefore produce an income for the property. The extra income might offset the costs of maintenance. Clear antigraffiti coatings might be needed to minimize maintenance. The white color might be attractive to corporations associated with the color white (i.e., sugar manufacturers, linen manufacturers, lamp manufacturers, etc.), and therefore may be more likely to lease at the location.

Threats: Graffiti, urban grime from vehicle exhaust, rust, and dirt could present problems. Surrounding buildings may copy the white finish, thereby diminishing the visual impact of the property.

Table 13.2 recaps the key points outlined in this example.

TABLE 13.2 SWOT analysis for building exterior changes

Strengths	Weaknesses
High water resistance.	Difficult to maintain.
Unique	Prone to dirt.
Made from recycled materials.	Requires more lighting and security to prevent vandalism and/or graffiti.
Reflective surface lowers air-conditioning costs.	Damaged by high-pressure washing, requires antigraffiti coatings.
Opportunities	**Threats**
Signage will be highly visible on white background.	
Surface could be attractive enough to lure new tenants.	Lack of local contractors for application and future repair.
	Graffiti, vehicle exhaust, rust stains, and dirt.
	Other properties may copy the idea.

Summary of Project One

Although rising energy costs can be a threat to the economic success of many projects, reducing energy consumption in the wake of rising costs accelerates project payback. Rising energy costs can provide opportunities not found in periods of stable pricing. The project in this example might also benefit by expanding its scope to include renewable energy sources. In any event, an economic analysis will have to be performed to show that project costs will be paid back prior to the projected downturn in the leasing market in 2025.

Example: Project Two: Sale of Commercial Property with Outdated Chiller Equipment

A commercial property will be for sale shortly. An environmental scan has uncovered regulatory issues with the refrigerant currently used for building air conditioning. The property has an electric, centrifugal, water-cooled chiller and a steam absorption unit to serve air-conditioning needs. The steam absorption unit can operate at a low load, and it is frequently used on spring days that require only a small amount of cooling to satisfy the temperature requirements of the building when **_free cooling_** (using outside air to cool the building) isn't enough. Both air-conditioning units are the same size (600 tons each), but the chiller is more economical to operate.

When the building was new, operating either the chiller or the absorption unit provided enough cooling to satisfy the building's air-conditioning needs. Cooling equipment efficiency decreased with normal machinery wear, and the new open-office configurations increased the number of tenants in the building. The open-office configuration, with multiple cubicles instead of private offices, created more heat with more people and office equipment. More people also increased ventilation requirements. So cooling load increased at a time when the cooling equipment had a diminished capacity to meet that load. Well into the summer, with outside temperatures over 80 degrees Fahrenheit, one unit wasn't able to handle the entire cooling load. The absorption unit was then used to increase cooling capacity, which led to a disproportionately high cost. Steam is relatively expensive compared with electricity; this expense combined with the low efficiency of the absorber in turn caused two cost factors to increase. The steam absorption unit returned water to the cooling tower that was hotter than the water that the chiller returned. As tower water sump temperature went up, the electric chiller efficiency went down because warmer water caused higher condenser pressures.

Eddy current testing and maintenance records reveal that the absorber was in better condition than the electric chiller. Both units are twenty years old, but there is more runtime on the electric chiller.

The FM rectifies the situation by performing a SWOT analysis for replacement of the existing chiller with a larger chiller that can carry the building load without using the steam absorber. In the following analysis, the steam absorber would not be replaced.

Strengths:	A new larger chiller will be more energy-efficient than the old chiller. The building won't have to run the steam absorber at all. Two reasons for running the steam absorption chiller are that, during high cooling demand, it could be added to the system to increase output. The second reason was that it could be run during low-load periods when the centrifugal chiller would trip on low load. The new chiller can also be run at a very low load and has an increased capacity, so the absorber won't be needed under most operating conditions. Cooling tower tonnage is more than sufficient to handle the larger chiller, so no retrofit is needed. The original pumps and piping can remain. The steam absorption unit can be used as a backup during electric chiller maintenance or when problems occur. Replacement refrigerant will not be a problem for the new machine. The steam absorption unit uses lithium bromide solution with water as the refrigerant. Although lithium bromide is highly corrosive to steel, no major environmental concerns exist because environmentally friendly corrosion inhibitors can be used to protect the steel. The new chiller will be less costly to operate and will provide an energy cost savings payback in seven years. So if the property isn't sold, the building can realize a cost savings over time. The new equipment will also drop the energy costs per square foot for the building, which is likely to be attractive to new and existing tenants.
Weaknesses:	Cost is high for a replacement chiller: The project cost is $350,000. The new chiller has a four-month lead time for purchase, and the unit will take one month to install. It is also difficult to get the full value of HVAC upgrades from a property buyer. Property buyers expect proper heating and air conditioning: Both are frequently viewed more as requirements for the sale rather than as major benefits.
Opportunities:	The new chiller is more energy-efficient and utilizes earth-friendly refrigerant, a fact that can be marketed to new and existing tenants. A more comfortable building can be touted as a marketing advantage.
Threats:	The energy savings for project payback will not be realized by the current owner, and the full value of the replacement will probably not be realized in the sale. High equipment lead time could cause a service disruption when the building needs to be operating to impress prospective buyers. Poor tenant survey results could lead to a significantly lower price for the building if service problems result from the replacement process.

Table 13.3 recaps the key points outlined in this example.

TABLE 13.3 SWOT analysis of sale of property with outdated chiller equipment

Strengths	Weaknesses
More energy efficient.	High initial cost.
Can handle the entire load.	Long lead time.
Absorber is not required for normal operations.	Unlikely to improve building resale value directly.
Original equipment can be reused, and the steam absorber can be used as a backup.	The old chiller is covered by insurance, so if it were to fail catastrophically, it would be replaced at a small cost to the facility. The new chiller is an elective purchase that is not covered.
The refrigerant is earth-friendly, available, and less expensive.	
Lower electric costs per square foot.	

Opportunities	Threats
Would be able to market a more comfortable building using an earth-friendly refrigerant and a lower cost per square foot.	Service disruption due to high lead time and one-month installation time requirements.
	Tenant dissatisfaction and poor tenant survey results.

Summary of Project Two

The chiller provided a seven-year payback that was unacceptable to the property owner because the property was for sale. To sell a property, it used to be more important for an office building to be in a good location with a beautiful lobby than it was for it to have the latest energy-efficient equipment. Personnel performing building acquisitions today have become more technically aware; they now understand the value of major mechanical improvements and their effect on the building's financial performance. The proof of this is the great success of the Leadership in Energy and Environmental Design, existing buildings: operations and maintenance (LEED® EBOM) rating, which can be an attractive feature in regard to the sale of a property. Even though a LEED rating is a highly attractive feature for any property, market research by the owner of the property, indicated that the high cost of a major HVAC system upgrade would not be recoverable by raising the building sale price above most of the properties in the market. An alternative solution was found by creating a mind map of the project. A mind map (discussed later in this chapter) was used to find alternatives. By improving tower efficiency, the efficiency of the existing chiller was improved, tenant satisfaction went up, and the building was sold. Instead of spending $350,000 on a new chiller, $28,000 was spent to improve tower efficiency by installing new spray nozzles and high-efficiency fill media. These retrofits achieved the same goal that a chiller replacement would at a fraction of the cost.

13.3 BRAINSTORMING

Brainstorming is a method to obtain project considerations and ideas. In a brainstorming session, ideas are quickly put forward, one after the other, for a specified time. Then all ideas are collected and reviewed by the team members. At this stage, some of the ideas can include seemingly impractical suggestions. The ideas might not be well thought out at this point, so no criticism is allowed. Criticism of any idea might shut down the flow of ideas, not only from the member making the suggestion, but by those who have an idea and might not like be embarrassed by stating it. Ideas, not judgments, are what is needed at this point in the process. Even seemingly poor ideas frequently lead to workable solutions. The goals of brainstorming are to consider the ideas and discover their benefits, or eliminate them from further consideration if the ideas are found to be impractical.

A frequent difficulty that occurs during brainstorming sessions is no one responds to the challenge, especially if several managers are present. An agenda can prevent this situation by providing a focus. The agenda needs to be reviewed by the participants prior to the meeting so they can think about their contributions in advance.

13.4 MIND MAPS*

Mind maps are drawings that bring about and record creative thoughts and new ideas. Mind maps are excellent tools for brainstorming sessions. The value of mind maps is that they help people discover topics that might normally be missed through conventional lists.

Mind maps are based on the theory that people think in a radiating pattern. From a central idea, a vast number of connecting ideas can flourish through associations made by the brain. Mind maps generally tend to be a personal affair, with the resulting diagram being most valuable to the person who drew it because of an intimate understanding of the drawing on the part of the creator.

Developing mind maps can also be a group project. The resulting ideas can be tabulated and used by all team members. By combining the ideas from the team members, a collaborative mind map can be produced. This sharing turns other team members into creators as well. The collaborative mind map then becomes a visual record of all the team members' contributions.

Creating a Mind Map

Mind mapping begins with a central topic that captures the main aspect or goal of a project. This starting point is like the hub of a wheel or the crown of a tree. It is a point of departure for ideas. It must be broad in scope so a fixed direction is not created at this early stage. From the center, other topics and ideas can branch out. A hand-drawn picture or even a simple shape, such as a circle, can be at the center with the topic written on it. Next the main branches are created. These branches can represent the major challenges or ideas that are part of the central topic. These subtopics are written separately on top of each branching line. Once again, the mapper should have broad subtopics. From these subtopics grow multiple branches—the ideas that become more and more specific. For the main branch to branch out, the mapper needs to think about all the important topics and subtopics that make up the branch. The mapper needs to think broadly about what is important about a topic and allow her or his mind to take the topic further and further outward, with the mind map growing more and more in specific points and generating new ideas, which in turn result in additional branching.

The branch lines drawn on the map can be curved for interest. They can start thick and become thinner as they branch out. Each main branch and subsequent smaller branches should have their own color. The length of the branch should only be as long as the single word on the branch. Multiple words that can represent different ideas may confuse the mapping process because the creator has to choose from which word to start branching. Words can be written in bold or in capitals for emphasis. To avoid breaking viewers' concentration, branches must be drawn and the words on them must be written so that viewers do not have to turn their heads and do not have to rotate the map to read the topic of the branch.

Mind maps should reflect the individual character and inner working of the mind of the creator. For example, pictures may be used to spur creativity; color might maintain interest; more compact or larger, spread-out drawings might keep the thought process moving forward.

Labeling the Branches

Figures 13.1 and 13.2 show the differences between using specific branch labels compared to using general branch labels. Pollution is the main branch in both figures. In Figure 13.1, it branches off into the two common types of pollution, namely, air and water. But Figure 13.2, shows that, by using more encompassing terms, a great deal more branching results. Although *air* and *water pollution* are common terms when thinking about pollution, more encompassing branch terms such as *earth*, *human*, and *atmospheric* open the door to additional branching. If the mind mapper uses only *air* and *water*, he or she could leave out soil contamination, noise and light pollution, electromagnetic waves, consumption issues, and much of the pollution resulting from the built environment. For every mind mapper, the most important point about the branch terms is that they motivate the individual's creativity.

---

*Adapted from information provided in *The Mind Map Book: How to Use Radiant to Maximize your Brain's Untapped Potential*, by Tony Buzan and Barry Buzan, Copyright 1993. A respected expert and proponent of mind maps is Tony Buzan. Mr. Buzan has written books on the subject, one of those being *The Mind Map Book*. The term *Mind Maps* is a trademark of the The Buzan Organization, Ltd.

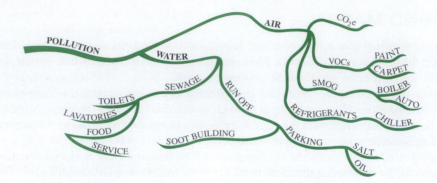

FIGURE 13.1 Pollution mind map with two major topics: air and water

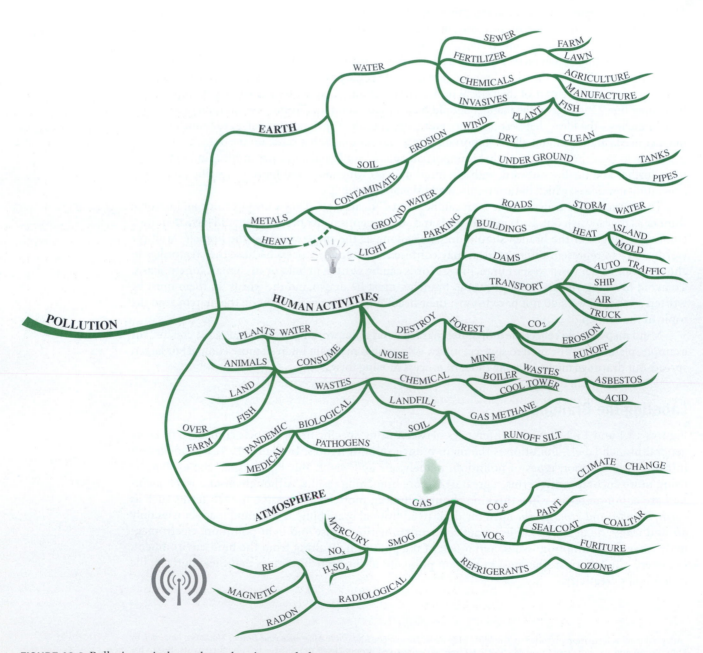

FIGURE 13.2 Pollution mind map branch using *earth*, *human*, and *atmospheric* toevoke more branching

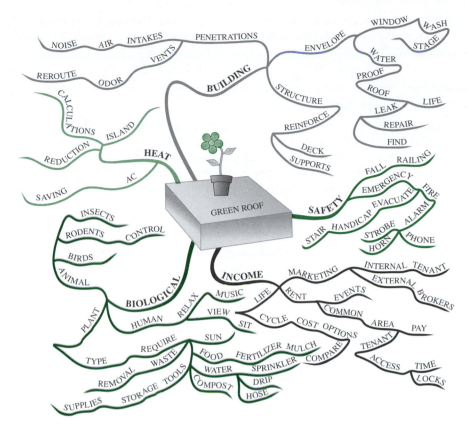

FIGURE 13.3 Mind map of the topics to consider for a green roof project that will allow tenants roof access

For the mind map of a green roof project in Figure 13.3, a viewer might point out that some considerations have been omitted. The green roof mind map shown in the figure is not complete because no mind map is ever complete. The goal is to generate ideas. Additional ideas can spring to mind at any time, thus creating more branching opportunities and, of course, more new ideas.

13.5   PROJECT FINANCIAL ANALYSIS

Projects of substantial size, cost or importance to a facility require a thorough analysis that is formulated into a report for decision makers. The report provides the justification necessary to get the project approved by upper management. A universal requirement for all reports is the price estimate to complete the project, which can be determined with a thorough economic cost-benefit analysis. This provides the needed justification for doing the work. Cost-benefit analyses include the following:

- Simple payback analysis
- Return on investment (ROI)
- Life-cycle cost analysis (LCCA)

The analysis should be included in the *executive summary* of the report, which is the section that most decision makers turn to first. The executive summary has all the information decision makers need to determine if further report reading is necessary to help in making a decision about undertaking the project.

Simple Payback Analysis

Simple payback is one of the easiest tools to convey project costs to upper management because it is so easy to understand. Simple payback answers the question, How long will it take for a project to produce benefits and/or savings equivalent to the project's cost? In fact, upper management may have a policy about the payback of new projects. For example, top managers may have established a minimum payback period for the initial screening of all projects. Their policy might be that no projects will be undertaken unless the project's payback period is less than 2.5 years. *Simple payback* considers a project's initial cost and the dollar payback for the savings created by the project over the life of the equipment.

Example: Simple Payback Analysis

A facility is considering the purchase of a new electric centrifugal chiller to replace an old chiller that has reached the end of its useful life but is still functioning. The new chiller will cost $200,000.Retro-commissioning and subsequent analysis estimates that the new chiller will save $15,000 per year in utility costs compared to the existing chiller. Operating and maintenance costs for both chillers are the same. The new unit has a twenty-year service life. To find the payback period, divide the cost of the unit by the savings per year:

Payback period = $200,000 ÷ $15,000 = 13.33 years to achieve project payback

The new unit has a twenty-year life, so it will continue provide savings for at least another 6.66 years. During this time, an additional $15,000 × 6.66 years = $99,900 in utility costs will be saved before it is time to consider chiller replacement. These additional savings will help to offset the cost of a replacement chiller.

Return on Investment

Return on investment (ROI) is used to define the success or failure for a project that is expected to turn a profit. ROI is also a quick scan on the merits of a project. It is frequently used in real estate transactions to determine whether to purchase a particular property. *Return on investment (ROI)* is a percentage of investment cost that is returned to the owner on a yearly basis. Therefore, ROI is:

Profit ÷ value of the asset

This equation can be stated differently:

ROI = net profit after taxes ÷ total asset value

Example: Return on Investment: Rental Income

A building for sale costs $20,000,000. It provides a net profit financial return to the current owner of $2,200,000 per year in the form of rental income. The ROI is:

$2.2M ÷ $20M = 11%

A real estate investment trust may use ROI to scan a potential property investment quickly. If the return on investment meets the investment trust's minimum required profit level, the investment may undergo further consideration for a potential investment.

Life-Cycle Cost Analysis

Life-cycle cost analysis (LCCA) provides an evaluation tool for determining which capital investment is the most cost-effective over time. Life-cycle cost analysis looks at the time value of money and its present worth. *Present worth (PW)* is the value of the money today that will either be received or paid at a known time in the future. The *time value of money* is a concept that is based on the fact that a dollar received today is worth more than a dollar received a year from now. If an investor receives a dollar today, the investor can immediately purchase something worth a dollar, or she or he can invest the dollar and begin receiving interest on the dollar. If the interest rate is 5 percent and inflation is 3 percent, an investor opting to receive the dollar in a year's time would have to receive an additional 8 percent, or $1.08, to equal the dollar received today. To find out today's value of a dollar that will be paid in one year, the single present worth table is used (see Table 13.4). Find year 1 in the table and move from left to right across the row to the 8 percent column. The factor to be multiplied by the $1.00 amount is 0.926, so $1.00 × 0.926 = $0.926. The dollar promised to be paid in a year's time has a value today of 92.6¢.

TABLE 13.4 Single Present Worth Factors

	Discount Rate										
Year	1%	2%	3%	4%	5%	6%	7%	8%	9%	10%	11%
1	0.990	0.980	0.971	0.962	0.952	0.943	0.935	0.926	0.917	0.909	0.901
2	0.980	0.961	0.943	0.925	0.907	0.890	0.873	0.857	0.842	0.826	0.812
3	0.971	0.942	0.915	0.889	0.864	0.840	0.816	0.794	0.772	0.751	0.731
4	0.961	0.924	0.888	0.855	0.823	0.792	0.763	0.735	0.708	0.683	0.659
5	0.951	0.906	0.863	0.822	0.784	0.747	0.713	0.681	0.650	0.621	0.593
6	0.942	0.888	0.837	0.790	0.746	0.705	0.666	0.630	0.596	0.564	0.535
7	0.933	0.871	0.813	0.760	0.711	0.665	0.623	0.583	0.547	0.513	0.482
8	0.923	0.853	0.789	0.731	0.677	0.627	0.582	0.540	0.502	0.467	0.434
9	0.914	0.837	0.766	0.703	0.645	0.592	0.544	0.500	0.460	0.424	0.391
10	0.905	0.820	0.744	0.676	0.614	0.558	0.508	0.463	0.422	0.386	0.352
11	0.896	0.804	0.722	0.650	0.585	0.527	0.475	0.429	0.388	0.350	0.317
12	0.887	0.788	0.701	0.625	0.557	0.497	0.444	0.397	0.356	0.319	0.286
13	0.879	0.773	0.681	0.601	0.530	0.469	0.415	0.368	0.326	0.290	0.258
14	0.870	0.758	0.661	0.577	0.505	0.442	0.388	0.340	0.299	0.263	0.232
15	0.861	0.743	0.642	0.555	0.481	0.417	0.362	0.315	0.275	0.239	0.209
16	0.853	0.728	0.623	0.534	0.458	0.394	0.339	0.292	0.252	0.218	0.188
17	0.844	0.714	0.605	0.513	0.436	0.371	0.317	0.270	0.231	0.198	0.170
18	0.836	0.700	0.587	0.494	0.416	0.350	0.296	0.250	0.212	0.180	0.153
19	0.828	0.686	0.570	0.475	0.396	0.331	0.277	0.232	0.194	0.164	0.138
20	0.820	0.673	0.554	0.456	0.377	0.312	0.258	0.215	0.178	0.149	0.124
21	0.811	0.660	0.538	0.439	0.359	0.294	0.242	0.199	0.164	0.135	0.112
22	0.803	0.647	0.522	0.422	0.342	0.278	0.226	0.184	0.150	0.123	0.101
23	0.795	0.634	0.507	0.406	0.326	0.262	0.211	0.170	0.138	0.112	0.091
24	0.788	0.622	0.492	0.390	0.310	0.247	0.197	0.158	0.126	0.102	0.082
25	0.780	0.610	0.478	0.375	0.295	0.233	0.184	0.146	0.116	0.092	0.074

Instead of getting a dollar today, the investor was promised a dollar ten years from now. The promise of a dollar ten years from now still has value, but certainly not as much value as a dollar promised only one year from now. Table 13.4 is used to discount costs, such as replacement costs or salvage values, that occur in a particular year. According to this table, the promised dollar would be worth only $0.46 today. The present worth can be obtained from a present worth table by multiplying the dollar by the factor 0.46 located in intersection of the ten-year row and the 8 percent column, or by using the single present worth (SPW) formula:

$$SPW = F/(1 + I)^y$$

$$SPW = \$1.00/(1 + 0.08)^{10} = \$0.46$$

where
PW = present worth
 F = the future sum of money
 I = the interest rate or discount rate
 y = the number of years

The SPW formula is easier to use then the table when the length of time is not a whole-year value. For example, if we are promised a dollar ten years and six months from now, we can find its present value by simply adjusting the years (y) to reflect that time. In this case, the factor y in the formula is 10.5 (the "point five" being the result of converting months to a decimal fraction of one year: six months divided by twelve months in one year equals 0.5).

Example: Decision Making with the Single Present Worth Formula

The single present worth formula can also be used in a basic analysis of two building purchase options:

- First option: Purchase a building for $200 million that can serve the needs now and for the next thirty years.

- Second option: Purchase a smaller building for $100 million that can serve the needs for fifteen years, then purchase another building for $125 million that can serve the needs for the remaining fifteen years.

Which is the better option if one assumes a 5 percent interest rate?

1. The PW (present worth, or purchase price) of the first option is $200 million.

2. The PW (combination of immediate purchase price and the purchase price fifteen years from now) of the second option is:

$$PW = (\text{amount paid today}) + (\text{amount paid in 15 years} \times \text{SPW factor [15 years at 5\%]})$$
$$= \$100M + \$125M \times \text{SPW at 5\% for 15 years}$$
$$= \$100M + \$125M \times 0.481$$
$$= \$100M + \$60.1M$$
$$PW = \$160.1M$$

Project Summary

Although the second option will eventually cost $25 million more than the first option, the value of the $125 million held for fifteen years provides a return that makes the second option the better choice. Stated another way, $125 million with a 5 percent discount rate and paid out fifteen years from now is worth $60.1 million today.

Life-cycle cost analysis looks at the total cost of owning a product over its useful life. It compares products by taking future costs and converting them to their present value. In this way, different products or projects with different costs, useful lives, and benefits can be evaluated equally. For an accurate evaluation, a discount factor is the one essential fact that needs to be known first. The *discount factor* is a percentage that describes the value of money to the organization.

Instead of spending money on a new project, an organization may opt to put that money in the bank to earn interest. In this case, the percentage that the money can earn in the bank is the discount factor. Different organizations may have different abilities to generate a return on their money, so the discount factor can vary from one organization to the next. Governmental organizations may establish a discount factor to be used in LCCAs performed by any branch office throughout the agency.

The formula for LCCA takes into consideration the following factors:

$$LCCA = C + M_{spw} + E_{upw} + R_{spw} - S_{spw}$$

where
C = initial cost
M = maintenance costs
E = energy costs
R = replacement cost
S = salvage value

Cost is the only factor in the LCCA formula that is not discounted because it is paid immediately. The other factors can be discounted by using formulas or tables, as we will see.

For fixed recurring costs, the formula for *uniform present worth (UPW)* is used to calculate present worth:

$$UPW = A(1 - [1 + I^{-y}) \div I$$

where
A = annual amount
I = discount rate
y = number of years

It has been found that energy costs increase faster than the rate of inflation. Therefore, an escalation factor is used to compensate for the additional costs. The escalation factor can be shown in equation form by using a modified UPW formula to include energy escalation:

Modified UPW that includes energy escalation

$$E_{UPW} = A[(1 + E)/(i - E)] \times 1 - [(1 + E)/(1 + i)]^n]$$

where
  P = present value
  E = escalation factor (percentage increase in energy)
  A = uniform series of consecutive payment amounts
  i = interest
  n = number of periods

Easier Method for Finding PW with Increasing Energy Costs

An easier method for finding present worth with increasing energy costs is to use the uniform present worth factor (see Table 13.5) and make an adjustment to the discount rate to compensate for the increasing energy costs. Uniform present worth factor (UPWF) is used to discount annually recurring costs such as maintenance or energy. For example, if the discount rate is 5 percent and the energy escalation factor is 2 percent, the FM can take the 5 percent discount rate and subtract the 2 percent energy escalation factor (5 percent − 2 percent = 3 percent). Three percent is then used as the discount factor in the uniform present worth table, as shown in Table 13.5.

TABLE 13.5 Uniform present worth factors

Year	\multicolumn{11}{c}{Discount Rate}
	1%	2%	3%	4%	5%	6%	7%	8%	9%	10%	11%
1	0.99	0.98	0.97	0.96	0.95	0.94	0.93	0.93	0.92	0.91	0.90
2	1.97	1.94	1.91	1.89	1.86	1.83	1.81	1.78	1.76	1.74	1.71
3	2.94	2.88	2.83	2.78	2.72	2.67	2.62	2.58	2.53	2.49	2.44
4	3.90	3.81	3.72	3.63	3.55	3.47	3.39	3.31	3.24	3.17	3.10
5	4.85	4.71	4.58	4.45	4.33	4.21	4.10	3.99	3.89	3.79	3.70
6	5.80	5.60	5.42	5.24	5.08	4.92	4.77	4.62	4.49	4.36	4.23
7	6.73	6.47	6.23	6.00	5.79	5.58	5.39	5.21	5.03	4.87	4.71
8	7.65	7.33	7.02	6.73	6.46	6.21	5.97	5.75	5.53	5.33	5.15
9	8.57	8.16	7.79	7.44	7.11	6.80	6.52	6.25	6.00	5.76	5.54
10	9.47	8.98	8.53	8.11	7.72	7.36	7.02	6.71	6.42	6.14	5.89
11	10.37	9.79	9.25	8.76	8.31	7.89	7.50	7.14	6.81	6.50	6.21
12	11.26	10.58	9.95	9.39	8.86	8.38	7.94	7.54	7.16	6.81	6.49
13	12.13	11.35	10.63	9.99	9.39	8.85	8.36	7.90	7.49	7.10	6.75
14	13.00	12.11	11.30	10.56	9.90	9.29	8.75	8.24	7.79	7.37	6.98
15	13.87	12.85	11.94	11.12	10.38	9.71	9.11	8.56	8.06	7.61	7.19
16	14.72	13.58	12.56	11.65	10.84	10.11	9.45	8.85	8.31	7.82	7.38
17	15.56	14.29	13.17	12.17	11.27	10.48	9.76	9.12	8.54	8.02	7.55
18	16.40	14.99	13.75	12.66	11.69	10.83	10.06	9.37	8.76	8.20	7.70
19	17.23	15.68	14.32	13.13	12.09	11.16	10.34	9.60	8.95	8.36	7.84
20	18.05	16.35	14.88	13.59	12.46	11.47	10.59	9.82	9.13	8.51	7.96
21	18.86	17.01	15.42	14.03	12.82	11.76	10.84	10.02	9.29	8.65	8.08
22	19.66	17.66	15.94	14.45	13.16	12.04	11.06	10.20	9.44	8.77	8.18
23	20.46	18.29	16.44	14.86	13.49	12.30	11.27	10.37	9.58	8.88	8.27
24	21.24	18.91	16.94	15.25	13.80	12.55	11.47	10.53	9.71	8.98	8.35
25	22.02	19.52	17.41	15.62	14.09	12.78	11.65	10.67	9.82	9.08	8.42

Example: Project Evaluation Using Both the Simple Payback Method and LCCA

An FM needs to replace a water-cooled centrifugal chiller because of it age. The old unit consumes $55,000 per year worth of electricity during the cooling season. The FM is reviewing three chiller options (A, B, and C). All three chillers have a capacity of 500 tons and have a useful life of twenty-three years.

> Chiller A is the least expensive, at $150,000, and it consumes $43,000 worth of electricity per year.
>
> Chiller B costs $175,000 and uses $38,000 worth of electricity per year.
>
> Chiller C is the most expensive and costs $220,000, but it uses $35,000 per year in electricity.

The building will be sold in twenty years. Maintenance and operating costs are the same for all three chillers; thus, they are not entered into the calculations. Which option is the most cost effective?

Part 1: Simple Payback Method

The cost in electricity to run the old chiller was $55,000 per year. Based on this dollar amount, the potential savings using each chiller is found by subtracting the new chiller cost from the old chiller cost, as follows:

> Chiller A saves $55,000 – $43,000 = $12,000 saved per year
>
> Chiller B saves $55,000 – $38,000 = $17,000 saved per year
>
> Chiller C saves $55,000 – $35,000 = $20,000 saved per year

To find the payback period, use the following equation:

$$\text{Initial price} \div \text{savings per year} = \text{payback period}$$

> Chiller A: $150,000 ÷ $12,000 = 12.5 years
>
> Chiller B: $175,000 ÷ $17,000 = 10.3 years
>
> Chiller C: $220,000 ÷ $20,000 = 11 years

According to simple payback, Chiller B has the shortest payback period and therefore should be chosen.

Part 2: LCCA

The FM decides to perform an LCCA of the project. Table 13.6 illustrates this analysis. The discount factor for the company is 4 percent, with 2 percent energy escalation. The period is twenty years. The chillers last for twenty-three years, but the organization will own them for only twenty years because the building will be sold.

TABLE 13.6 LCCA project analysis

	Chiller A	Chiller B	Chiller C
Initial cost	$150,000	$175,000	$220,000
Electricity used per year	$43,000	$38,000	$35,000
Uniform present worth of electricity used for 20 years, UPWF 2% = 16.35	$43,000 × **16.35*** = $703,000	$38,000 × **16.35*** = $621,300	$35,000 × **16.35*** = $572,250
Present worth = initial cost of equipment + E_{upw}	$150,000 + $703,050 = $853,050	$175,000 + $621,300 = $796,300	$220,000 + $572,250 = $792,000
Lowest life-cycle cost			Chiller C

*The values in bold type are taken from Table 13.5 by finding the intersection of the 2 percent column and the 20-year row.

According to LCCA, Chiller C is the best option because it has the lowest life-cycle cost. To show the value of LCCA, consider the simple payback method once again, but now evaluate the additional savings gained after the payback period, as shown in Table 13.7.

TABLE 13.7 Additional savings after payback period

	Chiller A	Chiller B	Chiller C
Payback period	12.5 years	10.3 years	11 years
Time remaining after payback	7.5 years	9.7 years	9 years
Yearly savings	$12,000	$17,000	$20,000
Additional savings after payback period	$90,000	$164,900	$180,000

It is clear that if project B were picked by the simple payback method, $15,100 worth of savings would be missed.

To make a thorough evaluation, LCCA is only the first step. Facility managers are not dealing simply with financial instruments that can be readily compared. FMs are dealing with the realities of the physical world. FMs making a capital equipment decision should look at all the other factors that are important to a facility too. The FM should collect the following information:

- Does the vendor of the equipment have a trained technical staff that can make repairs to equipment on short notice? What are the charges for these services?
- Does the vendor have design capability?
- Are parts readily available?
- What is the reputation of the local distributor?

Then the FM needs to:

- Benchmark the experiences of other facilities that have made a similar purchase.
- Put all the factors that are important to the facility into a decision matrix.
- Discover the positive factors that may weigh in favor of each vendor.
- Evaluate the negative factors to see if the savings are enough to compensate for these issues.

A final decision might not solely reflect the financial aspects of the project. The other factors should also be considered because they may outweigh direct financial benefits.

Master Plans

Master plans are frequently used by public higher education facilities to plan for the future. Sometimes these types of plane are set up as five-year plans that list several goals to be achieved within a five-year period. In developing a master plan, one of the most important elements is designing the plan to meet the mission of the institution. The master plan provides guidance for capital improvements, major maintenance tasks, building and site development, landscaping, pedestrian circulation, parking facilities, utilities, recreation, athletic facilities, energy, storm water management, and sustainability. Master plans may also include cost estimates for all budgetary items projected over a five-year period, including income projections. This information is very helpful for the state or local tax jurisdiction that might provide the funds for major cost elements of the plan.

Master plans have a huge impact on the financial status of an organization. Careful financial review prior to moving forward is essential. With a multiyear plan, subsequent reviews during project execution are necessary to ensure that the plan's financial aspects have not changed.

13.6 CHURN MANAGEMENT/SPACE PLANNING

Churn management is the handling of the space needs when personnel are moved into or out of existing spaces. The goal is to provide flexible workspaces that meet the current need at the lowest cost. Cost can frequently be reduced by having an efficient layout of office equipment, furniture, and cubicles that reduces the occupied square footage.

Space planning is an area of interior design in which the knowledge and abilities of professionals are extremely valuable. Compliance issues in space design must be evaluated thoroughly. These issues arise from the requirements of the Occupational Safety and Health Administration (OSHA), the Americans with Disabilities Act (ADA), building codes, and health and fire department requirements. A mistake in any of these areas can be costly. Interior design professionals have up-to-date knowledge of the latest office technology, lighting, day lighting, acoustical issues, and they have the ability to design a space that accommodates the needs of the organization.

Do not assume that the FM has no input in these compliance issues: far from it. The FM can use his or her knowledge and can access information to provide project leadership in many areas, including:

- Business operations
- HVAC, HVAC layout, and HVAC system capabilities
- Electric system capabilities
- Telephone and data systems
- Storage capacity and availability
- Elevator capabilities
- Exhaust ventilation for kitchens and office equipment
- Plan review
- Logistics of moving personnel

The FM generally has a good idea of how the organization operates as a whole. The FM can also bring all the department heads of the organization together to discuss how each department works within the organization and discover the flow of work throughout the organization. It is important to know how work moves through the organization, from department to department and down to persons within a department. This information is essential when arranging work layouts for personnel.

It is also important to know how the normal work of a person can affect the work of others around her or him. For example, even though sales and engineering work together on projects, it might not make sense for the departments to be right next to one another. Sales personnel answer the telephone, receive visits from clients, and make calls, while engineering may need a more quiet area to make calculations and think through issues. A solution may be to include some partition separation or to install sound-masking equipment.

For existing spaces, the FM can provide information on HVAC and lighting layouts, and whether personnel density can be accommodated by the existing ventilation system. Decisions about changing the layout or moving the equipment based on the building's ability to deliver a comfortable environment may have to be made.

13.7 COUNCIL HOUSE 2, MELBOURNE, AUSTRALIA

A government office building in Melbourne, Australia, known as Council House 2 is a project rated 6 Stars by the Green Building Council of Australia (GBCA) and has garnered many national and international awards for its sustainability and green initiatives. Constructed in 2006, Council House 2 is a $51 million building that is ten stories tall, with 134,940 square feet of floor space, and a staff of 540. The first floor provides retail space.

Approximately 23 percent of the building's cost went toward green initiatives. The building is a daring masterpiece of innovation that overcomes seemingly insurmountable difficulties by its creative use of existing technologies and environmental site conditions. The building is designed to work with the ecology and urban landscape around it. An entire book could easily be devoted to this one project. A brief summary of some of the major innovations are provided in this chapter.

Melbourne, the capital city of Victoria, has a population of approximately 4 million. Its weather is described as moderate oceanic. Its location in the southern hemisphere means that summers are from December to February and winters are from June to August. Snow is almost unknown. Temperatures in the summer can rise to 35 degrees Celsius, with the hottest temperature up to 41 degrees Celsius (105 degrees Fahrenheit). The temperature range is mainly from 25 to 33 degrees Celsius (77 to 91 degrees Fahrenheit). High temperatures usually occur during the day and generally drop as the wind shifts to the southwest. Temperatures can drop abruptly to 10 degrees Celsius (50 degrees Fahrenheit). The coldest days occur in July and August, when the temperature rises to only about 9 degrees Celsius. Temperatures may fall to below 4 degrees Celsius, but the average winter daily temperature is about 14 degrees Celsius (about 57 degrees Fahrenheit). These environmental conditions make heating the building a minor issue, but cooling the building is a major one.

FIGURE 13.4 Council House 2

Source: Photo courtesy of the City of Melbourne.

The success of the Council House 2 project lies in the integration of various technologies and consideration of the outdoor environment. The building, building mechanical systems, and surrounding environment are designed to work together. Rather than fight nature, as is the case with glass- and steel-encased skyscrapers, Council House 2 embraced it. Many of its systems would not work without this feature. Council House 2 was constructed using recycled concrete. Recycled timbers were used for its timber windows, which cover much of the building. As compared with similar buildings, Council House 2 has reduced its electrical consumption by 85 percent, its natural gas usage by 87 percent, and its water usage by 72 percent. It produces only 13 percent of the emissions (compared to similar buildings) into the surrounding air. Sometimes during the year, no supplemental heating or cooling is required by the building.

The major obstacle that needed to be overcome was lack of water. Australia has suffered a drought for over a decade, and water is at a premium. In spite of this, water was chosen as the source for much of the building's cooling, in addition to watering plants and flushing toilets.

The site presented another difficulty. Council House 2 is in an urban location and thus is surrounded by other buildings, which limited the orientation of the building on the site (see Figure 13.4). Proper orientation can take advantage of prevailing winds for cooling and the sun for heating, or it can limit unwanted solar gain by reducing the number of windows facing the sun.

Building Systems

To understand the synergies between building and environment, it is necessary to review the systems that make up Council House 2. So many systems are integrated by performing multiple functions that complement one another, so it is difficult to compartmentalize any one system into just one function. The major parts of the various systems will be discussed so that you can see how parts of the various systems work together to contribute to this sustainable building.

Ventilation Stacks   Large ventilation stacks are located on the north and south sides of the building. The south stacks are for supplying cool air through a vent system. The north stack vents air through a stack effect that is aided by wind-driven air turbines (see Figure 13.5). The north stacks receive a lot of solar radiation and are painted black, so they heat up and aid in the removal of building air. None of the air inside the building is recirculated; thus, the building operates on 100 percent outside air. The council members at Council House 2 imposed a minimum of 22.5 liters per second per person of fresh outside air, which is more than double the Australian national standard of 10 liters per second per person. This additional ventilation provides a healthy indoor environment. The wind turbines operate to exhaust air

FIGURE 13.5 Wind turbine–driven ventilation stacks on Council House 2, as seen from the roof
Source: Photo courtesy of the City of Melbourne.

only during the nighttime purge, which maintains a positive pressure in the building during occupied times. During the day, wind turbines generate electricity.

Ceiling Cooling   Ceilings in the building's interior are constructed of concrete in curved wavy shapes that maximize surface area and provide a greater thermal mass. This concrete mass absorbs interior heat during the day. This absorbed interior heat is removed through a night purge whereby cool night air is allowed to flow through the building from automatically opened windows, which are controlled by the building's building automation system (BAS). By controlling the opening and closing of windows on every floor, the ceiling temperature is allowed to cool to 20 degrees Celsius in the summer and 24 degrees Celsius in the winter. This prevents overcooling the space in the winter months. The windows on one side of the building can also be shut during high wind conditions. Wind-powered air turbines located on the roof maintain exhaust ventilation for purging the space. The ceiling design also provides a channel to direct hot air into ventilation stacks.

Radiant Cooling   Mounted on the ceiling is a passive, chilled beam, sensible heat cooling system. Heated air from the space below rises into the finned tube radiators of the chilled beam. Chilled water running through the tubes cools the air, which naturally circulates down to the area below.

Phase Change Cooling   Approximately 30,000 small stainless steel balls (the size of baseballs) are located in three large tanks in the basement. The balls are filled with a phase change material (PCM) comprised of a salt solution that freezes at 15 degrees Celsius (59 degrees Fahrenheit). During the winter months, the solution in the balls is frozen by operating the cooling towers at night when the outside air is cool. During the summer months, the temperatures are too high to freeze the PCM balls, so rooftop chill-

FIGURE 13.6 Close-up of a shower tower on the exterior of Council House 2
Source: Photo courtesy of the City of Melbourne.

ers are used to chill the water to a point where the PCM balls can freeze. The tanks of PCM balls act as storage tanks for cooling. The chilled water from the frozen balls is sent to the chilled beams. The water from the beams returns to the tank 2 to 3 degrees Celsius warmer. As long as the PCM balls are frozen, the balls can rechill the water. When the PCM balls remove much of the heat inside the building, chillers are used to supplement the building cooling by directly cooling the water sent to the chilled beams.

Evaporative Cooling   Council House 2 uses traditional cooling towers that operate with recycled water. The cooling towers provide cool water for the absorption chiller to satisfy part of the building's air-conditioning needs. Shower towers (see Figure 13.6), which are highly visible features on the building's exterior, operate with city water (called mains water in Australia). The shower towers are made of lightweight fabric and are 1.4 meters in diameter and 13 meters tall. The towers (see Figure 13.7) are located on the southern façade of the building. Outside air drawn into the top of the shower towers is cooled by the evaporation of water. The resulting cool air is used to cool retail spaces in the building. The remaining cool water is collected and used to precool the water returning from the chilled beams.

Absorptive Cooling   As discussed Chapter 9, absorptive chillers are not cost-effective when running on heat supplied by a power utility. Council House 2 uses waste heat from a microturbine that generates electricity. Waste heat that would normally be released into the atmosphere is instead used as a heat source for the absorptive chiller. The absorptive chiller is used to provide cool water for the building's chilled beam system.

Heating System   The building relies on a hydronic heating system located under the floor along the perimeter windows. Air supplied to the office space is heated to 20 degrees Celsius, so the radiant floor system keeps cold radiating from the windows from affecting the interior space temperature.

FIGURE 13.7 Shower towers on Council House 2

Source: Photo courtesy of the City of Melbourne.

Power Generation   Power generation for Council House 2 comes from many sources. A 60-kilowatt gas-fired microturbine generator produces enough electricity to meet up to 30 percent of the building's electrical needs and provides 105 kilowatts of heat. This heat is used to heat domestic water and to power an absorptive chiller. Twenty three photovoltaic cells on the roof generate 3.5 kilowatts of power. Hot water solar panels produce up to 60 percent of the building's domestic hot water needs. Even the elevators in the building generate power when braking.

Windows and Lighting   The windows on the north and south sides of the building are timber-framed and double-glazed. Wood is a low conductor of heat, so the timber frames add to the insulation of the exterior wall. Balconies also provide shading for the windows. Upward rolling interior fabric shades above the door leading out to the balcony also prevent unwanted glaring light from entering the interior space. The balconies have plantings in plastic containers that include trellises. Light is diffused by the leaves as it enters the building, which cuts down on glare. The exterior supply and return ducting also provide a reveal for the window. The windows themselves have an upper and lower shade. By closing the lower shade, solar gain can be limited while still getting the benefits of natural light. Chilled beams above the windows cool the air that is heated by solar gain off the windows, and hydronic underfloor heat next to the window provides heat when cool air is falling from the window during cold weather. Windows are opened automatically during the night purge. ***Light shelves*** constructed of fabric in a steel frame are on the north side of the building and reflect light into the interior space.

The west side of the building is covered by a series of timber louvers (see Figure 13.8) that pivot to optimize the amount of sunlight. They also shade the building from the harsh western sun. The

FIGURE 13.8 Automatic timber louvers on the west side of Council House 2

Source: Photo courtesy of the City of Melbourne.

louvers are opened and closed automatically by a hydraulic system in response to the amount of sunlight.

Water Conservation   Council House 2 overcame the water problem by sewer mining: 12,000 gallons per day of sewer water is extracted from the sewer main running in the street along with the gray and black water generated by the building. This sewage water is processed and treated, and then used for building cooling by sending it to the cooling tower. Treated water is also used for watering plants and the flushing of toilets and urinals. Council House 2 also collects and stores rainwater that is used in conjunction with the treated sewage water.

Council House 2 also collects potable water in a tank that is obtained when the fire pump is tested. Water that is used in sprinkler testing is contaminated and is therefore sent to the sewage-processing system. Other strategies include low-flow fixtures and dual-flush toilets.

Impact of Council House 2 Innovations

Council House 2 has many innovations that, considered individually, would be a great success on their own. Council House 2 takes these innovations further by implementing them in an aesthetic and artistic manner, synergistically melding the building, building equipment, the building's occupants, and the surrounding urban environment into a successful outcome. This achievement provides benefits not only for the occupants of Council House 2, but it also enhances living and working in the surrounding community. If projects like Council House 2 are undertaken around the globe, they could have far-reaching benefits for the world. Council House 2 doesn't fight the forces of nature or lock them out; it exists within the urban ecology for the betterment of the planet.

13.8 CONCLUSION

Sustainable facility management is the quest that successful facility managers everywhere have undertaken. These forward-thinking managers make everyday decisions with an eye toward sustainability that directly affects the success of their entire organization. Sustainability in facility management encompasses all the management topics that the job requires: management methods, maintenance systems, HVAC, construction and contract, environmental, human resources, financial management, housekeeping, landscaping, energy, risk management, disaster management and prevention, churn management and security, along with the green topics like waste reduction and recycling, alternative energy, carbon reduction, green cleaning and landscaping, nonchemical treatment methods, indoor air quality, renewable energy, and renewable materials. Study of these topics will provide you, the facility manager or aspiring facility manager, with the tools you need to undertake a lifelong journey toward sustainability.

REVIEW QUESTIONS

1. How does simple payback differ from life-cycle costing analysis?
2. What are the advantages of life-cycle costing analysis compared to simple payback?
3. How can mind maps be used in the brainstorming process?
4. What items in a SWOT analysis look mostly at the facility organization?
5. What items in a SWOT analysis look mostly outside the facility organization?
6. Name some reasons for drawing a mind map. What are some of the basic rules of drawing mind maps?
7. Draw a mind map to address the sustainability issues encountered in new building construction.
8. Draw a mind map for an elevator modernization project.
9. When are uniform present worth factors used?
10. Explain ROI.
11. What is churn management?

Essay Questions

12. What are the challenges that face an FM at a facility similar to Council House 2?
13. Could a Council House 2–type of building be constructed in a location with four seasons and temperature extremes? What systems might you want to augment with additional equipment?

ABBREVIATIONS

ABS	acrylonitrile butadiene styrene (black plastic pipe used for drains and vents)
ACM	asbestos-containing material
ADA	Americans with Disabilities Act
afd	adjustable frequency drive
AFE	Association for Facilities Engineering
AFUE	annual fuel utilization efficiency (measure of the amount of fuel energy that is converted to heat)
AHJ	authority having jurisdiction
AHRI	Air-Conditioning Heating and Refrigeration Institute
AHU	air-handling unit
AIA	American Institute of Architects
AOA	activity on arrow
AON	activity on node
ASHRAE	American Society of Heating Refrigerating and Air-Conditioning Engineers
BAS	building automation system
BATNA	best alternative to a negotiated agreement
bEQ	Building Energy Quotient (ASHRAE's green building rating system)
BHP	boiler horsepower
BIM	building information modeling
BOMA	Building Owners and Managers Association
BREEAM	Building Research Establishment Ltd. Environmental Assessment Method (BRE's green building rating system)
BTU	British thermal unit, the amount of heat necessary to raise the temperature of 1 pound of water 1 degree Fahrenheit. MMBTU is 1 million BTUs.
BUR	built-up roof, a roof comprised of alternate layers of asphalt or cold tar and felts
CCHP	combined cooling heat and power
CERCLA	Comprehensive Environmental Response, Compensation, and Liability Act
CFL	compact fluorescent lamp
cfm	cubic feet per minute
CFM	Certified Facility Manager
CHFM	Certified Healthcare Facility Manger
CHP	combined heat and power
CH2	Council House 2
CM	construction manager
CMMS	computerized maintenance management system
COBie	Construction Operations Building Information Exchange
COP	coefficient of performance
CO_2	carbon dioxide
CO_2e	carbon dioxide equivalent gas
CPM	critical path method
CRI	color rendering index (measure of how accurately colors appear under the light source); also Carpet and Rug Institute
DCV	demand-controlled ventilation
DDC	direct digital control
DfE	Designation for the Environment
DMAIC	defining, measuring, analyzing, improving and controlling
ECM	electronically commutated motor
EER	electrical efficiency rating (equal to BTUs out/watts of input energy)
EF	early finish (term on a network diagram using the critical path method)
EFSR	early suppression fast response
EFVM	electric field vector mapping (method of roof leak detection)
EIFS	exterior insulation finish system
EPDM	ethylene propylene diene monomer, better known as a single-ply rubber roofing material
ES	early start (term on a network diagram using the critical path method)
FM	facility manager

FMT	facility maintenance technician
fpm	feet per minute
FTE	full time equivalent
GBC	Green Building Council
GBI	Green Building Initiative
GC	general contractor
GHS	Globally Harmonized System
GPM	guaranteed maximum price
HCFC-22	hydrochlorofluorocarbon (a type of refrigerant phased out by the Montreal Protocol)
HFC-134a	hydrofluorocarbon or 1,1,1,2 tetrafluoroethane (a refrigerant)
HPS	high-pressure sodium (type of lamp used for outdoor lighting)
Hz	hertz or cycles per second (the frequency of AC power)
IAQ	indoor air quality
IE	infrared emittance
IEER	integrated energy efficiency rating
IFMA	International Facility Managers Association
IPM	integrated pest management
kWh	kilowatt hours
LCCA	life-cycle cost analysis
LED	light-emitting diode
LEED	Leadership in Energy and Environmental Design
LEED AP	Leadership in Energy and Environmental Design Accredited Professional
LF	late finish (term on a network diagram using the critical path method)
LS	late start (term on a network diagram using the critical path method)
MAU	makeup air unit
MSDS	material data safety sheet
MWBA	management by walking around
NESHAP	National Emissions Standard for Hazardous Air Pollutants
NFPA	National Fire Protection Association
OPIM	other potentially infectious material
OSHA	Occupational Safety and Health Administration
PACF	phosphoric acid fuel cell
PCM	phase change material
PCS	property condition survey
PET	polyethylene terephthalate
PEX	cross-linked polyethylene tubing
ph	potential hydrogen
PHAs	polycyclic aromatic hydrocarbons
PM	preventive maintenance
PMP	pressure maintenance pump or jockey pump
PMWO	preventive maintenance work order
PPE	personal protective equipment
PSI	pounds per square inch
PVC	polyvinylchloride (a type of single-ply roofing)
RA	return air
REIT	real estate investment trust
RH	relative humidity
RO	reverse osmosis
ROI	return on investment
RTD	resistance temperature detector
SDS	safety data sheet
SEER	seasonal energy efficiency rating
SfP™	sustainability facility professional
SLA	service-level agreement
SPM	sustainability project manager
SR	solar reflectance
SWOT	SWOT (strengths, weakness, opportunities and threats) analysis (a project analysis method)
T&M	time and materials
TCC	triclosan
TJC	The Joint Commission
TPO	thermoplastic olefin (a type of single-ply roofing)
TQM	total quality management
TUPE	Transfer of Undertakings

TWA	time-weighted average
UFAD	underfloor air distribution
USD	United States dollars
UVC	C-band of ultraviolet light
VAT	vinyl asbestos tile
VAV	variable air volume
VCT	vinyl composition tile
VFD	variable frequency drive
VOC	volatile organic compound
XRF	X-ray fluorescence

GLOSSARY

"24/72-hour work order" method—This method of measuring worker productivity differentiates between work orders that should be completed in a day and those that are more difficult and may take up to three days to complete. The rationale behind this method is that it is unfair to penalize the worker or the department for handling difficult and thus more time-consuming tasks.

"days to enter and close out all work orders" method—One of the simplest methods of measuring worker productivity, it sets the goal that all work orders are completed in a specified number of days.

"percentage of completed work orders" method—A calculation that provides a means of evaluating worker productivity by comparing the number of work orders completed to the number of work orders issued, and multiplying the difference by 100 to give a percentage.

"time to complete one work order" method—A method of measuring worker productivity that examines all work orders, considers them equal in all respects, and displays a number based on the average time (in hours) to complete one work order.

A

absorption chiller—An air-conditioning chiller that uses a vacuum to produce water vapor at a temperature of about 45 degrees Fahrenheit to be used to cool chilled water. Absorption chillers do not require a mechanically driven refrigerant compressor.

accrued—In accounting, it is a term used to describe funds reserved for later use.

acoustical contractor—A subcontracting firm that specializes in the installation of suspended ceilings using acoustical tile that is laid on a metal grid.

actuals—The amount(s) of money actually spent as compared to the amount budgeted.

additionally insured under general liability—Requirement that provides the owner with additional protection from claims against the work done or any negligence on the part of the contractor.

admixtures—Chemicals used to provide a variety of desirable properties in concrete.

aeration—A process where ½-inch diameter, 3-inch long plugs of soil and turf are removed from a lawn, allowing water, fertilizer, and oxygen to reach the turf's roots.

aggregates—Small stone pebbles, granulated sand, and stone chips added to cement to make concrete. Aggregates give concrete much of its strength and provide for up to 85 percent of the mass of the concrete.

agreement for service—A contract document in the form of a template that has been developed by the property owner's legal representation. This one contract can be used multiple times for most outsourcing situations involving the same contractor.

Americans with Disabilities Act—Legislation enacted by Congress in 1990 to protect the rights of persons with disabilities as defined by the act and to provide them access to goods and services.

approved equal—A construction term that is part of the languages of bid documents and allows the contractor to offer the owner a substitute product for one specified in the bid documents.

as-built drawing—A final construction plan set that shows all the deviations from the original construction drawings shown on a set of plans that were necessitated by the actual field conditions, and all the additional work resulting from change orders.

B

backward pass—The process of calculating the late finish times for activities on the network diagram.

benchmarking—Comparing the results of an activity at the facility manager's property with the results of leaders in the field.

beneficial occupancy—The owner's takeover action (moving into a space) at the completion of a construction project.

bid bond—A type of bond provided by the construction contractor to the owner to ensure that the contractor will do the work for the bid price.

bid form—A preprinted form given to contractors who are bidding on a service. The form ensures that all contractors provide the same bid pricing information.

biogenic carbon—Carbon given off by the combustion of biomass or wood. Because plants absorb carbon over their lifetime, the carbon emitted in combustion is not counted in the amount of CO_2 a facility produces.

blood-borne pathogens—Infectious viral agents found in the blood.

blowdown—The dumping of a portion of the water in a boiler or cooling tower water to reduce the concentration of chemicals and scale-forming materials by the addition of fresh water.

brainstorming—A method to obtain project considerations and ideas. Brainstorming is performed by a group of team members who exchange ideas on a project or issue in an informal matter. All ideas are considered and the idea is to generate new ideas quickly and uncover issues that may have been previously unknown.

breakdown—In contract terminology, a complete list of the prices of all the components in a bid amount.

budget—An item-by-item, projected cost breakdown of all the expenses a facility is projected to incur during the coming year.

budget variance report—A report that compares the actual money spent to the amount of money budgeted for an item in the budget.

building master key—A key that opens all the doors in just one building.

building standard materials—Chosen parts or materials that will be used throughout the building. Building standard materials cut down on the amount of spare parts required to be kept in inventory through standardization.

C

calcium carbonate ($CaCO_3$)—Along with magnesium carbonate, calcium carbonate is one of the two main scale-producing compounds in heat transfer equipment that utilizes water or steam.

capital budget—Funds budgeted for major equipment purchase or maintenance, large projects, and construction.

carbon calculator—An online resource from the United States Environmental Agency (U.S. EPA) that aids businesses in discovering their carbon footprints.

carbon footprint—The amount of greenhouse gases produced directly and indirectly by a facility's operation, construction, and maintenance. It is a measure of the amount of CO_2 and CO_2e produced or used by the facility.

centrifugal chiller—A device, not unlike a centrifugal pump, that compresses refrigerant used to reduce the temperature of water that is used for air conditioning.

chain drag—The process of dragging one or more chains across a concrete deck to find voids. When a void is found, the chain will vibrate differently, and this vibration is felt easily by the person dragging the chain.

change key—A term applied to a door that is on the master key system, but only one key beyond the master keys opens it.

change order—In construction projects, a modification or addition. Change orders frequently result in additional costs to the owner.

chief sustainability officer—The person in charge of sustainable initiatives for a corporation.

chilled beams—Fin tube water systems that provide latent cooling for interior spaces.

churn management—Handling the space needs of personnel when they are being moved into and out of existing spaces.

closed bid—A bidding process in which only a select few contractors are allowed to bid on a contract.

coanda effect—It is a physical property of flowing air or liquid that causes it to adhere to a smooth curved surface in the flow path.

color rendering index (CRI)—Measure that describes how accurately colors appear under a particular light source.

color temperature—A measure of the color of light given off by a lamp based on the light given off by a metal as it is heated. An incandescent lamp has a color temperature of 2,700 degrees Kelvin or 4,400 degrees Fahrenheit.

combination budget—A budget that employs a combination of vendor data budgets, formula-based budgets, zero-based budgets, and percentage budgets to achieve budget numbers that are as accurate as possible.

combined cooling heat and power (CCHP)—Utilization of waste heat from the equipment that generates electricity for cooling and heating purposes. Absorption cooling is frequently used because it requires heat to make chilled water for building cooling.

combined heat and power (CHP)—The utilization of waste heat from the equipment that generates electricity for heating purposes.

common area—Space accessible to the tenant that is outside the tenant's rented space. Common areas are places such as hallways, exercise rooms for tenant use, and parking areas.

common area maintenance—A charge shared by tenants for the maintenance of areas outside their tenant space and available for the tenants' and customers' use.

composting—The process of taking organic matter (leaves, lawn clippings, fruits, vegetables, etc.) and allowing it to decompose by microbial activity, thus producing a nutrient-rich material that can be worked into the soil.

compost tea—Compost mixed with water and sprayed around the plants, making sort of a tea. Compost teas provide nutrients that can be quickly taken up by the plants.

Comprehensive Environmental Response, Compensation, and Liability Act (CERCLA)—A law enacted in 1980 and better known as the Superfund, was designed to correct the decades of unregulated and improper disposal of hazardous wastes.

computerized maintenance management system (CMMS)—A software system that takes inputs from facility personnel, digital devices, control equipment, and other computer systems to provide for proper maintenance of facility equipment, management of parts inventory, work-order tracking, financial management, and worker productivity measurement.

condensing boilers—Boilers that drop the stack temperatures to a point where water vapor condenses out of the stack gas. This captures latent heat that would otherwise be lost. Condensing boilers can achieve combustion efficiencies of 90 percent to as high as 97 percent under ideal conditions.

cooling tower—A device that uses the process of evaporative cooling to cool the warm water that was used to liquefy the refrigerant gas in the condenser.

cool roof—A roof that is either made of a light-colored material or is coated with a light-colored paint or aluminized coating that reflects solar radiation.

critical path method (CPM)—A planning method that uses a network diagram to schedule project activities. The network diagram shows the tasks and the time it takes to complete them. It also displays the relationship between construction activities. The longest path through the diagram is the critical path and it controls the overall project duration.

cyberattack—The illegal disruption of an organization by the purposeful compromise of computer security to obtain information, damage computer equipment, or damage the software.

D

dashboard—A page on a computer screen that displays pertinent energy data in a format that allows the facility manager to extract the information easily.

demand—The average highest amount of energy or peak demand a facility registers on its meter over a fifteen-minute period.

demand charge—A fee charged to customers by a power company for the customers' peak demand (highest demand recorded over a fifteen-minute interval).

demising walls—Walls separating tenant spaces from each other and separating tenant spaces from common areas such as corridors. Demising walls are constructed from the floor to the underside deck of the floor above.

designation for the environment (DfE)—A United States Environmental Protection Agency (U.S. EPA) program that promotes partnerships for safer chemistry and developing products that are safer for people and the environment. Products labeled with DfE have been tested for various safety and green features. A list of these products can be found in the U.S. EPA's database.

desuperheater—A heat exchanger located at the outlet of the compressor; it cools the refrigerant gas to the saturation temperature without reducing refrigerant pressure.

dew point—Temperature at which water condenses out of the air onto the surface of a material.

direct expansion (DX) system—DX systems send refrigerant liquid directly to a refrigerant expansion valve located at the entrance to the evaporator coil.

discount factor—The rate of return.

dry back boiler—A fire tube boiler whose combustion chamber contains refractory brick or other types of refractory insulation to contain the high heat of the flue gas.

due diligence—The process of evaluating all the factors of a property that affects its economic viability prior to property purchase.

E

early finish (EF)—On a network diagram using the critical path method (CPM), it is the earliest day an activity can be completed.

early start (ES)—On a network diagram using the critical path method (CPM), it is the earliest day an activity can start.

emergency action plan—A plan that defines the steps to be taken during an emergency, such as fire, earthquake, the spill of a hazardous liquid, other hazards identified by risk analysis.

ENERGY STAR label—Rating achieve by commercial buildings from the United States Environmental Protection Agency (U.S. EPA) by initiating energy and water resource conservation measures.

engineering controls—Physical barriers or devices that prevent employee contact with blood-borne pathogens.

enthalpy—The heat content of moist air. It is measured in BTU per pound of dry air and is the combination of heat energy due to air temperature and moisture content of the air.

environmental management plan—A plan customized to meet the property's current use by tenants or in-house personnel if the property is owned and managed by the same entity. The plan must also reflect past use by previous owners and the use of surrounding properties, and it must be based on the property's current environmental condition.

environmental scan—Focused research performed on the major elements of the project or the important ideas under consideration. Environmental scans provide the topics, investigation, and backup data for the SWOT analysis.

equipment maintenance history—A summary of all the work done, equipment runtimes between maintenance events, the person(s) who performed the maintenance, and the parts used.

evapotranspiration—The process by which water evaporates from soil and transpires from plants. This form of evaporative cooling can lower a green roof deck temperature 50 degrees Fahrenheit.

executive summary—The section of a report that most decision makers turn to first. The executive summary has all the information decision makers need to determine if further report reading is necessary to help in making a decision about undertaking the project.

exposure control plan—A list of the steps that employers are taking to control workers' exposure to blood-borne pathogens, to keep the workers safe, and to keep the facility in compliance with regulations.

F

face and bypass dampers—A set of dampers placed in front of the cooling coil, and another set of dampers placed on top of the cooling coil, respectively. Opening and closing these dampers regulates the airflow through and around the cooling coil, which in turn regulates cooling capacity without having to change the cooling medium's temperature or volume.

first cousin keys—Keys that vary only slightly from the keys used to unlock other doors.

float—Extra time to complete a project task before the next task needs to begin. Also known as *slack time*.

floor master key—A key that opens all the doors on one particular floor in a multifloored building.

formula-based budget—Budgeting for items based on a formula such as dollars per square foot.

forward pass—The process of calculating the early start and early finish times for activities on the network diagram used in the critical path method (CPM).

free cooling—Process that causes a major reduction in the amount of energy needed to cool a building. Air economizers, strainer cycles, refrigerant migration, and phase change materials (PCMs) in building materials are all methods of free cooling.

fuel cell—An electrochemical, battery-like device that converts a hydrogen-containing fuel to electricity.

G

Gantt chart—A chart, similar to a bar chart, shows the start and end dates of project activities.

gateway—A device that allows digital equipment that utilizes one protocol to communicate with equipment using a different protocol.

geothermal loop—A closed water loop that may consist of thousands of feet of high-density polyethylene (HDPE) piping installed in the earth at depths of 10 to 300 feet.

global warming—the trapping of heat in the earth's atmosphere by carbon dioxide gas, a variety of other gases, and water vapor. Also known as the greenhouse effect.

globally harmonized system (GHS)—An international approach to hazard communication that replaces the material data safety sheet (MSDS) system.

green carpet—A carpet that produces low levels of volatile organic compounds (VOCs), is recyclable or made from recycled or renewable product, and is attached with a low-VOC adhesive.

green roof—A roof covered with vegetation that reduces the temperature of the roof.

green walls—Living vegetative systems growing on the outside of a building. Green walls can be used to cool exterior walls and shade window glass.

greenhouse effect—The trapping of heat in the earth's atmosphere by carbon dioxide gas, a variety of other gases, and water vapor. The greenhouse effect causes not only temperature changes but global weather instability whereby storms are more devastating and droughts last for longer periods. See also global warming.

gross square feet—Areas that are needed for access and mechanical spaces; examples include corridors, stairways, elevators, and lobbies.

gross square footage—The total interior space within a building. The area includes mechanical spaces, hallways, common areas, and tenant areas.

gypsum board—A wall material composed of gypsum sandwiched between a finish paper face and cardboard backing.

H

Hazard Communication Standard—An Occupational Safety and Health Administration (OSHA) standard found in the code of federal regulations (29CFR1910.1200). It gives employees the right to know the identity of the chemical hazards they are exposed to when working.

hazardous waste plan—A plan that defines the required frequency of removal and treatment of tenant-generated hazardous waste to the appropriate hazardous waste landfill or recycling facility.

heat recovery chiller—A chiller that has an additional heat exchanger so that the heat normally released to the atmosphere can be captured and used by the building.

human resources management—The art of managing personnel and dealing with situations involving people.

I

innocent landowner—Designation under the Comprehensive Environmental Response, Compensation, and Liability Act (CERCLA) (42 U.S.C. 9602-9675) that limits a new owner's liability should contamination be found on his or her property now or in the future.

inspector test valve—A valve on a wet-type fire sprinkler system that is used to simulate the flow of one sprinkler head. The inspector test valve is located inside the building, but frequently the fitting that simulates a sprinkler head is mounted outside the building. When the valve is opened, water sprays outside.

inspector test valve—A valve used to simulate flow from one sprinkler head.

integrated pest management (IPM)—Reducing insect pests through the use of knowledge and biological pest management techniques. The management process involves a constant surveillance of the pests on a property to identify them and monitor their numbers to decide if the level of pest activity warrants that action should be taken.

intumescent material—Material that expands when exposed to the heat of a fire.

J

janitorial mode—Performing mostly low-skill-level work such as cleaning, emptying trash, setting up rooms, changing lightbulbs, and so on.

jockey pump—A small pump installed to maintain pressure in a fire sprinkler system generally for a high-rise building.

K

keyed alike—Doors whose locks are pinned the same so that one key opens any one of the doors.

L

laser-scanning camera—A device used to create a raw point cloud of existing conditions to be converted into a 3D building information modeling (BIM) image.

late finish (LF)—On a network diagram using the critical path method (CPM), the last day an activity can be finished without delaying the project.

late start (LS)—On a network diagram using the critical path method (CPM), the last possible day that an activity can begin without delaying the project.

latent heat—The heat required to change the state of a substance (from a solid to a liquid, or a liquid to gas) but does not change its temperature.

lead time—Used in construction to describe the amount of time it takes to get a necessary part or material.

Leadership in Energy and Environmental Design (LEED)—A point system for evaluating the various sustainable aspects of a building.

Leadership in Energy and Environmental Design Accredited Professional (LEED AP)—A designation awarded by the United Stages Green Building Council that describes a professional knowledge of sustainability issues and the point system employed to certify buildings under the LEED rating system.

lien—A legal claim against a property asset. Liens must be paid when a property is sold.

lien waiver—A document by which a construction contractor waives the right to any claims against the property.

life-cycle cost analysis (LCCA)—An evaluation tool for determining which capital investment option is the most cost-effective over time.

light shelves—Reflective ledges that direct exterior sunlight into the interior of a building.

light shelves—Shelves constructed of fabric in a steel frame on the outside of a building that reflect light into the interior space.

load profile—The method for describing the operation of equipment: mostly part load operation, mostly full load, or a combination of the two.

M

major maintenance—A maintenance item that represent the expenditure of a significant amount of funds taken from the capital budget.

material adverse fact—Legal terminology used when there is a transfer of ownership of a property. This information could affect the value of a property and must be disclosed by the seller during the sale of a property.

material safety data sheet—A document required by OSHA that provides employees with hazard information for the chemicals they use in their jobs.

mechanical completion—For industrial construction projects, the point in a construction project when the machinery has been installed and tested according to code, and documentation has been turned over to commissioning personnel.

methane (CH_4)—A greenhouse gas that is twenty times more effective in trapping heat in the atmosphere than CO_2. Methane is the major constituent of natural gas.

microgrids—Electric power grids separate from the electric utility's primary power grid that give facilities the capability of generating and storing their own power.

microturbine—Small gas turbine used to generate electricity and frequently used in a combined heat and power (CHP) plant.

mind map—Graphic diagram that expands outward, helping its creator make mental connections between ideas and generate new ideas on a particular topic, problem, or project.

model code—Building code developed by a standards organization. Examples of such codes include the International Building Code and the National Electric Code.

mold—A fungus that requires moisture and food to live and grow. Mycotoxic mold contains toxins that, in elevated amounts, can be harmful to humans. Mold spores can cause allergic reactions.

mudjacking—A process whereby holes are drilled in the concrete slab and a cement mixture called a slurry is pressure-pumped under the slab to raise and level it hydraulically.

mulch—A covering over the soil placed around plantings to retain moisture and prevent weeds from sprouting. Mulch may consist of shredded hardwood or various sizes of pine bark nuggets, nut shells, stone, or rubber.

N

National Emissions Standard for Hazardous Air Pollutants (NESHAP)—These rules are federal regulations under the Clean Air Act that apply to facility owners and contractors who perform work in public and commercial buildings. Asbestos in buildings is one of the areas covered by NESHAP.

net zero energy—A condition whereby the energy used by the building is obtained from its surrounding environment (wind, solar, etc.).

network diagram—A flow chart that shows the interrelationship between project events.

net-zero energy—A condition whereby the energy used by the building is obtained from its surrounding environment (wind, solar, etc.).

node—A point on a network diagram that represents an important event in a project schedule.

nonconforming use—The use of a building for activities that differ from the buildings surrounding it. Also, the use of a building that may be the same as the buildings surrounding it but on a much different scale (larger or smaller). Zoning regulations protect owners of existing properties from construction that is nonconforming.

O

off-peak hours—The period of the day when power demand is low.

on-peak hours—The period of time when power demand is highest.

open bidding—The process by which unknown contractors bid on a contract. Open bidding may be required by law for federal or state contracts.

open purchase order—A standing purchase order assigned one account number for all future uses.

operating budget—Funds budgeted for the day-to-day maintenance operations, utilities, and wages.

other potentially infectious material (OPIM)—These materials contain blood or bodily fluids that harbor infectious pathogens.

outside and stem and yoke (OS&Y) valve—A valve whose screw on the valve stem moves outward through the valve handle, which remains attached to the top of the valve.

outsourcing—Hiring an outside contractor to provide maintenance or repair services.

overseeding—The process of applying additional grass seed to the established turf to increase the turf density. Overseeding can be accomplished by first aerating the soil and then seeding and topdressing with soil, or through the use of a device called a slit seeder.

owner's representative—A person who works for the owner or organization who owns the property in a construction project. The owner's representative must work for the best interests of the owner. Also known as *owner's rep*.

ozone—An unstable gas sometimes used to disinfect cooling tower water.

P

partition walls—Walls used to separate rooms within a single tenant's space. Partition walls end a short distance above the finished ceiling and are not required to go all the way up to the underside of the floor deck above.

payment bond—A bond given by the contractor to the owner to ensure that the contractor will pay his subcontractors and materials suppliers after the contractor receives payment from the owner.

percentage budget—A budget for the coming year that is created by adding a percentage increase to the previous year's budget.

performance bond—A bond the contractor gives to the owner to ensure that the work will be performed as specified in the contract documents. If the work is not completed as specified, the owner can use the bond to make the necessary corrections.

personal protective equipment (PPE)—Safety equipment worn by an employee that is required for the hazards presented by the job.

pervious pavement—Pavement that allows rainwater to drain through the pavement into the base and into the soil below.

phase change material—A material used to absorb or transfer heat at a specific temperature. Phase change material may be mixed with building materials to achieve a latent cooling or heating effect without using mechanical equipment.

phase one all appropriate inquiry (phase I AAI)—An environmental site assessment method by which environmental consultant(s) visit the site and review environmental records.

plan set—The entire grouping of all the construction documents for a project.

plenum return—The area above the ceiling that is used as a duct to return air to the air handler from the spaces in the building.

pneumatic variable air volume (VAV) system—A cylindrical can that has an actuating shaft protruding from one end. The actuating shaft is connected to the rod of the damper blade. A spring-loaded rubber air bladder, piston, or diaphragm is used to move the actuating shaft. The spring provides a force against the bladder or piston. The spring returns the shaft to the proper position when air pressure has changed. Air pressure is supplied from a pneumatic thermostat at a pressure that is proportional to space temperature.

pneumatic variable air volume system—A ventilation system that controls the airflow into a zone with a pneumatic actuator that moves a damper. The actuator moves proportionally to a signal received from a pneumatic thermostat.

point—A single item of information in a control system used in a data storage location. Points are either input or output, and they can provide digital or analog signals to the controller.

polyethylene terephthalate (PET) carpet—A carpet made from recycled plastic beverage containers.

porous pavement—Pavement that allows rainwater to go through the asphalt and into a recharge bed for later use.

post purging—Having the boiler's stack fan run long enough to achieve eight changes of air in the furnace.

postpurge—The process venting of combustible gases from a boiler's furnace after the burner flame has been turned off.

power factor—The efficiency of a facility's use of electricity. Power factor = real power ÷ apparent power = kW ÷ kVA. Unity power factor is 1.00. A power utility may charge a customer for having a low power factor.

power factor—The way that a facility uses power, and the resulting power system efficiency.

predecessor activity—An activity along the same path on a network diagram that occurs before the activity being reviewed.

present worth (PW)—The value of the money today that will either be received or paid at a known time in the future.

pressure differential—The difference between the entering pressure and the leaving pressure. Pressure differential is especially important when looking at the condition of a filter.

pressure maintenance pump—A small pump used to keep a fire sprinkler system pressurized when the larger main fire pump is not running. Also known as jockey pump.

progress payment—Payment made to the construction contractor or general contractor as a project is completed. Progress payments are generally required for large projects that extend over several months.

project milestones—Specific events in the construction process that must be achieved before other work can continue. The dates of project milestones indicate whether the project is on schedule.

protocol—A machine language that allows different pieces of digital equipment to communicate with one another.

punch list—A list of items requiring correction at the end of a construction project that is presented to the general contractor or subcontractor who performed the work.

purchase order (PO)—A guarantee of payment issued to the vendor by the facility making the purchase. Purchase orders are identified with a number.

R

radon—An odorless, colorless, radioactive gas that is produced by decaying radium in soils under and around a building. It is linked to lung cancer.

raw point cloud—The collection of millions of points captured using a laser scanning camera. These points can be used as a picture and/or to create 3D construction models using programs such as AutoCAD®, Revit®, and MicroStation®.

red-line drawings—Drawn-in modifications on the construction documents to show corrections that were made in the field by the various trades.

retainage—The holding back of a percentage of the promised funds in a construction project to ensure that the contractor performs the corrections outlined in a punch list.

retro-commissioning—Investigating the efficiency and operating characteristics of the plant equipment at a facility and comparing the current status to the original design parameters in order make system energy efficiency improvements.

return air fan—A fan that creates a negative pressure to bring air back to the air-handling unit (AHU) from the spaces in the building that were supplied with ventilation.

return on investment (ROI)—A percentage of investment cost that is returned to the owner on a yearly basis.

reverse osmosis—A process whereby water is forced through a membrane to remove dissolved minerals in the water.

right-to-know law—Hazard communication state legislation in place prior to the federal legislation requiring employees to be made aware of chemical hazards in the workplace.

risk analysis—The first step in defining the probable disasters that can occur and the likelihood of their happening.

runaround coil—A series of three cooling coils that provide water above the dew point for the chilled beam coil and dehumidification of the supply air.

S

seasonal energy efficiency rating (SEER)—A number that reflects the amount of energy consumed in watt-hours over a cooling season compared to the amount of cooling in BTU provided over the cooling season.

secondary rate—A higher rate per kWh consumed that does not consider power factor in customer billing or levy an additional charge should the power factor be low.

sensible heat—Heat that can be measured with a thermometer. Adding sensible heat doesn't physically change the material that is being heated; it just makes its temperature rise as its heat content increases.

service-level agreements—Contracts that concentrate more on the final outcome of the work performed rather than on specifying the methods and materials employed in achieving that outcome.

simple payback—A method of determining a project's economic feasibility that considers a project's initial cost and the dollar payback for the savings that the project creates over the life of the equipment.

sling psychrometer—A manual type of hygrometer (device that measures relative humidity).

smart building technology—The integration of intelligent building systems that can make decisions based on events taking place, for example, a tenant swiping an access card after hours to open a door on the front of the building. The door opens and the heating and cooling is activated in the space the tenant is authorized to occupy. The time that the tenant remains in the building is recorded for a chargeback of the costs associated with after-hours building use.

solar array—A group of photovoltaic solar cells designed to produce a certain amount of direct current voltage.

space planning—Designing the interior space to accommodate its efficient use by its occupants.

stack effect—Ventilation created by a tall masonry chimney. In buildings, warm air rises through a shaft or atrium area in the building, causing a suction draft that pulls fresh air in through vents at the bottom level of the building.

static air pressure—The weight of air in a ventilation duct. A static pressure in a ventilation duct that is greater than the static pressure in the interior space allows the air to flow out of the duct and into the space due to the difference in pressure.

steam trap—A device that holds steam back in a heat transfer device until the steam has condensed, and then allows the condensed water to drain to a condensate tank.

substantial completion—The point in a construction project when the owner can use the building as intended, and only commission work and punch list items remain. Contract language can alter when substantial completion occurs.

successor activity—The activity on a network diagram using the critical path method (CPM) that occurs after the activity being reviewed.

supercritical boiler—A boiler operating above 3,200 pounds per square inch gauge (psig) that is used to generate electricity via a steam turbine at a power plant.

surety bond—An agreement among owner, contractor, and surety company that guarantees that the contractor will perform the contract in accordance with the contract documents.

sustainability analyst—A person who gathers data from the results of sustainability and energy conservation efforts.

sustainability project manager (SPM)—A person in charge of a sustainable construction project that will be seeking Leadership in Energy and Environmental Design (LEED) or other green certification. The SPM ensures that the certification is achieved.

SWOT (strengths, weaknesses, opportunities, and threats) analysis—A project evaluation whereby the strengths, weaknesses, opportunities, and threats to the project are evaluated.

T

thatch—A layer of living and dead stems, leaves, and roots on top of the soil around the shoots of grass plants.

tilt-up construction—Reinforced concrete wall sections that are fabricated on site and are generally 6 to 8 inches thick.

time and materials (T&M)—A contract format whereby the owner pays for the labor, time, and the materials used. A profit factor is built into the labor rate, and there is a markup on materials.

time of use—Billing power utility customers on a rate that is based not only on how much electricity the customer consumes, but also on when they consume the electricity.

time of use dollar savings—Overall electric utility efficiency gains achieved by evening out the load and reducing the peak demand for power.

time value of money—A theory based on the fact that a dollar received today is worth more than a dollar received a year from now. If an investor receives a dollar today, the investor can immediately purchase something worth a dollar, or she or he can invest the dollar and begin receiving interest on the dollar.

ton of refrigeration—Amount of refrigeration equivalent to 12,000 BTU per hour.

top dressing—Spreading of a thin layer of soil that falls between the blades of grass of a lawn. Used as a cultural method to remove thatch by causing the thatch to decompose naturally rather than by using a dethatching machine. Top dressing is also performed prior to overseeding.

total quality management's (TQM's) four main processes—*Kaizen, atarimae hinshitsu, miryokuteki hinshitsu,* and *kansei.*

tower fill—Material that spreads out water flowing through a cooling tower to increase the material's surface area. Air blowing across the wetted fill material causes some of the water to evaporate, thus reducing the remaining water's temperature.

Transfer of Undertakings Protection of Employment (2006)—Law in the United Kingdom that protects employees from losing their jobs when another company or contractor is hired to do their work.

turn down ratio—The ability of a boiler to operate efficiently at various load conditions. A boiler with a turn down ratio of 3 can operate at 33 percent of full load output.

U

ultraviolet light C-band (UVC)—The band of ultraviolet light that is filtered out by the ozone in the upper atmosphere. It is a strong disinfectant and is used to kill bacteria in air and water.

underfloor air distribution—A method whereby ventilation air is brought into the space under a raised floor by using the air ducting above the ceiling.

uniform present worth (UPW)—A formula that is used to calculate present worth for fixed recurring costs.

unity power factor—A power factor equal to 1. The facility's electrical system is most efficient when the power factor is at 1.00 (unity). This condition occurs in an alternating current circuit when voltage and current are in phase.

unity power factor—Real power $\div$ apparent power = kW $\div$ kVA = 1.00

V

variance—Constructing or remodeling a building for a use that does not conform to the zoning requirements of a particular area.

vendor data budget—A budget that is constructed from cost data provided by a vendor for goods or a services that the vendor provides to the facility.

ventilation reset—A control scheme that resets the outside air damper to ensure that all spaces are properly ventilated.

ventilation—The provision of the correct amount of outside air to ensure that the building has achieved a high-quality indoor environment and has met all health and energy code requirements.

W

walk-through—The act of touring the facility in the area where the work will be done prior to bidding.

walls—*See* demising walls; partition walls.

water softener—A device or substance used to remove minerals from boiler or cooling tower makeup water.

wet back boiler—A boiler whose chamber contains refractory brick or other types of insulation to insulate against the high heat of the flue gas, and also has a water jacket to absorb heat.

work practice controls—Requirements that are part of an exposure control plan. The controls require that employees use the appropriate personal protective equipment (PPE) for the job.

X

xeriscaping—The use of native plants that, under normal conditions, do not require any supplemental irrigation.

Z

zero-based budget—A budget in which the facility manager must provide justification for all items and amounts.

BIBLIOGRAPHY

AHRI. "Performance Rating of Commercial and Industrial Unitary Air-Conditioning Equipment," October 2011, http://www.ahrinet.org/App_Content/ahri/files/standards%20pdfs/ANSI%20standards%20pdfs/ANSI%20AHRI%20Standard%20340-360-2007%20with%20Addenda%201%20and%202.pdf (accessed January 15, 2013).

Alabama Power. "Desuperheaters and Refrigeration Heat Reclaim Systems," http://www.alabamapower.com/business/services/architects-engineers/library/water-heating/rhr.asp (accessed January 2, 2013).

American Pacific Corporation (AMPAC). "Products," http://www.halotron.com (accessed December 8, 2012).

Aquascape. "Boerner Botanical Garden Project," http://www.aquascapeinc.com/retailers/index.php?page=articles&a_id=117 (accessed December 20, 2012).

Aquionics. "UV Disinfection of Cooling Tower Water," http://halmapr.com/news/aquionics/uv-disinfection-of-cooling-tower-water/ (accessed November 11, 2012).

ASHRAE. "ASHRAE'S Building Energy Quotient—ASHRAE'S Building Labeling Program," updated March 1, 2012, http://www.ashrae.org (accessed June 1, 2012).

Atkinson, Megan. "The Art of Explaining Power Factor," April 2011, http://www.energycareerist.com/2011/04/07power-factor/ (accessed January 1, 2013).

BRE Trust. "What Is BREEAM?" http://www.breeam.org (accessed May 31, 2012).

Brucker, Beth. "Building Information Modeling to Benefit Entire Facility Life Cycle," 2010, http://www.cecer.army.mil/td/tips/docs/BIM-LC.pdf (accessed November 11, 2012).

Bureau of Labor Statistics, U.S. Department of Labor, *Occupational Outlook Handbook, 2012–13 Edition,* "General Maintenance and Repair Workers," http://www.bls.gov/ooh/installation-maintenance-and-repair/general-maintenance-and-repair-workers.htm (visited January 12, 2013).

Buzan, Tony, and Barry Buzan. *The Mind Map Book: How to Use Radiant Thinking to Maximize Your Brain's Untapped Potential.* New York: Plume Book, Penguin Group, 1993.

California Department of Toxic Substance Control, Office of Pollution Prevention and Green Technology. "Evaluation of Non-Chemical Treatment Technologies for Cooling Towers at Select California Facilities," OPPGT Document No. 1220, February 2009, http://www.dtsc.ca.gov (accessed September 30, 2012).

City of Melbourne. "About CH2," http://www.melbourne.vic.gov.au/Sustainability/CH2/aboutch2/Pages/AboutCH2.aspx (accessed November 11, 2012).

Darling, David. "The Encyclopedia of Science," http://www.daviddarling.info/encyclopedia/C/cork.html (accessed January 19, 2013).

Department of Environmental Protection, Bureau of Remediation & Waste Management, State of Maine. "Operations and Maintenance Plan Requirements for Schools with Asbestos," http://www.maine.gov/dep/waste/asbestos/operplan.html (accessed January 25, 2013)).

Durkin, Thomas H. "Boiler System Efficiency," ASHRAE Journal, Vol. 48, July 2006, http://www.dvpe.net/articles/boilerefficiency.pdf (accessed October 26, 2012).

Eastman, Chuck. "What Is BIM?" August 2009, http://bim.arch.gatech.edu/?id=402 (accessed November 11, 2012).

E.I. du Pont de Nemours and Company. "Dupont™FM200® Waterless Fire Suppression," June 2012 (accessed December 8, 2012).

E.I. du Pont de Nemours and Company. "Dupont™Opteon™ YF Refrigerant for Mobile Air Conditioning,"2012, http://www2.dupont.com/Refrigerants/en_US/uses_apps/automotive_ac/SmartAutoAC/HFO-1234yf.html#.UMlLW-Si9nY (accessed December 12, 2012).

Equal Employment Opportunity Commission. "Federal Laws Prohibiting Job Discrimination Questions and Answers," November 2009. http://www.eeoc.gov/facts/qanda.html (accessed January 6, 2013).

Georg Fischer Signet LLC. "Chlorine Dioxide Control," 2012, http://www.gfsignet.com/go/935D6CF3199943D4416EBD131BAADE6C (accessed December 19, 2012).

Georgia-Pacific. "Safe-T-Guard™Door Tissue Dispensers," http://productcatalog.gp.com/Main.aspx?View=2&Cat=6241 (accessed January 11, 2013).

Goidich, Stephen J., Kenneth P. Melzer, Robert V. Roche, and William Bousquet. "Innovation in Supercritical Boiler Technology—The 750 MWe Longview Power Project," December 2008, http://www.fwc.com/publications/tech_papers/files/TP_PC_08_01.pdf (accessed December 19, 2012).

Goodman, Leonard S. "Facility Managers Using LEED-EBOM As Blueprint for Sustainability," April 2010, http://www.facilitiesnet.com/green/article/Facility-Managers-Using-LEEDEBOM-As-Blueprint-for-Sustainability—11621 (accessed November 11, 2012).

Government Services Administration (GSA). "Internal Revenue Service GSA Public Building Service Preventive Maintenance Guides," http://www.irs.gov/pub/irs-procure/technicalexhibit4_pm_guide_cards.pdf (accessed November 9, 2012).

Green Building Council Australia (GBCA). "Green Star Overview," http://www.gbca.org.au/green-star/green-star-overview (accessed June 1, 2012).

Green Building Initiative (GBI). "Green Globes® Overview," http://www.thegbi.org (accessed October 4, 2012).

Green Energy Ohio. "Hydropower," http://www.greenenergyohio.org/page.cfm?pageId=54 (accessed January 6, 2013).

Gross and Associates. "Can an EFSR Sprinkler System Keep You from Getting Soaked?" 2001, http://westburyfd.org/files/InstructorsSafety/ESFR_Sprinkler_1_.pdf (accessed November 11, 2012).

Hannagan, Tim. *Management Concepts and Practices*, 3rd ed. London: Pearson Education, 2002.

Harvey, Bill M. "The Jockey Pump, an Important Part of a Fire Pump System," August 2009, http://harveyandassociates.com (accessed June 5, 2010).

High School Guide. "Top 100 Largest High Schools in America," September 2011, http://highschoolguide.org/624/top-100-largest-high-schools-in-america/ (accessed October 28, 2012).

Illinois Department of Public Health. "Illinois Department of Public Health Guidelines for Indoor Air Quality," http://www.idph.state.il.us/envhealth/factsheets/indoorairqualityguide_fs.htm (accessed September 5, 2010).

Katsnelson, Alla. "Chilled Beam Ceiling Systems Promise Energy, Maintenance Savings," August 2007, http://www.facilitiesnet.com/hvac/article/Chilled-Beam-Ceiling-Systems-Promise-Energy-Maintenance-Savings-7219 (accessed January 6, 2013).

Khan, Md. Touhidul Alam. "Going Beyond Simple Productivity Improvement," September 2009, http://www.thefinancialexpress-bd.com/2009/09/24/79556.html (accessed October 22, 2012).

Lay, Richard M. (Contributing Expert), and Sustainable Building.com. "Natural Ventilation and Cooling," Copyright 1994–2004, http://www.sustainablebuilding.com (accessed September 15, 2010).

Lennerts, Kunibert, Jochen Abel, Uwe Pfrunder, and Vishal Sharma. Step-by-step process analysis for hospital facility management. An insight into the OPIK research project, 2005, http://www.emeraldinsight.com/journals.htm?articleid=1463897 (accessed October 22, 2005).

Lundquist, Ragnar, Andre Schrief, Pertti Kinnunen, Kari Myohanen, and Mani Seshamani. "A Major Step Forward: The Supercritical CFB Boiler," http://www.energy.siemens.com (accessed December 19, 2012).

Mills, Evan. "Building Commissioning: A Golden Opportunity for Reducing Energy Costs and Greenhouse Gas Eemissions in the United States," February 2011, http://evanmills.lbl.gov/pubs/pdf/cx-enef-mills.pdf (accessed January 1, 2013).

Milwaukee County. "Rainwater Harvesting," http://county.milwaukee.gov/BoernerBotanicalGard10113/Rainwater-Harvesting.htm?PrintPage=yes (accessed December 20, 2012).

Minnesota Department of Health. "Maintenance and Testing of Fire Sprinkler Systems," http://www.health.state.mn.us/divs/fpc/Firesprinklers2.pdf (accessed October 26, 2012).

National Asphalt Pavement Association (NAPA). "Porous Asphalt," 2010, http://www.Hotmix.org/index.php?option=com_content&task=view&id=359&Itemid=863 (accessed June 1, 2010),

New Jersey Department of Health and Senor Services. "Hazardous Substance Fact Sheet—Coal Tar Pitch," August 2009, http://nj.gov/health/eoh/rtkweb/documents/fs/0519.pdf (accessed January 19, 2013).

North Carolina Division of Pollution Prevention and Environmental Assistance. "Boiler Blowdown," August 2004, http://infohouse.p2ric.org/ref/34/33027.pdf (accessed November 11, 2012).

Nunn, Robert. *Water Treatment Essentials for Boiler Plant Operation.* New York: McGraw Hill, 1997.

NV Energy. "The Waterside Economizer (aka "Tower Free-Cooling"): A Cost Effective Design Option for Larger Chilled Water Cooling systems," https://www.nvenergy.com/business/saveenergy/incentives/surebet/documents/images/TechSheet_WaterSideEconomizer_JCY_01062009.pdf (accessed January 6, 2013).

Occupational Safety and Health Administration. "Evacuating High Rise Buildings" (OSHA Fact Sheet), 2003, http://www.osha.gov/OshDoc/data_General_Facts/evacuating-highrise-factsheet.pdf (accessed January 25, 2013).

Occupational Safety and Health Administration. "Hazard Communication/HazCom 2012 Final Rule," 2012, http://www.osha.gov/dsg/hascom/ghs-final-rule.html (accessed December 20, 2012).

Occupational Safety and Health Administration. "Model Plans for OSHA Bloodborne Pathogens and Hazard Communication Standard," 2003, http://www.osha.gov/dsg/hazcom/solutions.html (accessed December 20, 2012).

Pacific Gas and Electric Company®. "PG&E's Energy Management Solutions for Boilers," 2006, http://www.pge.com/includes/docs/pdfs/mybusiness/energysavingsrebates/incentivesbyindustry/fs_boilers.pdf (accessed December 11, 2011).

PCM Construction Inc. "Sample Safe Work Procedures (Templates)," http://www.pcmconstruction.com/pdf/safe_work_procedures.pdf (accessed December 20, 2012).

Peterson, Steven. *Pearson's Pocket Guide to Construction Management*. Upper Saddle River, NJ: Pearson Education, 2012.

Pound, William E., and John R. Street. "Thatch: The Accumulation in Lawns," http://ohioline.osu.edu/hyg-fact/4000/4008.html (accessed December 4, 2012).

Pyant, Richard P., and Bernard T. Lewis. *Facility Manager's Maintenance Handbook*, 2nd ed. New York: McGraw-Hill, 2007.

Reid, Rollo. "How to Make Buildings & Structures Earthquake Proof," http://www.reidsteel.com/information/earthquake_resistant_building.htm (accessed September 30, 2012).

Reynolds, Gary L., Paul F. Tabolt (Ed.), Harvey H. Kaiser, Jay W. Klingel, Richard S. Fowler, John D. Houck (Ed.), Warren Corman, Grace C. Kelley, Mohammad H. Qayoumi, John P. Sluis, Douglas Forsman, Matthew S. Manfredi, Bryan R. Hines, Norman H. Bedell (Ed.), J. Richard Swistock, Roy Peterman, J. Kirk Campbell, Phillip R. Melnick, William S. Rose (Ed.), Susan A. Kirkpatrick, John P. Harrod Jr., Ralph O. Allen, Loras Jager, and James E. Ziebold. *Facilities Management: A Manual for Plant Administration. Part II: Maintenance and Operations of Buildings and Grounds*, 3rd ed. Alexandria, VA: The Association for Higher Education Facilities Officers, 1997.

Rumsey, Peter, and John Weale. "Practical Implementation of Chilled Beams for Offices," January 2011, http://www.esmagazine.com/articles/94746-practical-implementation-of-chilled-beams-for-offices?v=preview (accessed January 6, 2013).

Schelmetic, Tracey. "Green Facts about New York's One World Trade Center," January 2012, http://news.thomasnet.com/green_clean/2012/01/03/green-facts-about-new-yorks-new-one-world-trade-center/ (accessed January 6, 2013).

Stanfield, Carter, and David Skaves. *Fundamentals of HVACR*. Upper Saddle River, NJ: Pearson Education, 2010.

Studer, Craig. "Integrating Fire Systems with HVAC Controls." July 2012. https://www.csemag.com/home/single-article/integrating-fire-systems-with-hvac-controls/bcd770dce96eb4d0d6073074d1b7da04.html (accessed December 25, 2012).

Surety Information Office (SIO). "How to Obtain Surety Bonds," 2007, http://suretyinfo.org/?wpfb_dl=57 (accessed September 30, 2012).

System Sensor. "Fire Sprinkler Systems Monitoring," 2007, http://www.systemsensor.com/pdf/A05-1057.pdf (accessed November 11, 2012).

Tao, William, and Richard Janis. *Mechanical and Electrical Systems in Buildings*. Upper Saddle River, NJ: Pearson Education, 2001.

3M Company. "3M™ Novec™Fire Protection Fluid," http://solutions.3M.com (accessed December 8, 2012).

UK Green Building Council (UKGBC). "BREEAM and LEED: How Do They Compare?" *GreenBuilding Magazine*, http://www.ukgbc.co.uk/leed.php (accessed May 31, 2012).

University of Tennessee. "Introduction to MACT or NESHAP Rules," http://www.epamact.tennessee.edu (accessed December 20, 2012).

U.S. EPA, "Fact Sheet, Greenhouse Gas Reporting Program: Proposed Amendments and Confidentiality Determinations for Subpart I," August 2012, http://www.epa.gov/ghsreporting/documents/pdf/2012/documents/subpart_I_factsheet.pdf (accessed December 26, 2012).

U.S. EPA, Laboratories for the 21st Century. "Best Practice Guide: Chilled Beams in Laboratories, Key Strategies to Ensure Effective Design, Construction, and Operation," June 2009, http://www.epa.gov/labs 21 century/ (accessed April 1, 2011).

U.S. EPA. "Managing Asbestos in Place," updated April 2010, http://www.epa.gov/asbestos/pubs/section4.html (accessed June 4, 2010).

U.S. Department of Energy, Energy Efficiency and Renewable Energy, Federal Energy Management Program. "Demand-Controlled Ventilation Using CO2 Sensors," document DOE/EE-0293, March 2004, http://www1.eere.energy.gov/femp/pdfs/fta_Co2.pdf (accessed September 30, 2012).

U.S. Department of Energy. "Heat Pump Water Heaters," May 4, 2012, http://www.energy.gov. (accessed May 2, 2010).

U.S. Equal Employment Opportunity Commission (EEOC). "Federal Laws Prohibiting Job Discrimination: Questions and Answers," http://www.eeoc.gov/facts/quanda.html (accessed May 20, 2010).

U.S. Green Building Council (USGBC). FAQ. "What Is LEED 2012?" http://www.usgbc.org/ShowFile.aspx?DocumentID=18558 (accessed October 4, 2012)

Washington State University Cooperative Extension Energy Program and the Northwest Energy Efficiency Alliance. "Reducing Power Factor Cost," February 2003, http://www.energyideas.org/documents/factsheets/reducing_pwr.pdf (accessed October 26, 2012).

INDEX